Universitext

Springer
New York
Berlin
Heidelberg
Barcelona
Budapest
Hong Kong
London
Milan
Paris
Santa Clara
Singapore
Tokyo

Universitext

Editors (North America): S. Axler, F.W. Gehring, and P.R. Halmos

Aksoy/Khamsi: Nonstandard Methods in Fixed Point Theory
Aupetit: A Primer on Spectral Theory
Booss/Bleecker: Topology and Analysis
Borkar: Probability Theory; An Advanced Course
Carleson/Gamelin: Complex Dynamics
Cecil: Lie Sphere Geometry: With Applications to Submanifolds
Chae: Lebesgue Integration (2nd ed.)
Charlap: Bieberbach Groups and Flat Manifolds
Chern: Complex Manifolds Without Potential Theory
Cohn: A Classical Invitation to Algebraic Numbers and Class Fields
Curtis: Abstract Linear Algebra
Curtis: Matrix Groups
DiBenedetto: Degenerate Parabolic Equations
Dimca: Singularities and Topology of Hypersurfaces
Edwards: A Formal Background to Mathematics I a/b
Edwards: A Formal Background to Mathematics II a/b
Foulds: Graph Theory Applications
Gardiner: A First Course in Group Theory
Gårding/Tambour: Algebra for Computer Science
Goldblatt: Orthogonality and Spacetime Geometry
Hahn: Quadratic Algebras, Clifford Algebras, and Arithmetic Witt Groups
Holmgren: A First Course in Discrete Dynamical Systems
Howe/Tan: Non-Abelian Harmonic Analysis: Applications of $SL(2, \mathbb{R})$
Howes: Modern Analysis and Topology
Humi/Miller: Second Course in Ordinary Differential Equations
Hurwitz/Kritikos: Lectures on Number Theory
Jennings: Modern Geometry with Applications
Jones/Morris/Pearson: Abstract Algebra and Famous Impossibilities
Kannan/Krueger: Advanced Real Analysis
Kelly/Matthews: The Non-Euclidean Hyperbolic Plane
Kostrikin: Introduction to Algebra
Luecking/Rubel: Complex Analysis: A Functional Analysis Approach
MacLane/Moerdijk: Sheaves in Geometry and Logic
Marcus: Number Fields
McCarthy: Introduction to Arithmetical Functions
Meyer: Essential Mathematics for Applied Fields
Mines/Richman/Ruitenburg: A Course in Constructive Algebra
Moise: Introductory Problems Course in Analysis and Topology
Morris: Introduction to Game Theory
Porter/Woods: Extensions and Absolutes of Hausdorff Spaces
Ramsay/Richtmyer: Introduction to Hyperbolic Geometry
Reisel: Elementary Theory of Metric Spaces
Rickart: Natural Function Algebras
Rotman: Galois Theory
Rubel/Colliander: Entire and Meromorphic Functions

(continued after index)

Anadijiban Das

The Special Theory of Relativity

A Mathematical Exposition

With 27 Illustrations

Springer

Anadijiban Das
Department of Mathematics and Statistics
Simon Fraser University
Burnaby, V5A 1S6 British Columbia
Canada

Library of Congress Cataloging-in-Publication Data
Das, Anadijiban.
The special theory of relativity: a mathematical exposition /
Anadijiban Das, author.
p. cm.
Includes bibliographical references and index.
ISBN 0-387-94042-1. --ISBN 3-540-94042-1
1. Special relativity (Physics)--Mathematics. 2. Mathematical
physics. I. Title.
QC173.65.D38 1993
530.1'1--dc20 93-10256

Printed on acid-free paper.

Production managed by Jim Harbison; manufacturing supervised by Vincent Scelta.
Typeset by Asco Trade Typesetting Ltd, Hong Kong.
Printed and bound by Edwards Brothers, Inc., Ann Arbor, MI.
Printed in the United States of America.

9 8 7 6 5 4 3 2 (Corrected second printing, 1996)

ISBN 0-387-94042-1 Springer-Verlag New York Berlin Heidelberg
ISBN 3-540-94042-1 Springer-Verlag Berlin Heidelberg New York SPIN 10528157

Dedicated to Sri Gadadhar Chattopadhyaya

Preface

The material in this book is presented in a logical sequence rather than a historical sequence. Thus, we feel obligated to sketch briefly the history of the special theory of relativity. The brilliant experiments of Michelson and Morley in 1887 demonstrated the astonishing fact that the speed of light is independent of the state of relative linear motion of the source of light and the observer of the light. This fact necessitates the modification of the usual Galilean transformation (between two relatively moving observers), which tacitly assumes that time and space are absolute.

Fitzgerald in 1889 and Lorentz in 1892 altered the Galilean transformation by introducing a length contraction in the direction of relative motion. This contraction explained the Michelson–Morley experiment, but it was viewed by both Fitzgerald and Lorentz as a mathematical trick only and *not* indicative of the nature of reality. In 1898 Larmor introduced a similar time dilation in an attempt to find the transformations which leave Maxwell's equations invariant. Lorentz also introduced the time dilation independently sometime before 1904. Poincaré in 1905 also discovered the Lorentz transformation and asserted that it was the fundamental invariance group of nature. Einstein in 1905 discovered the Lorentz transformation from physical considerations. Einstein, alone among these mathematical physicists, recognized the philosophical implications of the Lorentz transformation in that it rejected the commonly held notion that space and time were both absolute. He postulated the equivalence of all inertial frames of reference (moving with constant velocities relative to each other) with regard to the formulation of natural laws. Furthermore, he recognized and postulated that the speed of light is the maximum speed of propagation of any physical action. Therefore, the speed of light must be invariant for all inertial observers. Thus the Michelson-Moreley experiment was reconciled with theory. Minkowski, a mathematician, combined both physical postulates of Einstein into one mathematical axiom. This axiom is that "all natural laws must be expressible as tensor field equations on a (flat) absolute space–time manifold." Thus, in that there is no preferred inertial frame for the formulation of natural laws, a universal democracy is postulated to exist among all inertial observers. This axiom is called the Principle of Special Relativity. Many experiments involving atoms and subatomic particles have verified the essential validity of this principle.

In the first chapter we introduce axiomatically the four-dimensional Minkowski vector space. This vector space is endowed with a nondegenerate inner product which is *not* positive definite. Therefore, the concepts of the *norm* (or length) of a four-vector and of the *angle* between two four-vectors have to be *abandoned*. A Lorentz mapping is introduced as an inner product preserving linear mapping of Minkowski vector space into itself.

In Chapter 2 we introduce the flat Minkowski space–time manifold with a proper axiomatic structure. It is proved that the transformation from one Minkowski chart to another must be given by a Poincaré transformation (or an inhomogeneous Lorentz transformation). The *conceptual difference* between a Lorentz transformation of coordinate charts and a Lorentz mapping of the tangent (Minkowski) vector space is clearly displayed. Minkowski tensor fields on the flat space–time are also defined.

In the third chapter, by applications of a particular Lorentz transformation (the "boost"), length contraction, time retardation, and the composition of velocities are explained. The group structure of the set of all Lorentz transformations is demonstrated, and real representations of the Lorentz group are presented. The proper orthochronous subgroup is defined and discussed also.

The fourth chapter defines the spinor space (a two-dimensional complex vector space) and the properties of spinors. Bispinor space (a four-dimensional complex vector space) is also introduced. It is shown that a unimodular mapping of spinor space can induce a proper, orthochronous Lorentz mapping on Minkowski vector space. Furthermore, a unimodular mapping of bispinor space is shown to induce a general Lorentz mapping of Minkowski vector space.

In Chapter 5 prerelativistic mechanics is briefly reviewed. In the setting of prerelativistic mechanics in space and time, $\mathbb{E}_3 \times \mathbb{R}$, the momentum conjugate to the time variable turns out to be the negative of energy! After this, the relativistic mechanics is investigated. The Lagrangian is assumed to be a positive homogeneous function of degree one in the velocity variables (which makes the generalized Hamiltonian identically zero!). Examples from electromagnetic theory and the linearized gravitational theory of Einstein are worked out.

In Chapter 6 the relativistic (classical) field theory is developed. Noether's theorem (essential for the differential conservation laws) is rigorously proved. As examples of special fields, the Klein–Gordon scalar field, the electromagnetic tensor field, nonabelian gauge fields, and the Dirac bispinor field are presented. However, at the present level of treatment, gauge fields are *not* derived as connections in a fibre bundle over the base (Minkowski) manifold. In each chapter, examples and exercises of various degrees of difficulty are provided.

Chapter 7 deals with a research topic, namely, classical fields in the eight-dimensional extended (or covariant) phase space. Historically, Born and Yukawa advocated the extended phase space on the basis of the principle of *reciprocity* (covariance under the canonical transformation $\hat{p} = -q$, $\hat{q} = p$). In recent years, Caianello and others have considered the principle of *maximal proper acceleration* arising out of the extended phase space geometry. We ourselves have done some research on classical fields in the eight-dimensional phase space. We can obtain, in a certain sense, a unified meson field and a unification of fermionic fields. These fields, however, contain *infinitely* many modes or particles.

We have *changed* the usual notation for the Lorentz metric η_{ij} in favor of d_{ij} (since η_{ijkl} is used for the pseudotensor) and $\gamma \equiv (1 - v^2)^{-1/2}$ in favor of $\beta \equiv (1 - v^2)^{-1/2}$ (since γ is used to denote a curve).

This book has grown out of lectures delivered at Jadavpur University (Calcutta), University College of Dublin, Carnegie–Mellon University, and mostly at Simon Fraser University (Canada). The material is intended mainly for students at the fourth and the fifth year university level. We have taken special care to steer a *middle course* between abstruse mathematics and theoretical physics, so that this book can be used for courses in special relativity in both mathematics and physics departments. Furthermore, the material presented here is a suitable prerequisite for further study in either general relativity or relativistic particle theory.

In conclusion, I would like to acknowledge gratefully several people for various reasons. I was fortunate to learn the subject of special relativity from the late Professor S. N. Bose F.R.S. (of Bose–Einstein statistics) in Calcutta University. I also had the privilege for three years of being a research associate of the late Professor J. L. Synge F.R.S. at the Dublin Institute for Advanced Studies. Their influence, direct or indirect, is evident in the presentation of the material (although the errors in the book are solely due to me!). In preparation of the manuscript, I have been helped very much by Dr. Ted Biech, who typed the manuscript and suggested various improvements. Mrs. J. Fabricius typed the difficult Chapter 7. Mrs. E. Carefoot drew the diagrams. Dr. Shounak Das has suggested some literary improvements. I also owe thanks to many of my students for stimulating discussions during lectures.

I thank Dr. S. Kloster for the careful proof reading.

Finally, I thank my wife Mrs. Purabi Das for constant encouragement.

Contents

1
Four-Dimensional Vector Spaces and Linear Mappings

1.1. Minkowski Vector Space $\mathbf{V}_4$

The three-dimensional vectors in Newtonian physics are generalized into four-dimensional vectors in the theory of relativity. This four-dimensional vector space is called the *Minkowski vector space* and is denoted by $\mathbf{V}_4$. This vector space is over the real field $\mathbb{R}$. The mathematical axioms for addition and scalar multiplication of Minkowski vectors are as follows:

A1. $\mathbf{a} + \mathbf{b} \in \mathbf{V}_4$ for all $\mathbf{a}, \mathbf{b} \in \mathbf{V}_4$.

A2. $\mathbf{a} + \mathbf{b} = \mathbf{b} + \mathbf{a}$ for all $\mathbf{a}, \mathbf{b} \in \mathbf{V}_4$.

A3. $(\mathbf{a} + \mathbf{b}) + \mathbf{c} = \mathbf{a} + (\mathbf{b} + \mathbf{c})$ for all $\mathbf{a}, \mathbf{b}, \mathbf{c} \in \mathbf{V}_4$.

A4. There is $\mathbf{0} \in \mathbf{V}_4$ such that $\mathbf{a} + \mathbf{0} = \mathbf{a}$ for all $\mathbf{a} \in \mathbf{V}_4$.

A5. For all $\mathbf{a} \in \mathbf{V}_4$ there is $-\mathbf{a} \in \mathbf{V}_4$ so that $(-\mathbf{a}) + \mathbf{a} = 0$. (1.1.1)

M1. $\alpha\mathbf{a} \in \mathbf{V}_4$ for all $\alpha \in \mathbb{R}$, for all $\mathbf{a} \in \mathbf{V}_4$.

M2. $\alpha(\beta\mathbf{a}) = (\alpha\beta)\mathbf{a}$ for all $\alpha, \beta \in \mathbb{R}$, for all $\mathbf{a} \in \mathbf{V}_4$.

M3. $1\mathbf{a} = \mathbf{a}$ for all $\mathbf{a} \in \mathbf{V}_4$.

M4. $\alpha(\mathbf{a} + \mathbf{b}) = \alpha\mathbf{a} + \alpha\mathbf{b}$ for all $\alpha \in \mathbb{R}$, for all $\mathbf{a}, \mathbf{b} \in \mathbf{V}_4$.

M5. $(\alpha + \beta)\mathbf{a} = \alpha\mathbf{a} + \beta\mathbf{a}$ for all $\alpha, \beta \in \mathbb{R}$, for all $\mathbf{a} \in \mathbf{V}_4$.

We shall also assume the existence of an inner product for $\mathbf{V}_4$ satisfying the following axioms:

I1. $\mathbf{a} \cdot \mathbf{b} \in \mathbb{R}$ for all $\mathbf{a}, \mathbf{b} \in \mathbf{V}_4$.

I2. $\mathbf{a} \cdot \mathbf{b} = \mathbf{b} \cdot \mathbf{a}$ for all $\mathbf{a}, \mathbf{b} \in \mathbf{V}_4$.

I3. $(\lambda\mathbf{a} + \mu\mathbf{b}) \cdot \mathbf{c} = \lambda(\mathbf{a} \cdot \mathbf{c}) + \mu(\mathbf{b} \cdot \mathbf{c})$ for all $\lambda, \mu \in \mathbb{R}$, for all $\mathbf{a}, \mathbf{b}, \mathbf{c} \in \mathbf{V}_4$. (1.1.2)

I4. $\mathbf{a} \cdot \mathbf{x} = 0$ for all $\mathbf{x} \in \mathbf{V}_4$ if and only if $\mathbf{a} = \mathbf{0}$.

The axiom I4 is called the *axiom of nondegeneracy*. It is a weaker axiom than

I5. $\mathbf{a}\cdot\mathbf{a} \geq 0$ for all $\mathbf{a} \in \mathbf{V}_4$, and $\mathbf{a}\cdot\mathbf{a} = 0$ if and only if $\mathbf{a} = 0$. (1.1.3)

For a *positive definite* inner product axiom I5 replaces I4. In addition to these axioms we impose the *axiom of dimensionality* on Minkowski vector space:

D1. $\dim \mathbf{V}_4 = 4$.

Let $\{\mathbf{e}_1, \mathbf{e}_2, \mathbf{e}_3, \mathbf{e}_4\}$ be a basis set for $\mathbf{V}_4$. The *metric tensor components* relative to this basis are defined by

$$g_{ij} \equiv \mathbf{e}_i \cdot \mathbf{e}_j \quad \text{for } i,\, j \in \{1, 2, 3, 4\}. \tag{1.1.4}$$

From axiom I2 it follows that $g_{ji} = g_{ij}$ for all $i,\ j \in \{1, 2, 3, 4\}$. The four-dimensional unit matrix is denoted by $\mathbf{I} \equiv [\delta_{ij}]$. The eigenvalues of the matrix $[g_{ij}]$ are the roots of the characteristic equation

$$\det[g_{ij} - \lambda\delta_{ij}] = 0. \tag{1.1.5}$$

Since the matrix g_{ij} is symmetric, the roots of (1.1.5) are all real. By the axiom of nondegeneracy I4 it follows that all the eigenvalues of g_{ij} are nonzero. The signs of the eigenvalues of g_{ij} are determined by the *axiom of Lorentz signature*:

S1. $\lambda_1 > 0, \lambda_2 > 0, \lambda_3 > 0, \lambda_4 < 0$.

The vector space obeying the sixteen axioms A1–A5, M1–M5, I1–I4, D1, and S1 is called *Minkowski vector space* and is denoted by $\mathbf{V}_4$.

In $\mathbf{V}_4$, the two vectors $\mathbf{a}$, $\mathbf{b}$ are defined to be *Minkowski orthogonal* (or *M-orthogonal*) provided

$$\mathbf{a}\cdot\mathbf{b} = 0. \tag{1.1.6}$$

Theorem (1.1.1): *There exists an M-orthonormal basis* $\{\mathbf{e}_1, \mathbf{e}_2, \mathbf{e}_3, \mathbf{e}_4\}$ *for* $\mathbf{V}_4$ *such that*

$$g_{ij} = \mathbf{e}_i \cdot \mathbf{e}_j = d_{ij}, \tag{1.1.7}$$

where

$$D = [d_{ij}] \equiv \begin{bmatrix} 1 & 0 & 0 & 0 \\ 0 & 1 & 0 & 0 \\ 0 & 0 & 1 & 0 \\ 0 & 0 & 0 & -1 \end{bmatrix}.$$

The proof is rather involved and is omitted. The metric d_{ij} in (1.1.7) is called the *Lorentz metric*. The signature of d_{ij} is defined to be the trace of $[d_{ij}]$. We shall use a choice of $[d_{ij}]$ so that the signature is equal to 2. Note that some authors use the signature -2.

Now we shall explain the *Einstein summation convention*. In a mathematical expression, wherever two repeated Roman indices are present, the sum over the repeated index is implied. For example, we write

$$u^k v_k \equiv \sum_{k=1}^{4} u^k v_k = \sum_{l=1}^{4} u^l v_l = u^l v_l,$$

$$g_{ij} u^i v^j \equiv \sum_{i=1}^{4} \sum_{j=1}^{4} g_{ij} u^i v^j = \sum_{k=1}^{4} \sum_{l=1}^{4} g_{kl} u^k v^l = g_{kl} u^k v^l.$$

The summation indices are called dummy indices, since they can be replaced by other indices over the same range. In the summation convention, *never* use dummy indices that repeat more than twice. This is necessary in order to avoid wrong answers; for example,

$$u^k v_k u^k v_k \equiv \sum_{k=1}^{4} u^k v_k u^k v_k \neq \sum_{k=1}^{4} \sum_{l=1}^{4} u^k v_k u^l v_l = u^k v_k u^l v_l = (u^k v_k)^2.$$

Let $\{\mathbf{e}_1, \mathbf{e}_2, \mathbf{e}_3, \mathbf{e}_4\}$ be an M-orthonormal basis (or *tetrad*) for $\mathbf{V}_4$. For any vector $\mathbf{u} \in \mathbf{V}_4$, there exists a linear combination

$$\mathbf{u} = \sum_{l=1}^{4} u^l \mathbf{e}_l = u^l \mathbf{e}_l. \tag{1.1.8}$$

The unique numbers or scalars u^i are called the *Minkowski components* of the vector $\mathbf{u}$ relative to the basis $\{\mathbf{e}_1, \mathbf{e}_2, \mathbf{e}_3, \mathbf{e}_4\}$.

Theorem (1.1.2): *In terms of the Minkowski components, the inner product between vectors* $\mathbf{u}$, $\mathbf{v}$ *is given by*

$$\mathbf{u} \cdot \mathbf{v} = d_{ij} u^i v^j. \tag{1.1.9}$$

Proof: Choose an M-orthonormal basis $\{\mathbf{e}_1, \mathbf{e}_2, \mathbf{e}_3, \mathbf{e}_4\}$ such that

$$\mathbf{u} = u^i \mathbf{e}_i, \qquad \mathbf{v} = v^i \mathbf{e}_i.$$

By the axioms in (1.1.2) and (1.1.8) we have

$$\mathbf{u} \cdot \mathbf{v} = (u^i \mathbf{e}_i) \cdot (v^j \mathbf{e}_j) = d_{ij} u^i v^j. \quad \blacksquare$$

Note that from (1.1.9)

$$\mathbf{u} \cdot \mathbf{u} = d_{ij} u^i u^j = (u^1)^2 + (u^2)^2 + (u^3)^2 - (u^4)^2. \tag{1.1.10}$$

The above expression is *not* positive definite. Thus the concept of the *length* (or norm) of a vector in $\mathbf{V}_4$ *is abandoned*. Furthermore, if we define $\cos(\mathbf{u}, \mathbf{v}) \equiv (\mathbf{u} \cdot \mathbf{v})/\sqrt{(\mathbf{u} \cdot \mathbf{u})(\mathbf{v} \cdot \mathbf{v})}$, then we are led to contradictions. For example, if we choose $\mathbf{v}_n = \mathbf{e}_1 + [(n-1)/n]\mathbf{e}_4$ for $n \in \mathbb{Z}^+$ and let $\mathbf{u} = \mathbf{e}_1$, then $\cos(\mathbf{e}_1, \mathbf{v}_n) \equiv n/\sqrt{(2n-1)}$. Therefore, $1 \leq \cos(\mathbf{u}, \mathbf{v}_n)$ and $\lim_{n \to \infty} \cos(\mathbf{u}, \mathbf{v}_n) \to \infty$, which is absurd. That is why the concept of an *angle* between two vectors $\mathbf{u}, \mathbf{v} \in \mathbf{V}_4$ *is abandoned* as well. However, for a spatial vector subspace $\mathbf{V}_3 \equiv \{\mathbf{v} \in \mathbf{V}_4 : v^4 = 0\}$, the usual concept of the length and angle can be restored.

Since the expression (1.1.10) for $\mathbf{u} \cdot \mathbf{u}$ is indefinite, we can define three kinds of vectors in $\mathbf{V}_4$:

(i) a vector $\mathbf{u} \in \mathbf{V}_4$ that satisfies $\mathbf{u} \cdot \mathbf{u} > 0$ is called a *spacelike* vector;
(ii) a vector $\mathbf{u} \in \mathbf{V}_4$ for which $\mathbf{u} \cdot \mathbf{u} < 0$ is called a *timelike* vector;
(iii) a vector $\mathbf{u} \in \mathbf{V}_4$ for which $\mathbf{u} \cdot \mathbf{u} = 0$ is called a *null* vector.

Example: Let $\{\mathbf{e}_1, \mathbf{e}_2, \mathbf{e}_3, \mathbf{e}_4\}$ be an M-orthonormal basis for $\mathbf{V}_4$. By (1.1.9), $\mathbf{e}_1 \cdot \mathbf{e}_1 = d_{11} = 1$. Thus $\mathbf{e}_1$ is a spacelike vector. Similarly $\mathbf{e}_2$, $\mathbf{e}_3$ are spacelike vectors. But $\mathbf{e}_4 \cdot \mathbf{e}_4 = d_{44} = -1$, so $\mathbf{e}_4$ is a timelike vector. Set $\mathbf{u} = \mathbf{e}_1 + \mathbf{e}_4$ and observe that $\mathbf{u} \cdot \mathbf{u} = 0$, so we see that $\mathbf{u}$ is a null vector. □

The *separation number* is a generalization of the concept of length and is denoted by $\sigma(\mathbf{u})$. It is defined by

$$\sigma(\mathbf{u}) \equiv \sqrt{|\mathbf{u} \cdot \mathbf{u}|} \geq 0. \tag{1.1.11}$$

Thus for either timelike or spacelike vectors $\mathbf{u} \in \mathbf{V}_4$ we have $\sigma(\mathbf{u}) > 0$. But for a null vector $\mathbf{n}$ we have $\sigma(\mathbf{n}) = 0$. For an example choose $\mathbf{u} = (\mathbf{e}_1 - 2\mathbf{e}_4)/2$. Then $\sigma(\mathbf{u}) = \sqrt{|-3/4|} = \sqrt{3}/2$.

A vector $\mathbf{e}$ in $\mathbf{V}_4$ is called a *unit vector* if $\sigma(\mathbf{e}) = 1$. Subsequently we shall use only M-orthonormal bases for $\mathbf{V}_4$ unless mentioned otherwise. A spatial vector subspace relative to an M-orthonormal basis is defined as

$$\mathbf{V}_3 \equiv \{\mathbf{v} \in \mathbf{V}_4 \colon v^4 = 0\}. \tag{1.1.12}$$

Small Greek indices will take values in the set $\{1, 2, 3\}$, and *small Roman indices* will take values in the set $\{1, 2, 3, 4\}$. The appropriate summation convention will apply to each type of index.

Theorem (1.1.3) (Schwarz Inequality): *For any two vectors* $\mathbf{u}$, $\mathbf{v}$ *in* $\mathbf{V}_3$ *the following inequality holds*:

$$|u^\alpha v^\alpha| \leq \sqrt{u^\alpha u^\alpha v^\beta v^\beta}. \tag{1.1.13}$$

Equality holds if and only if $u^\alpha = \lambda v^\alpha$ *for some* $\lambda \in \mathbb{R}$.

Proof: Suppose that $u^\alpha \equiv 0$. Then (1.1.13) holds trivially. Now suppose $u^\alpha u^\alpha > 0$. Then for any $\lambda \in \mathbb{R}$ we have

$$(\lambda u^\alpha + v^\alpha)(\lambda u^\alpha + v^\alpha) = \lambda^2 u^\alpha u^\alpha + v^\alpha v^\alpha + 2\lambda u^\alpha v^\alpha \geq 0.$$

Setting the value $\lambda = -(u^\alpha v^\alpha)/(u^\gamma u^\gamma)$, we obtain

$$[-(u^\alpha v^\alpha)^2 + (u^\alpha u^\alpha)(v^\beta v^\beta)]/(u^\gamma u^\gamma) \geq 0.$$

From above (1.1.13) follows. The case of equality mentioned in the theorem is left as an exercise. ■

The M-orthogonality between two vectors in $\mathbf{V}_4$ is not always intuitively natural. We shall derive a few theorems on that topic now.

Theorem (1.1.4): *No two timelike vectors in* $\mathbf{V}_4$ *can be M-orthogonal.*

Proof: Let **u**, **v** be two timelike vectors. Thus we have

$$\mathbf{u}\cdot\mathbf{u} = d_{ij}u^iu^j = u^\alpha u^\alpha - (u^4)^2 < 0,$$
$$\mathbf{v}\cdot\mathbf{v} = d_{ij}v^iv^j = v^\alpha v^\alpha - (v^4)^2 < 0.$$

Combining these two inequalities we have

$$\sqrt{u^\alpha u^\alpha v^\beta v^\beta} < |u^4v^4|.$$

By the Schwarz inequality (1.1.13) we obtain

$$|u^\alpha v^\alpha| < |u^4v^4|. \tag{1.1.14}$$

Suppose, contrary to the conclusion of the theorem, that $\mathbf{u}\cdot\mathbf{v} = 0$. Then $|u^\alpha v^\alpha| = |u^4v^4|$. This last equality contradicts (1.1.14). ■

Corollary (1.1.1): *For two timelike vectors* **u**, **v** *such that* $u^4 > 0$, $v^4 > 0$, *we have* $\mathbf{u}\cdot\mathbf{v} < 0$.

Proof: $u^\alpha v^\alpha \leq |u^\alpha v^\alpha| \leq \sqrt{u^\alpha u^\alpha v^\beta v^\beta} < |u^4v^4| = u^4v^4$, so $\mathbf{u}\cdot\mathbf{v} < 0$. ■

Theorem (1.1.5) (Synge): *Let* $\hat{\mathbf{t}}$, **t** *be two timelike, future-pointing unit vectors (past-pointing may replace future-pointing). Then* $-\infty < \hat{\mathbf{t}}\cdot\mathbf{t} \leq -1$.

Proof: For definiteness assume that $\hat{\mathbf{t}}$, **t** are future-pointing timelike vectors. Thus

$$\sigma(\mathbf{t}) = 1, \quad t^\alpha t^\alpha - (t^4)^2 = -1, \quad t^4 > 0,$$
$$\sigma(\hat{\mathbf{t}}) = 1, \quad \hat{t}^\beta\hat{t}^\beta - (\hat{t}^4)^2 = -1, \quad \hat{t}^4 > 0.$$

We want to solve the above equations and inequalities. For that purpose consider two three-dimensional unit spatial vectors:

$$a^\alpha a^\alpha = 1, \quad \hat{a}^\alpha\hat{a}^\alpha = 1.$$

In spherical polar coordinates we can write

$$a^1 = \sin\theta\cos\phi, \quad \hat{a}^1 = \sin\hat{\theta}\cos\hat{\phi},$$
$$a^2 = \sin\theta\sin\phi, \quad \hat{a}^2 = \sin\hat{\theta}\sin\hat{\phi},$$
$$a^3 = \cos\theta, \quad \hat{a}^3 = \cos\hat{\theta},$$

where $0 \leq \theta \leq \pi$, $0 \leq \hat{\theta} \leq \pi$, $-\pi \leq \phi < \pi$, and $-\pi \leq \hat{\phi} < \pi$. Thus we have $\cos\psi = a^\alpha\hat{a}^\alpha = \cos\theta\cos\hat{\theta} + \sin\theta\sin\hat{\theta}\cos(\phi - \hat{\phi})$, where ψ is the angle between the two vectors a^ρ, $\hat{a}^\rho$ and $0 \leq \psi \leq \pi$. We can always express **t** and $\hat{\mathbf{t}}$ in the form

$$t^\alpha = (\sinh\chi)a^\alpha, \quad \hat{t}^\alpha = (\sinh\hat{\chi})\hat{a}^\alpha,$$
$$t^4 = \cosh\chi > 0, \quad \hat{t}^4 = \cosh\hat{\chi} > 0,$$

where $\chi, \hat{\chi} \in \mathbb{R}$. Therefore,

$$\begin{aligned} -\mathbf{t}\cdot\hat{\mathbf{t}} &= \cosh\chi\cosh\hat{\chi}[\cos^2(\psi/2) + \sin^2(\psi/2)] \\ &\quad - \sinh\chi\sinh\hat{\chi}[\cos^2(\psi/2) - \sin^2(\psi/2)] \\ &= \cosh(\chi - \hat{\chi})\cos^2(\psi/2) + \cosh(\chi + \hat{\chi})\sin^2(\psi/2) > 0. \end{aligned}$$

Making another transformation $x = \sin(\psi/2)$ so that $x \in [0, 1]$, we get $y \equiv -\mathbf{t}\cdot\hat{\mathbf{t}} = [\cosh(\chi + \hat{\chi}) - \cosh(\chi - \hat{\chi})]x^2 + \cosh(\chi - \hat{\chi}) \equiv f(x)$. Thus we can graph the function $f(x)$ over $x \in [0, 1]$. For the case $\cosh(\chi + \hat{\chi}) - \cosh(\chi - \hat{\chi}) > 0$, this graph is a portion of a parabola with $f'(x) \geq 0$ and $\cosh(\chi - \hat{\chi}) \leq y \leq \cosh(\chi + \hat{\chi})$. For the case $\cosh(\chi + \hat{\chi}) - \cosh(\chi - \hat{\chi}) < 0$, the graph is a portion of a parabola with $f'(x) \leq 0$ and $\cosh(\chi + \hat{\chi}) \leq y \leq \cosh(\chi - \hat{\chi})$. In the case $\cosh(\chi + \hat{\chi}) - \cosh(\chi - \hat{\chi}) = 0$, the graph is a portion of a straight line with $f'(x) \equiv 0$ and $y = \cosh(\chi + \hat{\chi}) = \cosh(\chi - \hat{\chi})$. In all three cases $\min\{\cosh(\chi + \hat{\chi}), \cosh(\chi - \hat{\chi})\} \leq y \leq \max\{\cosh(\chi + \hat{\chi}), \cosh(\chi - \hat{\chi})\}$. Since the function $\cosh(x)$ has the lower bound 1 and no upper bound, we see that $1 \leq y < \infty$, so $-\infty < \mathbf{t}\cdot\hat{\mathbf{t}} \leq -1$. ■

Corollary (1.1.2): *Let* $\mathbf{t}, \hat{\mathbf{t}}$ *be two timelike, future-pointing unit vectors such that* $\mathbf{t}\cdot\hat{\mathbf{t}} = -1$. *Then* $\mathbf{t} = \hat{\mathbf{t}}$.

Proof: From the proof of Theorem (1.1.5), it is clear that, for the case $\cosh(\chi + \hat{\chi}) > \cosh(\chi - \hat{\chi})$, the minimum value $y = 1$ is at $x = 0$ and the minimum value is $\cosh(\chi - \hat{\chi}) = 1$. Thus $\psi = 0$ and $\chi = \hat{\chi}$, so $\theta = \hat{\theta}$, $\phi = \hat{\phi}$. Thus $t^\alpha = \hat{t}^\alpha$, $t^4 = \hat{t}^4$; hence, $\mathbf{t} = \hat{\mathbf{t}}$. In the second case $\cosh(\chi + \hat{\chi}) < \cosh(\chi - \hat{\chi})$, and the minimum value is $\cosh(\chi + \hat{\chi}) = 1$ at $x = 1$. Therefore, $\psi = \pi$ and $\chi = -\hat{\chi}$ and so $\hat{\theta} = \pi - \theta$, $\hat{\phi} = \phi + \pi$. Thus $t^\alpha = \hat{t}^\alpha$, $t^4 = \hat{t}^4$, so $\mathbf{t} = \hat{\mathbf{t}}$. In the case $\cosh(\chi + \hat{\chi}) = \cosh(\chi - \hat{\chi})$, the minimum value is 1, which is attained by all $x \in [0, 1]$. Then $\cosh(\chi + \hat{\chi}) = \cosh(\chi - \hat{\chi}) = 1$; hence, $\chi = \hat{\chi} = 0$. So $t^\alpha = \hat{t}^\alpha = 0$, $t^4 = \hat{t}^4$, so $\mathbf{t} = \hat{\mathbf{t}}$. ■

Theorem (1.1.6): *A timelike vector cannot be M-orthogonal to a nonzero null vector.*

Proof: Suppose that $\mathbf{t}$ is a timelike vector and $\mathbf{n}$ is a nonzero null vector. In terms of their Minkowski components $t^\alpha t^\alpha < (t^4)^2$ and $n^\alpha n^\alpha = (n^4)^2$ with $|t^4| > 0$, $|n^4| > 0$. Combining these expressions and using the Schwarz inequality (1.1.13) we have

$$(t^\alpha n^\alpha)^2 \leq t^\alpha t^\alpha n^\beta n^\beta < (t^4 n^4)^2. \tag{1.1.15}$$

Suppose that $\mathbf{t}\cdot\mathbf{n} = 0$, or $(t^\alpha n^\alpha)^2 = (t^4 n^4)^2$. This contradicts (1.1.15). ■

Now we shall prove a *very counterintuitive theorem.*

Theorem (1.1.7): *Two nonzero null vectors are M-orthogonal if and only if they are scalar multiples of each other.*

Proof: (i) Assume that two null vectors $\mathbf{m}$, $\mathbf{n}$ are such that $\mathbf{m} = \lambda\mathbf{n}$ for some $\lambda \in \mathbb{R}$. Then $\mathbf{m}\cdot\mathbf{n} = \lambda(\mathbf{n}\cdot\mathbf{n}) = 0$.

(ii) Suppose that two nonzero null vectors $\mathbf{m}$, $\mathbf{n}$ are M-orthogonal. Then

$$\begin{aligned} \mathbf{m}\cdot\mathbf{m} &= m^\alpha m^\alpha - (m^4)^2 = 0, \\ \mathbf{n}\cdot\mathbf{n} &= n^\alpha n^\alpha - (n^4)^2 = 0, \\ \mathbf{m}\cdot\mathbf{n} &= m^\alpha n^\alpha - m^4 n^4 = 0, \qquad m^4 \neq 0, n^4 \neq 0. \end{aligned} \tag{1.1.16}$$

From the above expressions we obtain

$$(m^\alpha n^\alpha)^2 = (m^4 n^4)^2 = m^\alpha m^\alpha n^\beta n^\beta, \qquad |m^\alpha n^\alpha| = \sqrt{m^\alpha m^\alpha n^\beta n^\beta}. \tag{1.1.17}$$

The above equation is the case of equality in the Schwartz inequality (1.1.13). Therefore, $m^\alpha = \lambda n^\alpha$ for some scalar $\lambda \neq 0$. Since $n^4 \neq 0$, we have $m^4 = m^\alpha n^\alpha / n^4 = \lambda n^\alpha n^\alpha / n^4 = \lambda n^4$. Thus $\mathbf{m} = \lambda\mathbf{n}$. ■

It is hard to plot the Minkowski vectors, since the concepts of length of a vector and angle between two vectors do not exist. However, the *parallelogram law* of vector addition still holds. It is worthwhile to draw Minkowski vectors to gain some geometrical insight. We have to plot these vectors on a piece of paper, which is part of a Euclidean plane. Let us plot M-orthonormal vectors $\mathbf{e}_1, \mathbf{e}_4$ such that $\mathbf{e}_1\cdot\mathbf{e}_4 = 0$ and $\mathbf{e}_1\cdot\mathbf{e}_1 = -\mathbf{e}_4\cdot\mathbf{e}_4 = 1$. It is quite natural to plot these two vectors as $\mathbf{i}$ and $\mathbf{j}$ of the usual two-dimensional Cartesian basis vectors; see Figure 1. As we have drawn, the Euclidean lengths $\|\mathbf{e}_1\| = \|\mathbf{e}_4\| = 1$ and $\mathbf{e}_1, \mathbf{e}_4$ are Euclidean orthogonal. However, the vectors $\mathbf{e}_1 + \mathbf{e}_4$, $-\mathbf{e}_1 + \mathbf{e}_4$ have Euclidean lengths $\|\mathbf{e}_1 + \mathbf{e}_4\| = \|-\mathbf{e}_1 + \mathbf{e}_4\| = \sqrt{2}$, and $\mathbf{e}_1 + \mathbf{e}_4$, $-\mathbf{e}_1 + \mathbf{e}_4$ are Euclidean orthogonal. But $\sigma(\mathbf{e}_1 + \mathbf{e}_4) = 0$, $\sigma(-\mathbf{e}_1 + \mathbf{e}_4) = 0$, and $(\mathbf{e}_1 + \mathbf{e}_4)\cdot(-\mathbf{e}_1 + \mathbf{e}_4) = -2 \neq 0$. So we have to *use caution* in order to interpret any plot of Minkowski vectors.

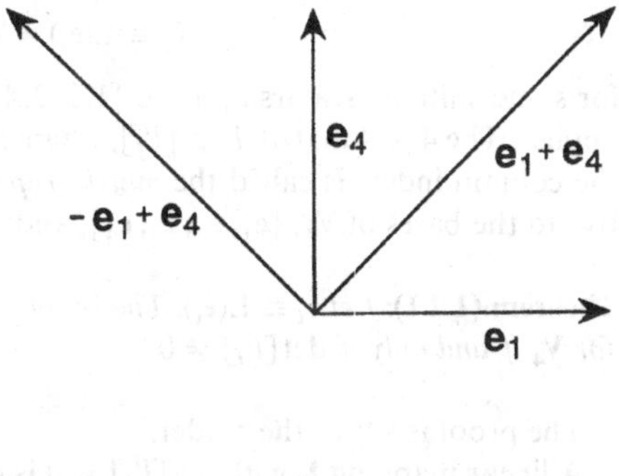

FIGURE 1. Minkowski vectors $\mathbf{e}_1$, $\mathbf{e}_4$, $\mathbf{e}_1 + \mathbf{e}_4$, $-\mathbf{e}_1 + \mathbf{e}_4$.

EXERCISES 1.1

1. Let $\{\mathbf{e}_1, \mathbf{e}_2, \mathbf{e}_3, \mathbf{e}_4\}$ be a Minkowski basis for $\mathbf{V}_4$. Let another basis be $\{\mathbf{a}, \mathbf{b}, \mathbf{c}, \mathbf{d}\}$ where

$$\mathbf{a} = \mathbf{e}_1 + \mathbf{e}_2 + \mathbf{e}_3 - \mathbf{e}_4, \qquad \mathbf{b} = \mathbf{e}_2 - \mathbf{e}_3 + \sqrt{2}\mathbf{e}_4,$$
$$\mathbf{c} = \mathbf{e}_3 - \mathbf{e}_4, \qquad \mathbf{d} = \mathbf{e}_4.$$

(i) Determine which of these are spacelike, timelike, or null vectors.
(ii) Obtain the separation numbers $\sigma(\mathbf{a})$, $\sigma(\mathbf{b})$, $\sigma(\mathbf{c})$, $\sigma(\mathbf{d})$.
(iii) Determine whether or not $\{\mathbf{a}, \mathbf{b}, \mathbf{c}, \mathbf{d}\}$ is a basis for $\mathbf{V}_4$.

2. Determine which of the following subsets of $\mathbf{V}_4$ is a vector subspace.
(i) The union of the subset of all spacelike vectors and $\{\mathbf{0}\}$.
(ii) The union of the subset of all timelike vectors and $\{\mathbf{0}\}$.
(iii) The subset of all null vectors.

3. Prove that the nondegeneracy axiom of the inner product in (1.1.2) implies that $\det[g_{ij}] \neq 0$.

4. Prove that for any two vectors $\mathbf{x}$, $\mathbf{y}$ in $\mathbf{V}_3$ that

$$|x^\alpha y^\alpha| = \sqrt{x^\alpha x^\alpha y^\beta y^\beta}$$

holds if and only if $\mathbf{x} = \lambda \mathbf{y}$ for some $\lambda \in \mathbb{R}$.

5. Let $\mathbf{s}$, $\mathbf{t}$ be two timelike future-pointing vectors in $\mathbf{V}_4$. Prove that $\sigma(\mathbf{s})\sigma(\mathbf{t}) \leq |\mathbf{s} \cdot \mathbf{t}|$. (This is called the *Reversed Schwarz Inequality* for timelike future-pointing vectors.) Does it hold for other cases, i.e., past-pointing or mixed orientation?

1.2. Lorentz Mappings of $\mathbf{V}_4$

A *linear mapping* $\mathbf{L}: \mathbf{V}_4 \to \mathbf{V}_4$ is defined to be such that

$$\mathbf{L}(\lambda \mathbf{a} + \mu \mathbf{b}) = \lambda \mathbf{L}(\mathbf{a}) + \mu \mathbf{L}(\mathbf{b}) \tag{1.2.1}$$

for all λ, $\mu \in \mathbb{R}$ and all $\mathbf{a}, \mathbf{b} \in \mathbf{V}_4$.

Suppose that $\{\mathbf{e}_1, \mathbf{e}_2, \mathbf{e}_3, \mathbf{e}_4\}$ is a basis for $\mathbf{V}_4$ that is not necessarily M-orthonormal. Since $\mathbf{L}(\mathbf{e}_1) \in \mathbf{V}_4$, we must have

$$\hat{\mathbf{e}}_j \equiv \mathbf{L}(\mathbf{e}_j) = l^i{}_j \mathbf{e}_i, \tag{1.2.2}$$

for some suitable scalars $l^i{}_j$, i, $j \in \{1, 2, 3, 4\}$, and the summation convention applies. The 4×4 matrix $L \equiv [l^i{}_j]$, where i stands for the row index and j for the column index, is called the *matrix representation* of the mapping $\mathbf{L}$ relative to the bases of $\mathbf{V}_4$, $\{\mathbf{e}_1, \mathbf{e}_2, \mathbf{e}_3, \mathbf{e}_4\}$, and $\{\hat{\mathbf{e}}_1, \hat{\mathbf{e}}_2, \hat{\mathbf{e}}_3, \hat{\mathbf{e}}_4\}$.

Theorem (1.2.1): *Let* $\hat{\mathbf{e}}_i \equiv \mathbf{L}(\mathbf{e}_i)$. *The set of vectors* $\{\hat{\mathbf{e}}_1, \hat{\mathbf{e}}_2, \hat{\mathbf{e}}_3, \hat{\mathbf{e}}_4\}$ *is also a basis for* $\mathbf{V}_4$ *if and only if* $\det[l^i{}_j] \neq 0$.

The proof is left to the reader.

A linear mapping $\mathbf{L}$ with $\det[l^i{}_j] \neq 0$ is called *invertible*.

Example: Let us consider a linear mapping $\mathbf{L}$ such that

$$\hat{\mathbf{e}}_1 = \mathbf{L}(\mathbf{e}_1) \equiv \sin(\pi/4)\mathbf{e}_1 + \cos(\pi/4)\mathbf{e}_2 = l^i{}_1 \mathbf{e}_i,$$

$$\hat{\mathbf{e}}_2 = \mathbf{L}(\mathbf{e}_2) \equiv -\cos(\pi/4)\mathbf{e}_1 + \sin(\pi/4)\mathbf{e}_2 = l^i{}_2 \mathbf{e}_i,$$

$$\hat{\mathbf{e}}_3 = \mathbf{L}(\mathbf{e}_3) \equiv \mathbf{e}_3 = l^i{}_3\mathbf{e}_i,$$
$$\hat{\mathbf{e}}_4 = \mathbf{L}(\mathbf{e}_4) \equiv \mathbf{e}_4 = l^i{}_4\mathbf{e}_i.$$

Therefore, the 4×4 matrix representation is

$$L = [l^i{}_j] = \begin{bmatrix} \sin(\pi/4) & -\cos(\pi/4) & 0 & 0 \\ \cos(\pi/4) & \sin(\pi/4) & 0 & 0 \\ 0 & 0 & 1 & 0 \\ 0 & 0 & 0 & 1 \end{bmatrix},$$

with $\det[l^i{}_j] = 1$. The linear mapping $\mathbf{L}$ is invertible. The basis $\{\hat{\mathbf{e}}_1, \hat{\mathbf{e}}_2, \hat{\mathbf{e}}_3, \hat{\mathbf{e}}_4\}$ is M-orthonormal whenever $\{\mathbf{e}_1, \mathbf{e}_2, \mathbf{e}_3, \mathbf{e}_4\}$ is M-orthonormal. □

Now we shall define the *Kronecker delta:*

$$\delta^i{}_j \equiv \begin{cases} 1 & \text{for } i = j, \\ 0 & \text{for } i \neq j. \end{cases} \tag{1.2.3}$$

These numbers are the entries of the 4×4 identity matrix I relative to the standard basis $\{\mathbf{e}_1, \mathbf{e}_2, \mathbf{e}_3, \mathbf{e}_4\}$. Similarly $\delta^{\alpha\beta}$ are the entries of the 3×3 identity matrix with respect to the standard basis $\{\mathbf{i}, \mathbf{j}, \mathbf{k}\}$. We shall work out some examples involving the Kronecker delta.

Example: Consider the sum

$$\begin{aligned} \delta^1{}_j u^j &= \delta^1{}_1 u^1 + \delta^1{}_2 u^2 + \delta^1{}_3 u^3 + \delta^1{}_4 u^4 \\ &= 1u^1 + 0u^2 + 0u^3 + 0u^4 = u^1. \end{aligned}$$

Similarly

$$\delta^i{}_j u^j = u^i, \tag{1.2.4}$$
$$\delta^i{}_j \delta^j{}_k = \delta^i{}_k. \quad □ \tag{1.2.5}$$

For an invertible linear mapping $\mathbf{L}$, we denote the inverse mapping by $\mathbf{A}$, so that the corresponding matrices satisfy

$$A = L^{-1}, \qquad AL = LA = I, \qquad a^i{}_j l^j{}_k = l^i{}_j a^j{}_k = \delta^i{}_k. \tag{1.2.6}$$

Example:

$$[l^i{}_j] = \begin{bmatrix} \sin(\pi/4) & \cos(\pi/4) & 0 & 0 \\ -\cos(\pi/4) & \sin(\pi/4) & 0 & 0 \\ 0 & 0 & 1 & 0 \\ 0 & 0 & 0 & 1 \end{bmatrix}$$

$$[a^i{}_j] = \begin{bmatrix} \sin(\pi/4) & -\cos(\pi/4) & 0 & 0 \\ \cos(\pi/4) & \sin(\pi/4) & 0 & 0 \\ 0 & 0 & 1 & 0 \\ 0 & 0 & 0 & 1 \end{bmatrix}. \quad □$$

Now we shall discuss the transformation properties of vector components under a change of basis sets.

Theorem (1.2.2): *Suppose that* $\{\mathbf{e}_1, \mathbf{e}_2, \mathbf{e}_3, \mathbf{e}_4\}$, $\{\hat{\mathbf{e}}_1, \hat{\mathbf{e}}_2, \hat{\mathbf{e}}_3, \hat{\mathbf{e}}_4\}$ *are two bases for* $\mathbf{V}_4$ *such that* $\hat{\mathbf{e}}_j = l^i{}_j \mathbf{e}_i$. *Then the corresponding components of a Minkowski vector* $\mathbf{u}$ *undergo the transformation*

$$\hat{u}^i = a^i{}_k u^k. \tag{1.2.7}$$

Proof: It is easy to see $\mathbf{u} = u^k \mathbf{e}_k = \hat{u}^j \hat{\mathbf{e}}_j = \hat{u}^j (l^k{}_j \mathbf{e}_k) = (l^k{}_j \hat{u}^j)\mathbf{e}_k$, so $(u^k - l^k{}_j \hat{u}^j)\mathbf{e}_k = \mathbf{0}$. The linear independence of the basis vectors $\mathbf{e}_k$ implies $u^k = l^k{}_j \hat{u}^j$, so $a^i{}_k u^k = a^i{}_k l^k{}_j \hat{u}^j = \delta^i{}_j \hat{u}^j = \hat{u}^i$. ■

Suppose that a linear mapping $\mathbf{L}: \mathbf{V}_4 \to \mathbf{V}_4$ satisfies $\mathbf{L}(\mathbf{a}) \cdot \mathbf{L}(\mathbf{b}) = \mathbf{a} \cdot \mathbf{b}$ for every pair $\mathbf{a}$, $\mathbf{b}$. Such a mapping is called a *Lorentz mapping*. A Lorentz mapping preserves inner products of pairs of vectors in $\mathbf{V}_4$. The corresponding matrix representation $L \equiv [l^i{}_j]$, relative to an M-orthonormal basis, is called a *Lorentz matrix*.

Theorem (1.2.3): *A Lorentz matrix L must satisfy the matrix equation*

$$L^{\mathsf{T}} D L = D, \tag{1.2.8}$$

where L^{T} *denotes the transpose for L and D is the Lorentz metric defined in* (1.1.7).

Proof: From the definition of a Lorentz mapping it follows that

$$\mathbf{L}(\mathbf{e}_i) \cdot \mathbf{L}(\mathbf{e}_j) = \mathbf{e}_i \cdot \mathbf{e}_j = d_{ij}, \tag{1.2.9}$$

where $\{\mathbf{e}_1, \mathbf{e}_2, \mathbf{e}_3, \mathbf{e}_4\}$ is an M-orthonormal basis. Since $\mathbf{L}(\mathbf{e}_i) = l^k{}_i \mathbf{e}_k$, Equation (1.2.9) yields $(l^k{}_i \mathbf{e}_k) \cdot (l^m{}_j \mathbf{e}_m) = d_{ij}$, so

$$l^k{}_i l^m{}_j (\mathbf{e}_k \cdot \mathbf{e}_m) = l^k{}_i l^m{}_j d_{km} = l^k{}_i d_{km} l^m{}_j = d_{ij}. \tag{1.2.10}$$

Thus $L^{\mathsf{T}} D L = D$. ■

Corollary (1.2.1): *The determinant of a Lorentz matrix L must satisfy*

$$\det L = \det[l^i{}_j] = \pm 1. \tag{1.2.11}$$

Proof: Taking the determinant of both sides of (1.2.8) we have

$$\det L^{\mathsf{T}} \cdot \det D \cdot \det L = \det D. \tag{1.2.12}$$

Recalling $\det L^{\mathsf{T}} = \det L$ and $\det D = -1$ we have from (1.2.12) that

$$(\det L)^2 = 1. \tag{1.2.13}$$

Thus equation (1.2.11) follows. ■

A Lorentz mapping $\mathbf{L}$ with $\det L = 1$ is called a *proper Lorentz mapping.* A Lorentz mapping with $\det L = -1$ is called an *improper Lorentz mapping.* A Lorentz mapping $\mathbf{L}$ such that $l^4{}_4 > 0$ when represented in an M-orthonormal basis is called an *orthochronous* Lorentz mapping. Now we shall give some elementary examples of Lorentz mappings. In all cases we use an M-orthonormal basis for $\mathbf{V}_4$.

Examples: (i) The *identity mapping* given by $\mathbf{I}(\mathbf{v}) = \mathbf{v}$ for all $\mathbf{v} \in \mathbf{V}_4$. The corresponding matrix $[\delta^i{}_j]$ shows that $\mathbf{I}$ is a proper, orthochronous Lorentz mapping.

(ii) The *space reflection* $\mathbf{P}$ is given by the mapping $\mathbf{P}(\mathbf{e}_\alpha) = -\mathbf{e}_\alpha$ and $\mathbf{P}(\mathbf{e}_4) = \mathbf{e}_4$. The matrix of $\mathbf{P}$ is

$$P = [p^i{}_j] = \begin{bmatrix} -1 & 0 & 0 & 0 \\ 0 & -1 & 0 & 0 \\ 0 & 0 & -1 & 0 \\ 0 & 0 & 0 & 1 \end{bmatrix}. \tag{1.2.14}$$

$\mathbf{P}$ is an improper orthochronous Lorentz mapping.

(iii) The time reversal $\mathbf{T}$ is given by $\mathbf{T}(\mathbf{e}_\alpha) = \mathbf{e}_\alpha$ and $\mathbf{T}(\mathbf{e}_4) = -\mathbf{e}_4$. The corresponding matrix is

$$T = [t^i{}_j] = \begin{bmatrix} 1 & 0 & 0 & 0 \\ 0 & 1 & 0 & 0 \\ 0 & 0 & 1 & 0 \\ 0 & 0 & 0 & -1 \end{bmatrix}. \tag{1.2.15}$$

$\mathbf{T}$ is an improper Lorentz mapping that is not orthochronous.

(iv) A plane rotation is given by the mapping $\mathbf{L}$ such that

$$\begin{aligned} &\mathbf{L}(\mathbf{e}_1) = \cos\theta\, \mathbf{e}_1 + \sin\theta\, \mathbf{e}_2 = \hat{\mathbf{e}}_1, \\ &\mathbf{L}(\mathbf{e}_2) = -\sin\theta\, \mathbf{e}_1 + \cos\theta\, \mathbf{e}_2 = \hat{\mathbf{e}}_2, \\ &\mathbf{L}(\mathbf{e}_3) = \mathbf{e}_3 = \hat{\mathbf{e}}_3, \qquad \mathbf{L}(\mathbf{e}_4) = \mathbf{e}_4 = \hat{\mathbf{e}}_4, \qquad \text{where } -\pi < \theta < \pi. \end{aligned} \tag{1.2.16}$$

The matrix of $\mathbf{L}$ is

$$L = [l^i{}_j] = \begin{bmatrix} \cos\theta & -\sin\theta & 0 & 0 \\ \sin\theta & \cos\theta & 0 & 0 \\ 0 & 0 & 1 & 0 \\ 0 & 0 & 0 & 1 \end{bmatrix}.$$

The Lorentz mapping $\mathbf{L}$ is proper and orthochronous. The plane rotation is shown in Figure 2.

(v) The *boost mapping* is a very important Lorentz mapping and is defined

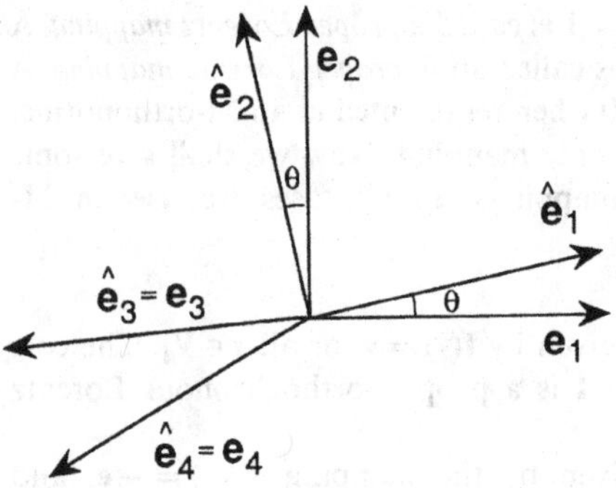

FIGURE 2. Plane rotation by the angle θ.

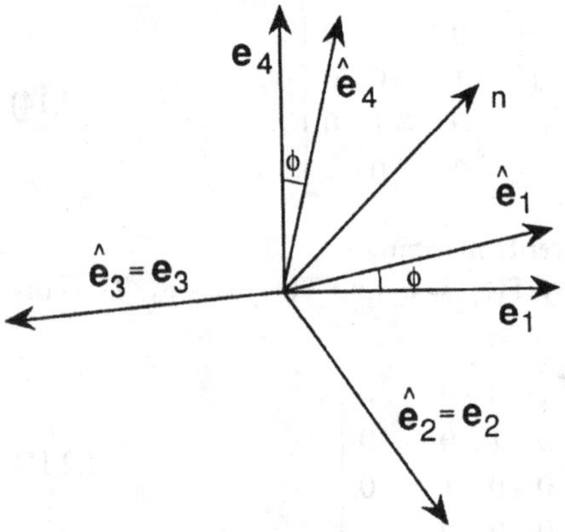

FIGURE 3. Boost mapping associated with a moving observer.

by

$$
\begin{aligned}
\mathbf{L}(\mathbf{e}_1) &= \beta(\mathbf{e}_1 + v\mathbf{e}_4) = \hat{\mathbf{e}}_1, \\
\mathbf{L}(\mathbf{e}_2) &= \mathbf{e}_2 = \hat{\mathbf{e}}_2, \\
\mathbf{L}(\mathbf{e}_3) &= \mathbf{e}_3 = \hat{\mathbf{e}}_3, \\
\mathbf{L}(\mathbf{e}_4) &= \beta(v\mathbf{e}_1 + \mathbf{e}_4) = \hat{\mathbf{e}}_4, \qquad |v| < 1, \ \beta \equiv (1 - v^2)^{-1/2}.
\end{aligned}
\tag{1.2.17}
$$

The matrix of the boost $\mathbf{L}$ is

$$
L = [l^i{}_j] = \begin{bmatrix} \beta & 0 & 0 & v\beta \\ 0 & 1 & 0 & 0 \\ 0 & 0 & 1 & 0 \\ v\beta & 0 & 0 & \beta \end{bmatrix}.
$$

The boost $\mathbf{L}$ is proper and orthochronous. This mapping is physically associated with a moving observer (see Chapter III). The transformation (1.2.17) is shown in Figure 3.

In Fig. 3, $\tan\phi \equiv v$, $|\phi| < \pi/4$. The null vector $\mathbf{n} \equiv (\mathbf{e}_1 + \mathbf{e}_4)/\sqrt{2}$ is symmetrically situated among the basis vectors $\mathbf{e}_1, \mathbf{e}_4$, and $\hat{\mathbf{e}}_1, \hat{\mathbf{e}}_4$. As the parameter v tends to 1, both the vectors $\hat{\mathbf{e}}_1$, $\hat{\mathbf{e}}_4$ tend to collapse into $\mathbf{n}$. The range of v is restricted to $|v| < 1$ in (1.2.17) to avoid this situation. □

EXERCISES 1.2

1. Show explicitly that the mapping $\mathbf{L}$ in equation (1.2.17) has a matrix that satisfies $L^{\mathrm{T}}DL = D$, $\det L = 1$, and $l^4{}_4 > 0$.

2. Let $\mathbf{L}$ be a Lorentz mapping. Prove that $\mathbf{L}^{-1}$ and $\mathbf{L}^{\mathrm{T}}$ are also Lorentz mappings by examination of their matrices.

3. Consider the mapping $\mathbf{L}$ whose matrix is

$$L = [l^i{}_j] = \begin{bmatrix} 0 & -1 & -1 & 1 \\ -1 & 0 & -1 & 1 \\ -1 & -1 & 0 & 1 \\ -1 & -1 & -1 & 2 \end{bmatrix}.$$

(i) Show that $\mathbf{L}$ is an improper Lorentz mapping.
(ii) Show that the Lorentz mapping takes the lattice vectors $n^i\mathbf{e}_i$ into the lattice vectors $\hat{n}^j\hat{\mathbf{e}}_j$ where both $n^i, \hat{n}^i \in \mathbb{Z}$.

4. Prove that, for the entry $l^4{}_4$ in the matrix L of a Lorentz mapping $\mathbf{L}$, the inequality $1 \le |l^4{}_4|$ holds.

1.3. The Minkowski Tensors

In case a Lorentz mapping $\mathbf{L}$ is given by $\hat{\mathbf{e}}_j = \mathbf{L}(\mathbf{e}_j) = l^k{}_j\mathbf{e}_k$, the components of a Minkowski vector $\mathbf{u}$ (a *contravariant vector*) change via the transformation rules

$$\hat{u}^j = a^j{}_i u^i, \tag{1.3.1}$$

$$a^k{}_j l^j{}_i = l^k{}_j a^j{}_i = \delta^k{}_i. \tag{1.3.2}$$

A *Minkowski covector* $\tilde{\mathbf{w}}$ (a *covariant vector*) is defined to be a linear mapping from $\mathbf{V}_4$ into $\mathbb{R}$. The components of $\tilde{\mathbf{w}}$ are defined by $w_i \equiv \tilde{\mathbf{w}}(\mathbf{e}_i)$.

Theorem (1.3.1): *The components w_i of a covector $\tilde{\mathbf{w}}$, under a Lorentz mapping $\hat{\mathbf{e}}_j = \mathbf{L}(\mathbf{e}_j)$, transform as*

$$\hat{w}_j = l^k{}_j w_k. \tag{1.3.3}$$

Proof: $\hat{w}_j = \tilde{\mathbf{w}}(\hat{\mathbf{e}}_j) = \tilde{\mathbf{w}}(l^k{}_j\mathbf{e}_k) = l^k{}_j\tilde{\mathbf{w}}(\mathbf{e}_k) = l^k{}_j w_k$. ■

Corollary (1.3.1): *Let u^i, w_i be components of a vector $\mathbf{u}$ and a covector $\tilde{\mathbf{w}}$ respectively. The sum $u^i w_i$ remains invariant under a Lorentz mapping.*

Proof: $\hat{u}^i\hat{w}_i = (a^i{}_k u^k)(l^m{}_i w_m) = (l^m{}_i a^i{}_k)u^k w_m = \delta^m{}_k u^k w_m = u^k w_k$. ■

The number $u^i w_i$, which is invariant under Lorentz mappings, is called a *Lorentz* (or *Minkowski*) *scalar*. We shall often simply call them scalars.

Consider a mapping $\mathbf{T}: \mathbf{V}_4 \times \mathbf{V}_4 \to \mathbb{R}$. Let $\mathbf{T}$ be a *bilinear mapping* so that we have

$$\mathbf{T}(\lambda\mathbf{a} + \mu\mathbf{b}, \mathbf{c}) = \lambda\mathbf{T}(\mathbf{a}, \mathbf{c}) + \mu\mathbf{T}(\mathbf{b}, \mathbf{c}),$$
$$\mathbf{T}(\mathbf{a}, \lambda\mathbf{b} + \mu\mathbf{c}) = \lambda\mathbf{T}(\mathbf{a}, \mathbf{b}) + \mu\mathbf{T}(\mathbf{a}, \mathbf{c}),$$

for all scalars λ, μ and all vectors $\mathbf{a}$, $\mathbf{b}$, $\mathbf{c}$. We call $\mathbf{T}$ a *second-order covariant Minkowski tensor* and define its components relative to the basis $\{\mathbf{e}_1, \mathbf{e}_2, \mathbf{e}_3, \mathbf{e}_4\}$ as $\tau_{ij} \equiv \mathbf{T}(\mathbf{e}_i, \mathbf{e}_j)$. The components τ_{ij} undergo the following transformation under a change of basis:

$$\begin{aligned}\hat{\tau}_{ij} &= \mathbf{T}(\hat{\mathbf{e}}_i, \hat{\mathbf{e}}_j) = \mathbf{T}(l^m{}_i \mathbf{e}_m, l^k{}_j \mathbf{e}_k) = l^m{}_i l^k{}_j \mathbf{T}(\mathbf{e}_m, \mathbf{e}_k)\\ &= l^m{}_i l^k{}_j \tau_{mk}.\end{aligned} \tag{1.3.4a}$$

The components τ^{ij} of a second-order *contravariant Minkowski tensor* are assumed to undergo the following transformation:

$$\hat{\tau}^{ij} = a^i{}_m a^j{}_k \tau^{mk}. \tag{1.3.4b}$$

Furthermore, the components $\tau^i{}_j$ of a second-order *mixed Minkowski tensor* are assumed to transform as

$$\hat{\tau}^i{}_j = a^i{}_m l^k{}_j \tau^m{}_k. \tag{1.3.4c}$$

An *antisymmetric* second-order covariant tensor is defined as a tensor α whose components satisfy

$$\alpha_{ij} = -\alpha_{ji}. \tag{1.3.5}$$

A *symmetric* second-order covariant tensor is defined as a tensor σ whose components satisfy

$$\sigma_{ij} = \sigma_{ji}. \tag{1.3.6}$$

The number of linearly independent components of an antisymmetric Minkowski tensor α_{ij} is six. Since $\alpha_{ij} = -\alpha_{ji}$, we have the following relationships on the components of α_{ij}:

$$\begin{aligned}&\alpha_{11} = -\alpha_{11} = 0, \qquad \alpha_{22} = -\alpha_{22} = 0,\\ &\alpha_{33} = -\alpha_{33} = 0, \qquad \alpha_{44} = -\alpha_{44} = 0,\\ &\alpha_{ij} = -\alpha_{ji} \quad \text{for } i \neq j.\end{aligned} \tag{1.3.7}$$

Thus, there are six independent nondiagonal components of α_{ij}. Similarly, for a symmetric tensor $\sigma_{ij} = \sigma_{ji}$, there are four independent diagonal and six independent nondiagonal components. Thus there are ten independent components of a second-order symmetric Minkowski tensor.

Example: Suppose that F_{ij} is an antisymmetric second-order Minkowski tensor such that $F_{12} = F_{23} = F_{31} = F_{14} = 0$. Under the Lorentz mapping given in (1.2.17), the transformed components are

$$\begin{aligned}
\hat{F}_{14} &= l^m{}_1 l^k{}_4 F_{mk} = l^1{}_1 l^4{}_4 F_{14} + l^4{}_1 l^1{}_4 F_{41} = 0,\\
\hat{F}_{24} &= l^m{}_2 l^k{}_4 F_{mk} = l^2{}_2 l^4{}_4 F_{24} + l^2{}_2 l^1{}_4 F_{21} = \beta F_{24},\\
\hat{F}_{34} &= l^m{}_3 l^k{}_4 F_{mk} = l^3{}_3 l^4{}_4 F_{34} = \beta F_{34}\\
\hat{F}_{23} &= l^m{}_2 l^k{}_3 F_{mk} = l^2{}_2 l^3{}_3 F_{23} = 0,\\
\hat{F}_{31} &= l^m{}_3 l^k{}_1 F_{mk} = l^3{}_3 l^4{}_1 F_{34} = v\beta F_{34},\\
\hat{F}_{12} &= l^m{}_1 l^k{}_2 F_{mk} = l^4{}_1 l^2{}_2 F_{42} = -v\beta F_{24},
\end{aligned}$$

where β is defined as in (1.2.17). This example is relevant in the electromagnetic theory. □

We can generalize the transformation properties (1.3.4*a*), (1.3.4*b*), (1.3.4*c*) to higher-order tensors. The 4^{r+s} components of an $(r + s)$*-order* (or *rank*) (r indices contravariant, s indices covariant) *Minkowski tensor* is assumed to transform as:

$$\hat{\tau}^{i_1 \cdots i_r}_{j_1 \cdots j_s} = a^{i_1}{}_{k_1} \cdots a^{i_r}{}_{k_r} l^{m_1}{}_{j_1} \cdots l^{m_s}{}_{j_s} \tau^{k_1 \cdots k_r}_{m_1 \cdots m_s}. \tag{1.3.8}$$

Note that, for multi-index tensors ($r + s \geq 3$), we shall write the contravariant indices above the covariant indices rather than offsetting them as we do for second-order tensors.

Example: Suppose that the components of a $(r + s)$-order Minkowski tensor $\tau^{i_1 \cdots i_r}_{j_1 \cdots j_s}$ have the following values:

$$\tau^{4 \cdots 4}_{4 \cdots 4} = -1, \qquad \tau^{i_1 \cdots i_r}_{j_1 \cdots j_s} = 0 \quad \text{otherwise.}$$

We want to compute $\hat{\tau}^{1 \cdots 1}_{1 \cdots 1}$ under the Lorentz mapping (1.2.17), so

$$\begin{aligned}
\hat{\tau}^{1 \cdots 1}_{1 \cdots 1} &= a^1_{k_1} \cdots a^1_{k_r} l^{m_1}{}_1 \cdots l^{m_s}{}_1 \tau^{k_1 \cdots k_r}_{m_1 \cdots m_s}\\
&= a^1{}_4 \cdots a^1{}_4 l^4{}_1 \cdots l^4{}_1 \tau^{4 \cdots 4}_{4 \cdots 4}\\
&= (-1)^{r+1} (\beta v)^{r+s}. \quad \square
\end{aligned}$$

In case some Minkowski tensors are given, there are ways to produce *new* tensors out of them. The following three theorems deal with the construction of these new tensors.

Theorem (1.3.2): *Given two $(r + s)$-order Minkowski tensors with components $\sigma^{i_1 \cdots i_r}_{j_1 \cdots j_s}$, $\tau^{i_1 \cdots i_r}_{j_1 \cdots j_s}$ the linear combination*

$\lambda \sigma^{i_1 \cdots i_r}_{j_1 \cdots j_s} + \mu \tau^{i_1 \cdots i_r}_{j_1 \cdots j_s}$ yields the components of another $(r + s)$-order Minkowski tensor.

Proof: Scalars transform as $\hat{\lambda} = \lambda$, $\hat{\mu} = \mu$. Thus by (1.3.8) we have

$$\hat{\lambda}\hat{\sigma}^{i_1\cdots i_r}_{j_1\cdots j_s} + \hat{\mu}\hat{\tau}^{i_1\cdots i_r}_{j_1\cdots j_s}$$
$$= \lambda a^{i_1}{}_{k_1}\cdots a^{i_r}{}_{k_r} l^{m_1}{}_{j_1}\cdots l^{m_s}{}_{j_s}\sigma^{k_1\cdots k_r}_{m_1\cdots m_s} + \mu a^{i_1}{}_{k_1}\cdots a^{i_r}{}_{k_r} l^{m_1}{}_{j_1}\cdots l^{m_s}{}_{j_s}\tau^{k_1\cdots k_r}_{m_1\cdots m_s}$$
$$= a^{i_1}{}_{k_1}\cdots a^{i_r}{}_{k_r} l^{m_1}{}_{j_1}\cdots l^{m_s}{}_{j_s}(\lambda\sigma^{k_1\cdots k_r}_{m_1\cdots m_s} + \mu\tau^{k_1\cdots k_r}_{m_1\cdots m_s}).$$

Comparing the above with (1.3.8), the theorem is proved. ■

Theorem (1.3.3): *Let the components of two Minkowski tensors be given by* $\sigma^{i_1\cdots i_r}_{j_1\cdots j_s}$ *and* $\tau^{k_1\cdots k_p}_{m_1\cdots m_q}$. *The tensor (or outer) product of the two tensors is given by the components* $\sigma^{i_1\cdots i_r}_{j_1\cdots j_s}\tau^{k_1\cdots k_p}_{m_1\cdots m_q}$ *and these form the components of an* $(r + p) + (s + q)$*-order Minkowski tensor.*

Proof: Using (1.3.8) we see that

$$\hat{\sigma}^{i_1\cdots i_r}_{j_1\cdots j_s}\hat{\tau}^{k_1\cdots k_p}_{m_1\cdots m_q} = (a^{i_1}{}_{c_1}\cdots a^{i_r}{}_{c_r} l^{h_1}{}_{j_1}\cdots l^{h_s}{}_{j_s}\sigma^{c_1\cdots c_r}_{h_1\cdots h_s})$$
$$\times (a^{k_1}{}_{d_1}\cdots a^{k_p}{}_{d_p} l^{t_1}{}_{m_1}\cdots l^{t_q}{}_{m_q}\tau^{d_1\cdots d_p}_{t_1\cdots t_q}),$$

so after a regrouping of terms the result follows. ■

Suppose that the components of a $(r + s)$-order Minkowski tensor are given by $\tau^{i_1\cdots i_r}_{j_1\cdots j_s}$. A *single contraction* of the tensor in terms of its components is defined by $\tau^{i_1\cdots i_{k-1}ci_{k+1}\cdots i_r}_{j_1\cdots j_{h-1}cj_{h+1}\cdots j_s}$. Similarly other single contractions can be defined. Multiple contractions may also be defined.

Theorem (1.3.4): *A single contraction* $\tau^{i_1\cdots i_{k-1}ci_{k+1}\cdots i_r}_{j_1\cdots j_{h-1}cj_{h+1}\cdots j_s}$ *of the components of an* $(r + s)$*-order Minkowski tensor yields the components of a* $[(r - 1) + (s - 1)]$*-order Minkowski tensor.*

The proof is left as an easy exercise for the reader.

Example: The Kronecker delta $\delta^i{}_j$, defined in equation (1.2.3), are the components of a mixed second-order Minkowski tensor (the identity tensor). The single contraction is given by the scalar $\delta^c{}_c = \delta^1{}_1 + \delta^2{}_2 + \delta^3{}_3 + \delta^4{}_4 = 4$. A double contraction of the tensor product of the identity tensor with itself is given by the scalar $\delta^c{}_k\delta^k{}_c = \delta^c{}_c = 4$. The other double contraction is $\delta^c{}_c\delta^k{}_k = 16$. □

Recall the metric tensor components d_{ij} in (1.1.7). The *contravariant metric tensor* components $d^{ij} = d^{ji}$ are defined to be the entries of the inverse of the metric such that

$$d^{ij}d_{jk} = d_{kj}d^{ji} = \delta^i{}_k. \tag{1.3.9}$$

In an inner-product vector space, such as $\mathbf{V}_4$, a contravariant tensor has a corresponding covariant tensor and vice-versa. This correspondence is brought about by the metric tensor in the following rules for *lowering and raising indices*:

$$\begin{aligned}
\tau^{i_1\cdots i_{k-1}i_{k+1}\cdots i_r}_{j_1\cdots j_s j} &\equiv d_{ji_k}\tau^{i_1\cdots i_{k-1}i_k i_{k+1}\cdots i_r}_{j_1\cdots j_s},\\
\tau^{i_1\cdots i_{k-1}i_k i_{k+1}\cdots i_r}_{j_1\cdots j_s} &\equiv d^{ji_k}\tau^{i_1\cdots i_{k-1}i_{k+1}\cdots i_r}_{j_1\cdots j_s j},\\
\tau^{i_1\cdots i_r i}_{j_1\cdots j_{h-1}j_{h+1}\cdots j_s} &\equiv d^{ij_h}\tau^{i_1\cdots i_r}_{j_1\cdots j_{h-1}j_h j_{h+1}\cdots j_s},\\
\tau^{i_1\cdots i_r}_{j_1\cdots j_{h-1}j_h j_{h+1}\cdots j_s} &\equiv d_{ij_h}\tau^{i_1\cdots i_r i}_{j_1\cdots j_{h-1}j_{h+1}\cdots j_s}.
\end{aligned} \tag{1.3.10}$$

Examples: $u_i = d_{ij}u^j$, so $u_1 = u^1, u_2 = u^2, u_3 = u^3, u_4 = -u^4$;

$$\begin{aligned}
u^i &= d^{ij}u_j;\\
\tau^k{}_l &= d_{lj}\tau^{kj} = d^{kj}\tau_{jl};\\
\tau_l{}^k &= d_{lj}\tau^{jk} = d^{kj}\tau_{lj}. \quad \square
\end{aligned} \tag{1.3.11}$$

There exist some special *numerical Minkowski tensors* whose components *do not* change values under Lorentz mappings. We shall list these tensors now.

(i) The $(r+s)$-order zero tensor has for its components $0^{i_1\cdots i_r}_{j_1\cdots j_s} \equiv 0$. These values do not change under any Lorentz mapping.

(ii) The components of the identity tensor are $\delta^i{}_j$. The transformed components are $\hat{\delta}^i_j = a^i_k l^m_j \delta^k_m = a^i_m l^m_j = \delta^i_j$ and thus remain unchanged.

(iii) The metric tensor components undergo the transformation $\hat{d}_{ij} = l^a{}_i l^b{}_j d_{ab} = d_{ij}$ by equation (1.2.10). Similarly d^{ij}, and $d^i{}_j \equiv \delta^i{}_j$ remain unchanged.

(iv) The tensor products of the above tensors produce other numerical tensors of higher-order with similar properties. For example, the $[(r+l)+(s+l)]$-order numerical tensor with components $d^{a_1a_2}\cdots d^{a_{r-1}a_r}d_{b_1b_2}\cdots d_{b_{s-1}b_s}\delta^{k_1}{}_{j_1}\cdots\delta^{k_l}{}_{j_l}$ can be mentioned.

These components do not change under any Lorentz mapping. There exists another important numerical symbol. It is the totally antisymmetric *Levi-Civita permutation symbol* defined by

$$\varepsilon_{mnrs} \equiv \begin{cases} 1 & \text{for } (m\ n\ r\ s) \text{ an even permutation of } (1\ 2\ 3\ 4),\\ -1 & \text{for } (m\ n\ r\ s) \text{ an odd permutation of } (1\ 2\ 3\ 4),\\ 0 & \text{otherwise.} \end{cases} \tag{1.3.12}$$

There are 256 components of this permutation symbol of which only twenty-four have nonzero values. Some components are explicitly shown below

$$\begin{aligned}
\varepsilon_{1234} = -\varepsilon_{2134} = -\varepsilon_{1243} = \varepsilon_{3412} &= 1,\\
\varepsilon_{1124} = \varepsilon_{1112} = \varepsilon_{4234} &= 0.
\end{aligned}$$

For the transformation properties of the permutation symbol ε_{mnrs}, we first notice that

$$(\det[l^i{}_j])\varepsilon_{mnrs} = l^a{}_m l^b{}_n l^c{}_r l^d{}_s \varepsilon_{abcd}.$$

Therefore, if we assign the transformation rule

$$\hat{\varepsilon}_{mnrs} = [\det[l^i{}_j]]^{-1} l^a{}_m l^b{}_n l^c{}_r l^d{}_s \varepsilon_{abcd}$$
$$= (\pm 1) l^a{}_m l^b{}_n l^c{}_r l^d{}_s \varepsilon_{abcd}, \tag{1.3.13a}$$

the ε_{mnrs} are components of a *numerical relative tensor*. We shall assume (1.3.13a) and define the totally antisymmetric *Levi–Civita pseudotensor* and its transformation properties by

$$\eta_{mnrs} \equiv [-\det[d_{ij}]]^{1/2} \varepsilon_{mnrs} = \varepsilon_{mnrs},$$
$$\hat{\eta}_{mnrs} = \operatorname{sgn}[\det[l^i{}_j]] l^a{}_m l^b{}_n l^c{}_r l^d{}_s \eta_{abcd}. \tag{1.3.13b}$$

(Although in a Minkowski coordinate system there is no difference between ε_{abcd} and η_{abcd}, these do differ in a curvilinear coordinate system of $\mathbf{M}_4$ in Chapter 2.) In case of a proper Lorentz mapping $\hat{\eta}_{mnrs} = \eta_{abcd}$, whereas in case of an improper Lorentz mapping $\hat{\eta}_{mnrs} = -\eta_{abcd}$. Now, for the raising of the indices of a pseudotensor, we follow the usual rules to get

$$\eta^{1234} = d^{1m} d^{2n} d^{3r} d^{4s} \eta_{mnrs} = d^{11} d^{22} d^{33} d^{44} \eta_{1234} = -\eta_{1234},$$
$$\eta^{abcd} = -\eta_{abcd}. \tag{1.3.14}$$

Example: Suppose that $F_{ji} = -F_{ij}$, which implies that $F^{ji} = -F^{ij}$. We define the components of the *Hodge dual pseudotensor* as

$$F^*{}_{ij} \equiv (1/2)\eta_{ijkl} F^{kl}. \tag{1.3.15}$$

We have by (1.3.12)

$$F^*{}_{12} \equiv (1/2)\eta_{1234} F^{34} + (1/2)\eta_{1243} F^{43} = F^{34},$$
$$F^*{}_{23} = F^{14}, \quad F^*{}_{31} = F^{24}, \tag{1.3.16}$$
$$F^*{}_{14} = F^{23}, \quad F^*{}_{24} = F^{31}, \quad F^*{}_{34} = F^{12}.$$

Raising and lowering indices in (1.3.16) we get

$$F_{34} = -F^{*12}, \quad F_{14} = -F^{*23}, \quad F_{24} = -F^{*31},$$
$$F_{23} = -F^{*14}, \quad F_{31} = -F^{*24}, \quad F_{12} = -F^{*34}.$$

These equations can be neatly summarized into

$$F_{ij} = -(1/2)\eta_{ijkl} F^{*kl}. \tag{1.3.17}$$

This example is relevant to the electromagnetic field theory. □

EXERCISES 1.3

1. Consider the lowering of the index

$$\tau^{i_1 \cdots i_{k-1}}{}_j{}^{i_{k+1} \cdots i_r} \equiv d_{ji_k} \tau^{i_1 \cdots i_{k-1} i_k i_{k+1} \cdots i_r}.$$

Prove *explicitly* that the above components transform as those of a $[(r-1)+1]$-order tensor.

2. (i) Consider the components τ_{ij} of an arbitrary second-order covariant tensor. Show that there exists a unique decomposition $\tau_{ij} = s_{ij} + a_{ij}$ such that $s_{ij} = s_{ji}$, $a_{ij} = -a_{ji}$.

(ii) Consider a symmetric second-order contravariant tensor and an antisymmetric second-order covariant tensor with components s^{ij} and a_{ij} respectively. Obtain the numerical scalar value of the double contraction $s^{ij}a_{ij}$.

3. Consider the numerical tensor defined by the components $d_{adbc} \equiv d_{ab}d_{dc} - d_{ac}d_{db}$. Prove that:

(i) $d_{dabc} = -d_{adbc} = -d_{dacb} = d_{bcda}$,

(ii) $d_{dabc} + d_{dbca} + d_{dcab} = 0$,

(iii) $d^{ad}{}_{bc} = (-1/2)\eta^{admn}\eta_{bcmn}$.

4. Consider an antisymmetric second-order covariant Minkowski tensor with components a_{ij}. Let a mixed tensor be defined by the components $\theta^i{}_j \equiv a^{ik}a_{jk} - (1/4)\delta^i{}_j a_{mn}a^{mn}$.

(i) Prove that $\theta^i{}_i \equiv 0$.

(ii) Show that $\theta^i{}_j = (1/2)(a^{ik}a_{jk} + a^{*ik}a^*{}_{jk})$, where $a^*{}_{ij} \equiv (1/2)\eta_{ijkl}a^{kl}$.

References

1. P. R. Halmos, *Finite dimensional vector spaces*, Van Nostrand, Princeton, NJ, 1958. [pp. 1, 12, 19]
2. W. Noll, Notes on Tensor Analysis prepared by C. C. Wang, The Johns Hopkins University, Mathematics Department, 1963.
3. J. L. Synge, *Relativity: The special theory*, North-Holland, Amsterdam, 1964. [pp. 6, 17]

2
Flat Minkowski Space–Time Manifold $\mathbf{M}_4$ and Tensor Fields

2.1. A Four-Dimensional Differentiable Manifold

The space–time of events in special relativity is *assumed* to be a *flat differentiable manifold* $\mathbf{M}_4$. Therefore, we shall go briefly through the definitions of a four-dimensional manifold.

First, it is assumed that $\mathbf{M}_4$ is a *topologized set*, endowed with a Hausdorff topology. That is, the question of whether a subset of $\mathbf{M}_4$ is open or not can be answered. Furthermore, for two distinct points $p, q \in \mathbf{M}_4$ there exist two disjoint open subsets containing p and q respectively.

A *chart* (χ, U) or local coordinate system is a pair consisting of an open set $U \subseteq \mathbf{M}_4$ together with a continuous, one-to-one mapping $\chi: U \to D \subseteq \mathbb{R}^4$, where D is an open subset of $\mathbb{R}^4$ in the usual topology. For a $p \in U$ we have $x \equiv (x^1, x^2, x^3, x^4) = \chi(p) \in D$. The coordinates (x^1, x^2, x^3, x^4) are the coordinates of p in the chart (χ, U) (see Figure 4).

Each of the coordinates x^i can be obtained by the projection mappings $\pi^i: D \to \mathbb{R}$, defined by $\pi^k(x) = \pi^k(x^1, x^2, x^3, x^4) \equiv x^k$.

Consider two charts (χ, U) and $(\hat{\chi}, \hat{U})$, such that the point p is in the non-empty intersection $U \cap \hat{U}$.

From Figure 5 we obtain

$$\hat{x} = (\hat{\chi} \circ \chi^{-1})(x),$$
$$x = (\chi \circ \hat{\chi}^{-1})(\hat{x}).$$

By projection of these points we get

$$\begin{aligned} \hat{x}^k &= [\pi^k \circ \hat{\chi} \circ \chi^{-1}](x) \equiv \hat{X}^k(x) = \hat{X}^k(x^1, x^2, x^3, x^4), \\ x^k &= [\pi^k \circ \chi \circ \hat{\chi}^{-1}](\hat{x}) \equiv X^k(\hat{x}) = X^k(\hat{x}^1, \hat{x}^2, \hat{x}^3, \hat{x}^4), \end{aligned} \tag{2.1.1}$$

where $x \in D_s$ and $\hat{x} \in \hat{D}_s$.

The above equations define a *general coordinate transformation* between two local coordinate systems. We shall assume that the eight functions X^k, $\hat{X}^k$ are continuously thrice differentiable or $\hat{X}^k \in \mathscr{C}^3(D_s)$ and $X^k \in \mathscr{C}^3(\hat{D}_s)$. We can also describe this fact by saying that the charts (χ, U) and $(\hat{\chi}, \hat{U})$ are $\mathscr{C}^3$-

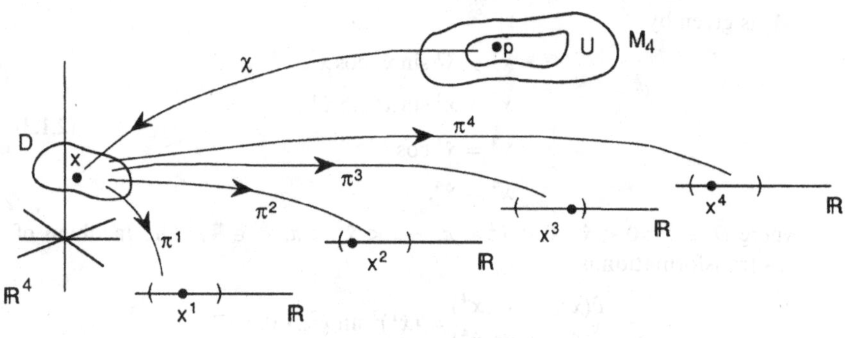

FIGURE 4. A chart (χ, U) and projection mappings.

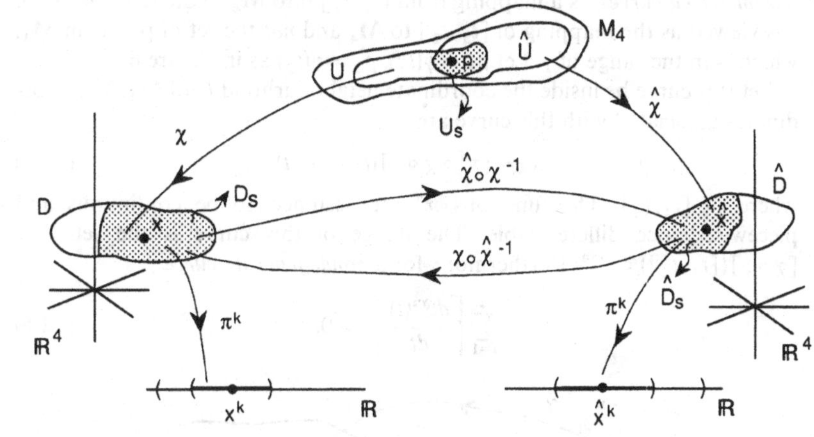

FIGURE 5. Two charts in $\mathbf{M}_4$ and coordinate transformations.

related. From the assumption of differentiability, we can prove that if the *Jacobian* of the transformation (2.1.1) satisfies

$$\frac{\partial(\hat{x}^1, \hat{x}^2, \hat{x}^3, \hat{x}^4)}{\partial(x^1, x^2, x^3, x^4)} \equiv \det\left[\frac{\partial \hat{X}^i(x)}{\partial x^j}\right] \neq 0 \tag{2.1.2}$$

for all $x \in D_s$, then the mapping is one-to-one.

A *subatlas* of class $\mathscr{C}^3$ is a collection of charts (χ_h, U_h), such that the union of all U_h covers $\mathbf{M}_4$ and the charts are mutually $\mathscr{C}^3$-related. A maximal collection of $\mathscr{C}^3$-related charts for $\mathbf{M}_4$ is an *atlas* of class $\mathscr{C}^3$. Finally, a *four-dimensional differentiable manifold* of class $\mathscr{C}^3$ is the set $\mathbf{M}_4$ together with an atlas of class $\mathscr{C}^3$.

Example: Suppose that an inverse coordinate transformation in a domain of

$\mathbf{M}_4$ is given by

$$\begin{aligned} x^1 &= \hat{x}^1 \sin \hat{x}^2 \cos \hat{x}^3, \\ x^2 &= \hat{x}^1 \sin \hat{x}^2 \sin \hat{x}^3, \\ x^3 &= \hat{x}^1 \cos \hat{x}^2, \\ x^4 &= \hat{x}^4, \end{aligned} \tag{2.1.3}$$

where $\hat{D}_s \equiv \{\hat{x}: 0 < \hat{x}^1, 0 < \hat{x}^2 < \pi, -\pi < \hat{x}^3 < \pi, \hat{x}^4 \in \mathbb{R}\}$. The Jacobian of this transformation is

$$\frac{\partial(x^1, x^2, x^3, x^4)}{\partial(\hat{x}^1, \hat{x}^2, \hat{x}^3, \hat{x}^4)} = (\hat{x}^1)^2 \sin \hat{x}^2 > 0. \quad \square$$

Now we shall define a curve in $\mathbf{M}_4$. Let $[t_1, t_2]$ be a closed interval in $\mathbb{R}$. A *parametrized curve* γ is a mapping from $[t_1, t_2]$ into $\mathbf{M}_4$. Note that the curve γ is viewed as the mapping of $[t_1, t_2]$ to $\mathbf{M}_4$ and *not* the set of points in $\mathbf{M}_4$, which is in the range of γ. Let $p_1 = \gamma(t_1)$, $p_2 = \gamma(t_2)$ as in Figure 6.

Let this curve be inside the coordinate neighbourhood U of $\mathbf{M}_4$. The coordinates associated with this curve are

$$x^i = [\pi^i \circ \chi \circ \gamma](t) \equiv \mathscr{X}^i(t), \tag{2.1.4}$$

where $t \in [t_1, t_2]$. The functions $\mathscr{X}^i$ are assumed to be continuous and piecewise twice differentiable. The image of this curve is the set $\Gamma \equiv [\chi \circ \gamma]([t_1, t_2]) \subset \mathbb{R}^4$. Furthermore, for a *nondegenerate curve* γ,

$$\sum_{i=1}^{4} \left[\frac{d\mathscr{X}^i(t)}{dt}\right]^2 > 0, \tag{2.1.5}$$

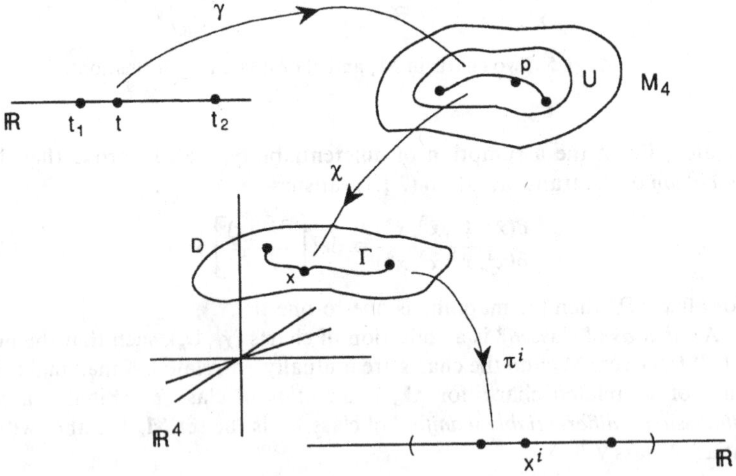

FIGURE 6. A curve γ in $\mathbf{M}_4$.

at each point of differentiability. The components of the *tangent vector* $\mathbf{t}_p$ (relative to the coordinate basis) to the curve γ at $p = \gamma(t)$ are defined to be $d\mathscr{X}^i(t)/dt$. The tangent vector $\mathbf{t}_p$ of γ at $p \in \mathbf{M}_4$ can be visualized as a directed line segment emanating from p tangential to γ. There is an *intrinsic* way of defining a tangent vector $\mathbf{t}_p$ as the *directional derivative*

$$\mathbf{t}_p = \mathbf{t}_{\gamma(t)} \equiv \frac{d\mathscr{X}^i(t)}{dt} \frac{\partial}{\partial x^i}\bigg|_{p=\gamma(t)}. \tag{2.1.6}$$

Let us consider all possible differentiable curves passing through p. The set of all possible tangent vectors, emanating from p, span a vector space. This vector space is called the *tangent vector space* $T_p\mathbf{M}_4$. It is clear that this vector space is four-dimensional. For the differentiable manifold $\mathbf{M}_4$, it is *assumed* that $T_p\mathbf{M}_4$ is isomorphic to the Minkowski vector space $\mathbf{V}_4$ for each $p \in U \subset \mathbf{M}_4$.

Let the components of the tangent vector $\mathbf{t}_p$ of a differentiable curve γ be $d\mathscr{X}^i(t)/dt$ in a coordinate system. Let

$$\hat{x}^k = \hat{X}^k(x) = \hat{X}^k(x^1, x^2, x^3, x^4),$$
$$x^k = X^k(\hat{x}) = X^k(\hat{x}^1, \hat{x}^2, \hat{x}^3, \hat{x}^4)$$

indicate a general coordinate transformation. The curve γ can be represented in the hatted coordinate system by

$$\hat{x}^i = [\pi^i \circ \hat{\chi} \circ \gamma](t) \equiv \hat{\mathscr{X}}^i(t). \tag{2.1.7}$$

The components of the tangent vector $\mathbf{t}_p$ in the hatted coordinate system are given by $d\hat{\mathscr{X}}^i(t)/dt$. Using the chain rule, the transformation between the two sets of components are

$$\begin{aligned} \frac{d\hat{\mathscr{X}}^i(t)}{dt} &= \left[\frac{\partial \hat{X}^i(x)}{\partial x^j}\right]\bigg|_{x=\mathscr{X}(t)} \frac{d\mathscr{X}^j(t)}{dt}, \\ \frac{d\mathscr{X}^j(t)}{dt} &= \left[\frac{\partial X^j(\hat{x})}{\partial \hat{x}^i}\right]\bigg|_{\hat{x}=\hat{\mathscr{X}}(t)} \frac{d\hat{\mathscr{X}}^i(t)}{dt}. \end{aligned} \tag{2.1.8}$$

The above equations imply that the transformation between the two sets of components of a tangent vector field $\mathbf{u}(p)$ must be

$$\begin{aligned} \hat{u}^i(\hat{x}) &= \frac{\partial \hat{X}^i(x)}{\partial x^j} u^j(x), \\ u^j(x) &= \frac{\partial X^j(\hat{x})}{\partial \hat{x}^i} \hat{u}^i(\hat{x}). \end{aligned} \tag{2.1.9}$$

Example: Let a parametrized, differentiable curve γ be given by

$$x^1 = \mathscr{X}^1(t) \equiv t^2,$$
$$x^2 = \mathscr{X}^2(t) \equiv t,$$

$$x^3 = \mathscr{X}^3(t) \equiv e^t,$$
$$x^4 = \mathscr{X}^4(t) \equiv 1,$$

where $t \in [0, 2]$ and we have used a coordinate chart (χ, U). For this curve $\sum_{i=1}^{4} [d\mathscr{X}^i(t)/dt]^2 = 4t^2 + 1 + e^{2t} > 0$ satisfying (2.1.5). Let another chart $(\hat{\chi}, \hat{U})$ be defined by the transformation

$$\hat{x}^1 = \hat{X}^1(x) \equiv x^1 + x^2,$$
$$\hat{x}^2 = \hat{X}^2(x) \equiv x^1 - x^2,$$
$$\hat{x}^3 = \hat{X}^3(x) \equiv 2x^3 + x^4,$$
$$\hat{x}^4 = \hat{X}^4(x) \equiv x^3 - x^4/2,$$
$$\frac{\partial(\hat{x}^1, \hat{x}^2, \hat{x}^3, \hat{x}^4)}{\partial(x^1, x^2, x^3, x^4)} = 4 \neq 0.$$

Therefore, the hatted components of the tangent vector, by equation (2.1.8), are given by

$$\frac{d\hat{\mathscr{X}}^1(t)}{dt} = (1)\frac{d\mathscr{X}^1(t)}{dt} + (1)\frac{d\mathscr{X}^2(t)}{dt} = 2t + 1,$$
$$\frac{d\hat{\mathscr{X}}^2(t)}{dt} = (1)\frac{d\mathscr{X}^1(t)}{dt} + (-1)\frac{d\mathscr{X}^2(t)}{dt} = 2t - 1,$$
$$\frac{d\hat{\mathscr{X}}^3(t)}{dt} = (2)\frac{d\mathscr{X}^3(t)}{dt} + (1)\frac{d\mathscr{X}^4(t)}{dt} = 2e^t,$$
$$\frac{d\hat{\mathscr{X}}^4(t)}{dt} = (1)\frac{d\mathscr{X}^3(t)}{dt} + (-1/2)\frac{d\mathscr{X}^4(t)}{dt} = e^t. \quad \square$$

EXERCISES 2.1

1. Consider a coordinate transformation given by

$$\hat{x}^1 = \sqrt{(x^1)^2 + (x^2)^2};$$

$$\hat{x}^2 = \operatorname{arc}(x^1, x^2) \equiv \begin{cases} \operatorname{Arctan}(x^2/x^1), & x^1 > 0, \\ (\pi/2)\operatorname{sgn}(x^2), & x^1 = 0 \text{ and } x^2 \neq 0, \\ \operatorname{Arctan}(x^2/x^1) + \pi \operatorname{sgn}(x^2), & x^1 < 0 \text{ and } x^2 \neq 0; \end{cases}$$

$$\hat{x}^3 = x^3;$$

$$\hat{x}^4 = x^4,$$

where $-\pi/2 < \operatorname{Arctan}(x^2/x^1) < \pi/2$ and the chart $(\chi, U) = (\chi, \mathbf{M}_4)$.

(i) Obtain $\hat{D} \equiv \hat{\chi}(\hat{U}) \subseteq \mathbb{R}^4$.

(ii) This transformation is $\mathscr{C}^r$-related. Obtain the maximal r-value.

(iii) Obtain the domain $\hat{D}_0 \subseteq \mathbb{R}^4$ such that the Jacobian

$$\frac{\partial(\hat{x}^1, \hat{x}^2, \hat{x}^3, \hat{x}^4)}{\partial(x^1, x^2, x^3, x^4)} > 0.$$

2. A family of differentiable curves in $\mathbf{M}_4$ (using the coordinate chart $(\chi, \mathbf{M}_4)$) satisfy the differential equations

$$\left[\frac{d\mathscr{X}^i(t)}{dt}\right]^2 + [\mathscr{X}^i(t)]^2 = 1$$

for all $t \in \mathbb{R}$. Obtain this family of curves explicitly.

3. Consider a coordinate transformation

$$\begin{aligned} x^1 &= \hat{x}^1 \sin \hat{x}^2 \cos \hat{x}^3, \\ x^2 &= \hat{x}^1 \sin \hat{x}^2 \sin \hat{x}^3, \\ x^3 &= \hat{x}^1 \cos \hat{x}^2, \\ x^4 &= \ln |\hat{x}^4|, \end{aligned}$$

where $\hat{D}_s \equiv \{\hat{x}\colon 0 < \hat{x}^1, 0 < \hat{x}^2 < \pi, -\pi < \hat{x}^3 < \pi, \hat{x}^4 \neq 0\}$. The components of a tangent vector field $\mathbf{u}(p)$ are given as

$$\begin{aligned} \hat{u}^1(\hat{x}) &= (1/\hat{x}^1) \cos \hat{x}^2 \cos 2\hat{x}^3, \\ \hat{u}^2(\hat{x}) &= (1/\hat{x}^1) \cos \hat{x}^2 \sin 2\hat{x}^3, \\ \hat{u}^3(\hat{x}) &= (1/\hat{x}^1) \sin \hat{x}^2, \\ \hat{u}^4(\hat{x}) &= 1/\hat{x}^4. \end{aligned}$$

Compute $u^i(x)$ at $x = (1, 0, 0, 0)$.

2.2. Minkowski Space–Time $\mathbf{M}_4$ and the Separation Function

A four-dimensional differentiable manifold $\mathbf{M}_4$ is *flat* (or *Minkowski*) provided it admits a global *Minkowski coordinate* chart $(\chi, \mathbf{M}_4)$. For such a chart the coordinate basis vectors (in the tangent space $T_p\mathbf{M}_4$) are assumed to be M-orthonormal:

$$[\mathbf{e}_i(p) \cdot \mathbf{e}_j(p)] = [d_{ij}] = \begin{bmatrix} 1 & 0 & 0 & 0 \\ 0 & 1 & 0 & 0 \\ 0 & 0 & 1 & 0 \\ 0 & 0 & 0 & -1 \end{bmatrix} \tag{2.2.1}$$

for every $p \in \mathbf{M}_4$. [In case the tangent space has an inner product, the manifold is called *Riemannian* or *pseudo-Riemann* (which need not be flat).]

The space–time universe of special relativity is *assumed* to be the flat Minkowski manifold $\mathbf{M}_4$. A point $p \in \mathbf{M}_4$ represents an *idealized event*. An event p has four coordinates with respect to a chart (χ, U), $\chi(p) = (x^1, x^2, x^3, x^4)$. If (χ, U) is a Minkowski coordinate system, then (x^1, x^2, x^3) represents the *spatial Cartesian* coordinates and x^4 represents the *time* of the event. The *history* of a point particle consists of a continuous locus of events in space–time. This history can be represented by a continuous curve γ in $\mathbf{M}_4$. Such a curve γ is also called the *world-line* of the particle. Usually a world-line is a twice differentiable curve. The image $\Gamma \subset \mathbb{R}^4$ of the world-line γ will be plotted (see Fig. 6) instead of the image of γ in $\mathbf{M}_4$. *Subsequently we shall restrict ourselves to the use of Minkowski coordinate systems on* $\mathbf{M}_4$ *unless explicitly noted otherwise.*

The world-line of a particle is locally classified by the value of

$$\mathbf{t}_p \cdot \mathbf{t}_p = \mathbf{t}_{\gamma(t)} \cdot \mathbf{t}_{\gamma(t)} = d_{ij} \frac{d\mathscr{X}^i(t)}{dt} \frac{d\mathscr{X}^j(t)}{dt}.$$

In the case of a *massive* particle it is *assumed* that

$$d_{ij} \frac{d\mathscr{X}^i(t)}{dt} \frac{d\mathscr{X}^j(t)}{dt} < 0.$$

Thus, the tangent vector to the world-line of a massive particle is timelike. For a massless *photon* (electromagnetic field quanta) or other massless particles such as neutrinos it is *assumed* that

$$d_{ij} \frac{d\mathscr{X}^i(t)}{dt} \frac{d\mathscr{X}^j(t)}{dt} = 0.$$

Thus, the world-line of a massless particle has a null tangent vector at each point. For the case of the hypothetical *tachyon* the tangent vector is assumed to be spacelike. For a *free* particle (not acted upon by any force) the world-line is determined by the differential equations

$$\frac{d^2\mathscr{X}^i(t)}{dt^2} = 0. \tag{2.2.2}$$

The assumptions made in this paragraph are compatible with the experimental results of physics. (Chapter 5 explores the details of motion curves in space–time.)

The general solutions of (2.2.2) are given by

$$x^k = \mathscr{X}^k(t) = x_0^k + v^k t, \tag{2.2.3}$$

where x_0^k, v^k are constants of integration. For known free particles $d_{ij}v^i v^j \leq 0$.

The *null-cone* (*light-cone*), which has its vertex at $x_0 \in \mathbb{R}^4$, corresponds to the set of points

$$N_{x_0} \equiv \{x \colon d_{ij}(x^i - x_0^i)(x^j - x_0^j) = 0\}, \tag{2.2.4}$$

where x^i are the Minkowski coordinates (see Fig. 7). The circles shown in Figure 7 represent (suppressed) spherical surfaces.

The events in space-time can be classified according to their corresponding positions relative to N_{x_0}. The events that correspond to the points in the

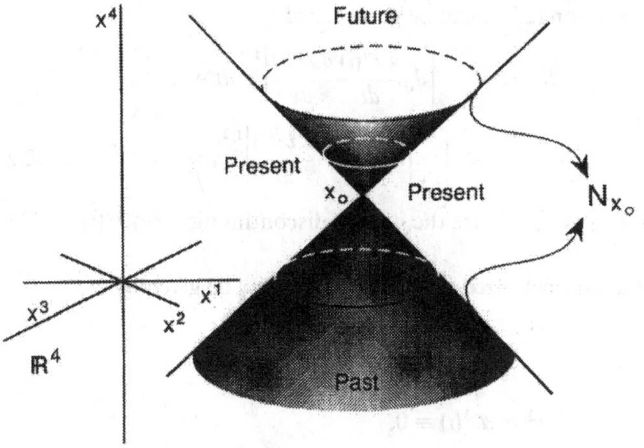

FIGURE 7. The null-cone N_{x_0}.

subset

$$\{x: d_{ij}(x^i - x_0^i)(x^j - x_0^j) \le 0, x_0^4 < x^4\}$$

are the *future events* relative to the event corresponding to x_0.

The events corresponding to the points in the subset

$$\{x: d_{ij}(x^i - x_0^i)(x^j - x_0^j) \le 0, x^4 < x_0^4\}$$

are the *past events* relative to x_0. The events in the subset

$$\{x_0\} \cup \{x: d_{ij}(x^i - x_0^i)(x^j - x_0^j) > 0\}$$

are called the *present events* relative to x_0. This concept of the present events generalizes the Newtonian concept of present events (which are called *simultaneous events* in relativity). This classification of events into future, past, and present subsets is independent of any observer.

The events on N_{x_0}, for which $x^4 \neq x_0^4$, satisfy

$$[(x^1 - x_0^1)^2 + (x^2 - x_0^2)^2 + (x^3 - x_0^3)^2]^{1/2}/|x^4 - x_0^4| = 1. \qquad (2.2.5)$$

The physical meaning of the above expression is that *the speed of light is* 1 *in our units.* Thus the points on N_{x_0} represent the history of a spherical wave front of light converging to and then emanating from the event x_0.

The space–time *separation* along a differentiable (nondegenerate, rectifiable) world-line $\gamma: [a, b] \to \mathbf{M}_4$ is defined by the *functional* S,

$$s = S(\gamma) \equiv \int_a^b \sigma[\mathbf{t}_{\gamma(t)}]\, dt \equiv \int_a^b \left| d_{ij} \frac{d\mathscr{X}^i(t)}{dt} \frac{d\mathscr{X}^j(t)}{dt} \right|^{1/2} dt. \qquad (2.2.6)$$

The above equation is the generalization of the concept of arc-length of a curve. In the case where the world-line γ is continuous but piecewise differentiable, the equation (2.2.6) can be generalized to

$$s = S(\gamma) \equiv \int_a^{t_1^-} \left| d_{ij} \frac{d\mathscr{X}^i(t)}{dt} \frac{d\mathscr{X}^j(t)}{dt} \right|^{1/2} dt + \cdots + \int_{t_n^+}^{b} \left| d_{ij} \frac{d\mathscr{X}^i(t)}{dt} \frac{d\mathscr{X}^j(t)}{dt} \right|^{1/2} dt, \tag{2.2.7}$$

where $a < t_1 < \cdots < t_n < b$ are the n jump discontinuities of $d\mathscr{X}^i(t)/dt$.

Example: Let a parameterized differentiable curve γ be given by

$$x^1 = \mathscr{X}^1(t) \equiv t^3,$$
$$x^2 = \mathscr{X}^2(t) \equiv 0,$$
$$x^3 = \mathscr{X}^3(t) \equiv 0,$$
$$x^4 = \mathscr{X}^4(t) \equiv \sqrt{3}t^3, \qquad t \in [0, 1].$$

The corresponding separation is

$$s = \int_0^1 |9t^4 + 0 + 0 - 27t^4|^{1/2}\, dt = \sqrt{2}. \quad \square$$

A straight timelike world-line satisfying (2.2.2) can represent an idealized point observer who is not subjected to any net external force. Using the equation $x^k = x_0^k + v^k t$, the separation along a straight world-line is

$$s = \int_0^t |d_{ij} v^i v^j|^{1/2}\, dt = |d_{ij}(x^i - x_0^i)(x^j - x_0^j)|^{1/2}. \tag{2.2.8}$$

An idealized point observer is assumed to follow a continuous, piecewise twice differentiable, timelike world-line that is not necessarily straight. He may carry a *standard clock*, which is an idealized point clock that runs perfectly. The elapsed time between two events of the history of the observer as measured by the standard clock is called the *proper time* between the events. The proper time elapsed between two events, $\gamma(t_1)$ and $\gamma(t_2)$, is *assumed* to be the separation

$$s = S(\gamma) = \int_{t_1}^{t_2} \left| -d_{ij} \frac{d\mathscr{X}^i(t)}{dt} \frac{d\mathscr{X}^j(t)}{dt} \right|^{1/2} dt. \tag{2.2.9}$$

Example: Let us confront the problem of the *twin paradox* with help of the preceding assumptions. Consider an identical twin at the same place on the earth. One of the twins leaves in a rocket at $t = t_1$ with a constant velocity v along the x^1-axis. After a while he reverses his velocity at $t = t_2$ and returns to the earth at $t = t_3$.

The paradox is whether or not the returning twin is at that instant younger than the twin that stayed at home. The piecewise straight timelike world-lines

FIGURE 8. The space-time diagram for the twin paradox.

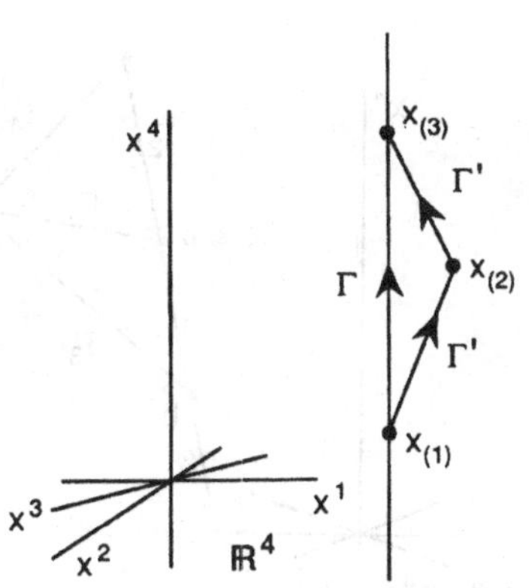

γ, γ' for the twins (see Figure 8) are given by

$$x^1 = c; \quad x^2 = 0; \quad x^3 = 0; \quad x^4 = t;$$

$$x'^1 = \begin{cases} c + v(t - t_1), & t_1 \le t \le t_2, \\ c + 2vt_2 - v(t + t_1), & t_2 < t \le t_3, \end{cases} \quad 0 < |v| < 1; \quad (2.2.10)$$

$$x'^2 = 0; \quad x'^3 = 0; \quad x'^4 = t.$$

It is easily seen that $t_3 = 2t_2 - t_1$. The proper time measured by the twin brothers are given by (2.2.6), (2.2.7) so that

$$s = S(\gamma) = \int_{t_1}^{t_3} |-1|^{1/2}\,dt = t_3 - t_1,$$

$$s' = S(\gamma') = \int_{t_1}^{t_2^-} |v^2 - 1|^{1/2}\,dt + \int_{t_2^+}^{t_3} |v^2 - 1|^{1/2}\,dt \quad (2.2.11)$$

$$= (1 - v^2)^{1/2}(t_3 - t_1).$$

The above equation clearly shows that $s' < s$ and the returning twin will be younger! The asymmetry of the aging processes is revealed in the asymmetry of the world-lines of the twin brothers. □

Now we shall obtain spacelike separations from timelike separations.

Theorem (2.2.1): *Let points a, b, q define a triangle in* $\mathbb{R}^4$ *(see Figure 9). Let p be an arbitrary intermediate point on the straight line joining a to b. Denoting the straight lines from a to p; b to p; p to q; q to a; and b to q by* $\Gamma_2, \Gamma_1, \Gamma, \Gamma_4$, *and* Γ_3 *respectively and assuming that* γ_1, γ_2 *are timelike,* γ_3, γ_4 *are null,* γ *is spacelike, one must have*

$$[S(\gamma)]^2 = [S(\gamma_1)][S(\gamma_2)]. \quad (2.2.12)$$

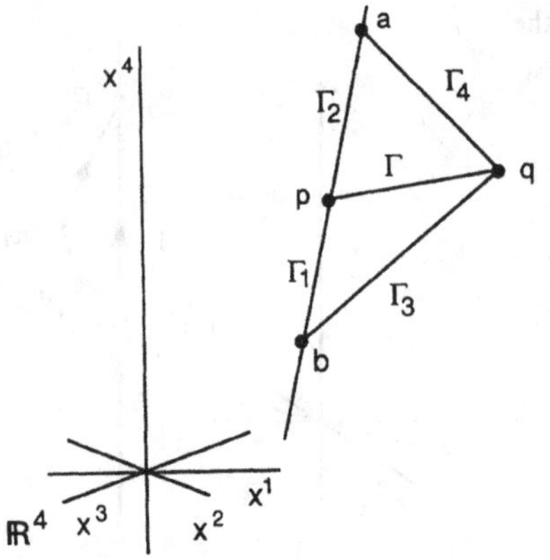

FIGURE 9. Measurement of a spacelike separation.

Proof: From the assumptions and equation (2.2.8), it follows that

$$d_{ij}(q^i - a^i)(q^j - a^j) = d_{ij}(q^i - b^i)(q^j - b^j) = 0.$$

We can write

$$q^i - a^i = (q^i - p^i) - (a^i - p^i),$$
$$q^i - b^i = (q^i - p^i) + (p^i - b^i),$$
$$p^i - b^i = \lambda(a^i - p^i), \qquad \lambda > 0.$$

Putting these equations into the conditions for null separations we obtain

$$\lambda d_{ij}[(q^i - p^i)(q^j - p^j) + (a^i - p^i)(a^j - p^j) - 2(q^i - p^i)(a^j - p^j)] = 0,$$
$$d_{ij}[(q^i - p^i)(q^j - p^j) + \lambda^2(a^i - p^i)(a^j - p^j) + 2\lambda(q^i - p^i)(a^j - p^j)] = 0.$$

Adding the above equations and cancelling the factor $(1 + \lambda)$ we get

$$d_{ij}(q^i - p^i)(q^j - p^j) = -\lambda d_{ij}(a^i - p^i)(a^j - p^j),$$

or $[S(\gamma)]^2 = \lambda[S(\gamma_2)]^2 = [S(\gamma_1)][S(\gamma_2)]$. ■

The equation (2.2.12) suggests an *operational method* of measuring a spacelike separation by a standard clock and emission, reflection, and absorption of photons.

Now we shall define the M-orthogonal intersection between two straight world-lines γ_1, γ_2 in $\mathbf{M}_4$. Suppose that the corresponding straight lines Γ_1, Γ_2 in $\mathbb{R}^4$ intersect at a (see Figure 10). The *intersection* at a is defined to be *M-orthogonal* provided

$$d_{ij}(b^i - a^i)(c^j - a^j) = 0, \tag{2.2.13}$$

where b, c are arbitrary points on Γ_1, Γ_2 respectively and all three points are distinct.

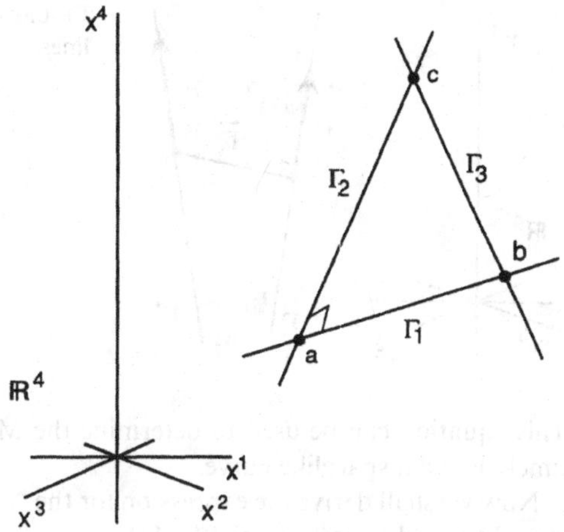

FIGURE 10. M-Orthogonal intersection.

In the case where the straight world-lines γ_1, γ_2 are both spacelike we have

$$d_{ij}(b^i - a^i)(b^j - a^j) > 0,$$
$$d_{ij}(c^i - a^i)(c^j - a^j) > 0.$$

Using (2.2.13) we derive

$$\begin{aligned} d_{ij}&(c^i - b^i)(c^j - b^j) \\ &= d_{ij}[(c^i - a^i) - (b^i - a^i)][(c^j - a^j) - (b^j - a^j)] \\ &= d_{ij}(c^i - a^i)(c^j - a^j) + d_{ij}(b^i - a^i)(b^j - a^j) > 0; \end{aligned}$$

that is, the straight world-line γ_3 is spacelike. Furthermore,

$$[S(\gamma_3)]^2 = [S(\gamma_1)]^2 + [S(\gamma_2)]^2,$$

which is the generalized theorem of Pythagoras. In case one of the straight world-lines, say γ_1, is spacelike and the other is timelike, we have

$$d_{ij}(b^i - a^i)(b^j - a^j) > 0,$$
$$d_{ij}(c^i - a^i)(c^j - a^j) < 0.$$

Using the M-orthogonality condition (2.2.13) we again get

$$d_{ij}(c^i - b^i)(c^j - b^j) = d_{ij}(c^i - a^i)(c^j - a^j) + d_{ij}(b^i - a^i)(b^j - a^j),$$

or

$$\pm[S(\gamma_3)]^2 = [S(\gamma_1)]^2 - [S(\gamma_2)]^2.$$

Choosing γ_3 to be a null straight world-line we have $S(\gamma_3) = 0$ and

$$S(\gamma_1) = S(\gamma_2). \tag{2.2.14}$$

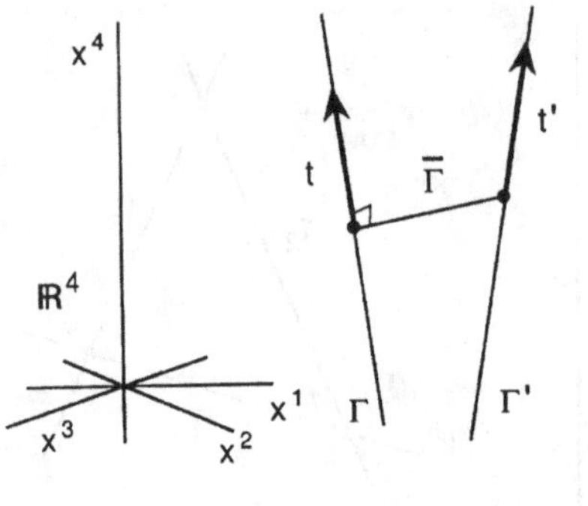

FIGURE 11. Distance between two straight lines.

This equation can be used to determine the M-orthogonal intersection of a timelike and a spacelike curve.

Now we shall derive the expression for the "distance" between two timelike straight world-lines γ, γ' (see Fig. 11).

In terms of proper time parameters s and s', the equations of two straight world-lines γ and γ' are

$$\begin{aligned} x^i &= \mathscr{X}^i(s) = t^i s + b^i, \qquad d_{ij}t^i t^j = -1, \qquad t^4 > 0, \\ x'^i &= \mathscr{X}'^i(s') = t'^i s' + b'^i, \qquad d_{ij}t'^i t'^j = -1, \qquad t'^4 > 0, \end{aligned} \tag{2.2.15}$$

The two straight world-lines γ, γ' are defined to be *parallel* if $t^i = t'^i$. (In general, parallel lines can be defined *only* for a *flat* differentiable manifold.) Assuming that the straight lines $\gamma, \bar{\gamma}$ intersect M-orthogonally, we have

$$\begin{aligned} d_{kj}(x'^k - x^k)t^j &= d_{kj}(t'^k s' + b'^k - t^k s - b^k)t^j = 0, \\ s' &= (d_{kj}t^j t'^k)^{-1}[d_{ac}t^a(b^c - b'^c) - s]. \end{aligned} \tag{2.2.16}$$

We can recall Theorem (1.1.4) to note that $d_{kj}t^j t'^k \neq 0$ and thus conclude that s' always exists. From (2.2.15), (2.2.16) we can express

$$\begin{aligned} x'^k - x^k &= -[t^k + (d_{ij}t^i t'^j)^{-1} t'^k]s + \beta^k, \\ \beta^k &\equiv b'^k - b^k + (d_{ij}t^i t'^j)^{-1}[d_{ac}t^a t'^k(b^c - b'^c)]. \end{aligned} \tag{2.2.17}$$

We can obtain the expression for the *distance* $D(s)$ along the straight world-line $\bar{\gamma}$ by the equation

$$\begin{aligned} [D(s)]^2 \equiv [S(\bar{\gamma})]^2 &= d_{kj}(x'^k - x^k)(x'^j - x^j) \\ &= [1 - (d_{ij}t^i t'^j)^{-2}]s^2 \\ &\quad - \{2d_{ij}\beta^i[t^j + (d_{kl}t^k t'^l)^{-1}t'^j]\}s + d_{ij}\beta^i\beta^j, \end{aligned} \tag{2.2.18}$$

which is a second-degree polynomial in the variable s.

Theorem (2.2.2) (Synge): *Let γ and γ' be two timelike straight world-lines in $\mathbf{M}_4$, given by equation* (2.2.15). *Then the world-lines γ and γ' are parallel if and only if $D(s)$ is a constant.*

Proof: (i) Assume that γ and γ' are parallel. Then $t^k = t'^k$ and by (2.2.15) and (2.2.18) we have $d_{kl}t^k t'^l = -1$, $1 - (d_{ij}t^i t'^j)^{-2} = 0$, $t^j + (d_{kl}t^k t'^l)t'^j = 0$, and $[D(s)]^2 = d_{ij}\beta^i\beta^j$. Thus $D(s)$ is a constant.

(ii) Assume that $D(s)$ is a constant. Then equation (2.2.18) gives

$$\frac{d^2}{ds^2}\{[D(s)]^2\} = 0, \qquad 2[1 - (d_{ij}t^i t'^j)^{-2}] = 0, \qquad d_{ij}t^i t'^j = \pm 1.$$

Recalling Theorem (1.1.5) we must have $d_{ij}t^i t'^j = -1$. Then by Corollary (1.1.2) we have $t'^i = t^i$. Thus γ and γ' are parallel. ■

Let $\gamma: [t_1, t_2] \to \mathbf{M}_4$ be a straight world-line. Among all the coordinate charts, a Minkowski coordinate chart $(\chi, \mathbf{M}_4)$ will give the separation $S(\gamma)$ along the straight world-line γ by the simple formula [see equation (2.2.8)]

$$[S(\gamma)]^2 = |(x^1_{(2)} - x^1_{(1)})^2 + (x^2_{(2)} - x^2_{(1)})^2 + (x^3_{(2)} - x^3_{(1)})^2 - (x^4_{(2)} - x^4_{(1)})^2|, \tag{2.2.19}$$

where $x^i_{(1)}$ are the coordinates of $\gamma(t_1)$ and $x^i_{(2)}$ are the coordinates of $\gamma(t_2)$.

A physicist, informed of the *mathematical existence* of such a Minkowski coordinate chart, would like to know how to assign such coordinates to events by the process of *physical measurements*. This is the problem of operational aspects of the *Minkowski chronometry*. We can assign coordinates of an event by emission, reflection, and absorption of photons and the measurements of proper times on a standard clock. The equations (2.2.12), (2.2.14), the generalized Pythagoras theorem, and Theorem (2.2.2) were all derived for the purpose of measurements in Minkowski chronometry. These equations and theorems tell us how to measure a spacelike separation by measurements of proper times; how to conclude M-orthogonality of intersections of straight world-lines by measurements of separations; and how to find that two straight world-lines are parallel. We shall use all these results to construct an *operational* Minkowski coordinate chart. We take an observer, who is not subjected to any net external force and thus has a straight timelike world-line. At an arbitrary event e (see Figure 12) on his history, he sets his standard clock

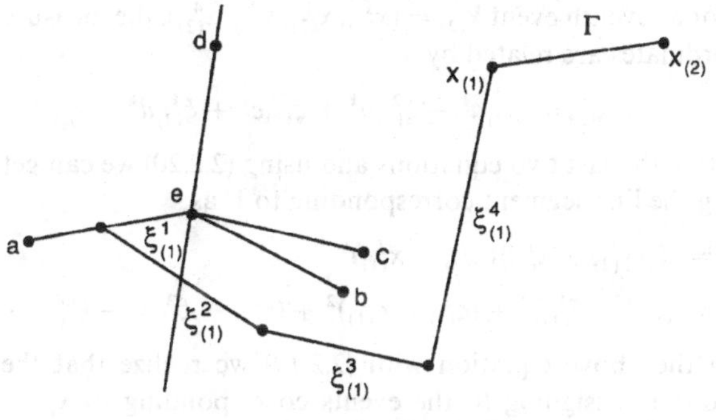

FIGURE 12. Operational construction of Minkowski coordinates.

to zero and his spatial coordinates to $(0, 0, 0)$. After a while, his standard clock reads unity and that event is denoted by d. By the measurements of proper times and by using equations (2.2.12), (2.2.14) and the Pythagoras theorem, the observer can identify three other events corresponding to a, b, c such that each of the line segments connecting e to a, e to b, e to c, and e to d has unit separation and all of them intersect M-orthogonally at e. The events a, b, c may almost be chosen arbitrarily. Let the events a, b, c, d, e have coordinates a^i, b^i, c^i, d^i, e^i with respect to a Minkowski coordinate chart. These coordinates exist in principle and may be unknown to the observer. But from the measurements, the separations along four line segments must satisfy the M-orthonormality conditions

$$\begin{aligned} d_{ij}(a^i - e^i)(a^j - e^j) &= d_{ij}(b^i - e^i)(b^j - e^j) \\ &= d_{ij}(c^i - e^i)(c^j - e^j) \\ &= -d_{ij}(d^i - e^i)(d^j - e^j) = 1, \\ d_{ij}(a^i - e^i)(b^j - e^j) &= d_{ij}(a^i - e^i)(c^j - e^j) = d_{ij}(a^i - e^i)(d^j - e^j) \\ &= d_{ij}(b^i - e^i)(c^j - e^j) = d_{ij}(b^i - e^i)(d^j - e^j) \\ &= d_{ij}(c^i - e^i)(d^j - e^j) = 0. \end{aligned} \tag{2.2.20}$$

In the above equations there are ten equations for twenty unknowns. Thus there are ten degrees of freedom in the solutions for a^i, b^i, c^i, d^i, e^i. There are four degrees of freedom for the event e and six degrees of freedom for the events a, b, c, d. Now we choose a sixth event corresponding to $x_{(1)}$ (see Fig. 12). We can go from e to $x_{(1)}$ by going from e toward a, then going *parallel* to e to b, e to c, and e to d in turn. (Going parallel is a *measurable* process, since it involves keeping the "distance" or separation to a *constant*.) The separations in these four steps are measurable in chronometry, and let their values be $\xi^1_{(1)}, \xi^2_{(1)}, \xi^3_{(1)}, \xi^4_{(1)}$. By virtue of the construction just carried out, the measured separations $\xi^k_{(1)}$'s are related by the equations

$$x^k_{(1)} = \xi^1_{(1)} a^k + \xi^2_{(1)} b^k + \xi^3_{(1)} c^k + \xi^4_{(1)} d^k.$$

Similarly, for a seventh event $x_{(2)} = (x^1_{(2)}, x^2_{(2)}, x^3_{(2)}, x^4_{(2)})$, the measured and unknown coordinates are related by

$$x^k_{(2)} = \xi^1_{(2)} a^k + \xi^2_{(2)} b^k + \xi^3_{(2)} c^k + \xi^4_{(2)} d^k.$$

By subtracting the last two equations and using (2.2.20) we can get the separation along the line segment corresponding to Γ as

$$\begin{aligned} [S(\gamma)]^2 &= |d_{ij}(x^i_{(2)} - x^i_{(1)})(x^j_{(2)} - x^j_{(1)})| \\ &= |(\xi^1_{(2)} - \xi^1_{(1)})^2 + (\xi^2_{(2)} - \xi^2_{(1)})^2 + (\xi^3_{(2)} - \xi^3_{(1)})^2 - (\xi^4_{(2)} - \xi^4_{(1)})^2|. \end{aligned}$$

Comparing the above equation with (2.2.19) we realize that the observer has succeeded in assigning to the events corresponding to $x_{(1)}$, $x_{(2)}$, other Minkowski coordinates, $\xi^k_{(1)}$ and $\xi^k_{(2)}$, obtained by measurements in Minkowski chronometry.

EXERCISES 2.2

1. A continuous, piecewise differentiable curve γ in $\mathbf{M}_4$ is given by the Minkowski coordinates

$$x^1 = \mathscr{X}^1(t) \equiv 0;$$

$$x^2 = \mathscr{X}^2(t) \equiv 0;$$

$$x^3 = \mathscr{X}^3(t) \equiv \begin{cases} t, & 0 \leq t \leq 2, \\ 2, & 2 < t \leq 3; \end{cases}$$

$$x^4 = \mathscr{X}^4(t) \equiv \begin{cases} 0, & 0 \leq t \leq 1, \\ t - 1, & 1 < t \leq 3. \end{cases}$$

Obtain the separation $s = S(\gamma)$.

2. Let γ, γ' be two timelike straight world-lines given by the equations

$$x^k = t^k s + b^k, \qquad d_{ij}t^i t^j = -1, \qquad t^4 > 0,$$

$$x'^k = t'^k s' + b'^k, \qquad d_{ij}t'^i t'^j = -1, \qquad t'^4 > 0.$$

Obtain all the possible conditions on t^k, t'^k, b^k, b'^k for the intersection of γ and γ', by equating the distance function $D(s)$ to zero.

2.3. Flat Submanifolds of Minkowski Space–Time $\mathbf{M}_4$

A differentiable *submanifold* $\sigma \subset \mathbf{M}_4$ is a proper open subset of $\mathbf{M}_4$ such that the assumptions of a differentiable manifold are satisfied for σ. If furthermore σ is flat, it is called a *flat submanifold*. We shall now define three flat submanifolds.

(i) Using a Minkowski coordinate system, a *3-flat* $\sigma_3 \subset \mathbf{M}_4$ is defined to be a subset corresponding to

$$\Sigma_3 \equiv \{x: A_i x^i + B = 0, \delta^{ij} A_i A_j > 0\} \subset \mathbb{R}^4. \tag{2.3.1}$$

Here, A_i's are constants.

(ii) A *2-flat* σ_2 is defined to be a subset corresponding to

$$\Sigma_2 \equiv \{x: A_i x^i + B = 0, A'_i x^i + B' = 0, \delta^{ij} A_i A_j > 0, \delta^{ij} A'_i A'_j > 0, A'_i \neq \lambda A_i\}. \tag{2.3.2}$$

(iii) A *1-flat* σ_1 is defined to be a subset corresponding to

$$\Sigma_1 \equiv \{x: A_i x^i + B = 0, A'_i x^i + B' = 0, A''_i x^i + B'' = 0\}. \tag{2.3.3}$$

It is furthermore assumed that the rank of the matrix

$$\begin{bmatrix} A_1 & A_2 & A_3 & A_4 \\ A'_1 & A'_2 & A'_3 & A'_4 \\ A''_1 & A''_2 & A''_3 & A''_4 \end{bmatrix}$$

is three. Therefore, solving the defining equations in (2.3.3), we get

$$\begin{aligned} x^1 &= c - At, \\ x^2 &= c' - A't, \\ x^3 &= c'' - A''t, \\ x^4 &= t \in \mathbb{R}^4 \end{aligned} \tag{2.3.4}$$

for some suitable constants c, c', c'', A, A', A'' so that $(A^2 + A'^2 + A''^2 > 0)$. Thus the subset σ_1 can always be identified with a straight world-line.

In the case where Σ_3, Σ_2 contain the origin $x = 0$, the defining equations simplify to

$$\begin{aligned} &\Sigma_3 \equiv \{x\colon A_i x^i = 0, \delta^{ij} A_i A_j > 0\} \\ &\Sigma_2 \equiv \{x\colon A_i x^i = 0, A'_i x^i = 0, \delta^{ij} A_i A_j > 0, \delta^{ij} A'_i A'_j > 0, A'_i \neq \lambda A_i\}. \end{aligned} \tag{2.3.5}$$

Let us solve the defining equations of Σ_2, namely,

$$A_i x^i = 0, \qquad A'_i x^i = 0. \tag{2.3.6}$$

Since the rank of the coefficient matrix is assumed to be two, we have solutions in the form

$$\begin{aligned} x^1 &= a_1 x^3 + b_1 x^4, \\ x^2 &= a_2 x^3 + b_2 x^4 \end{aligned} \tag{2.3.7}$$

for suitable constants a_1, a_2, b_1, b_2 such that

$$\begin{aligned} a_1^2 + a_2^2 > 0, \qquad b_1^2 + b_2^2 > 0, \\ (a_1, a_2) \neq \lambda(b_1, b_2). \end{aligned}$$

We can rewrite the solutions as

$$\begin{aligned} x^1 &= a_1 \bar{u} + b_1 \bar{v}, \\ x^2 &= a_2 \bar{u} + b_2 \bar{v}, \\ x^3 &= \bar{u}, \\ x^4 &= \bar{v}, \end{aligned} \tag{2.3.8}$$

where $(\bar{u}, \bar{v}) \in \mathbb{R}^2$. Since the solution space in (2.3.8) is two-dimensional, we can also write the symmetrical form of the parametric representation

$$\Sigma_2 \equiv \{x\colon x^i = a^i u + b^i v, \delta^{ij} a_i a_j > 0, \delta^{ij} b_i b_j > 0, a_i \neq \lambda b_i, (u, v) \in \mathbb{R}^2\}. \tag{2.3.9}$$

A straight world-line γ is *M-orthogonal* to a flat submanifold σ, provided every straight world-line in σ that intersects γ is M-orthogonal to γ.

Theorem (2.3.1): *Let $\Sigma_3 \equiv \{x\colon A_i x^i = 0, \delta^{ij} A_i A_j > 0\}$. Then the straight world-line γ, given by the equations $x^i = A^i t$, $t \in \mathbb{R}$ $(A^i \equiv d^{ij} A_j)$, intersects the corresponding 3-flat σ_3 M-orthogonally.*

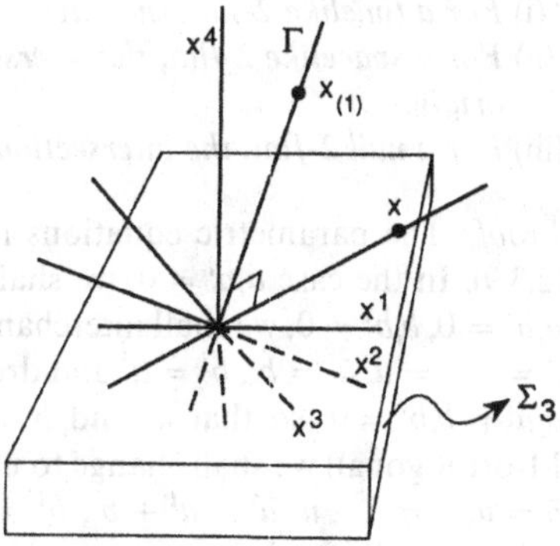

FIGURE 13. M-Orthogonal intersection of γ with σ_3.

Proof: Let $x^i_{(1)} = A^i t_{(1)}$, as in Figure 13. Therefore, we have $d_{ij}x^i x^j_{(1)} = d_{ij}x^i A^j t_{(1)} = (A_i x^i) t_{(1)} = 0$. ■

Example 1: Consider the straight world-line γ corresponding to the x^4-axis. It is given by the parametric equations

$$x^\alpha = A^\alpha t = 0, \qquad x^4 = A^4 t = t.$$

The corresponding M-orthogonal 3-flat σ_3 is given by

$$\Sigma_3 = \{x\colon x^4 = 0, (x^1, x^2, x^3) \in \mathbb{R}^3\}. \quad \square$$

Example 2: A 2-flat σ_2 is given by the corresponding subset Σ_2 with parametric equations

$$x^k = a^k u + b^k v,$$
$$a^i = \delta^i_1, \qquad b^i = \delta^i_4.$$

The same subset Σ_2 is also given by $\Sigma_2 = \{x\colon x^2 = x^3 = 0\}$. □

We can classify 3-flats corresponding to the value of $d^{ij}A_iA_j$. A 3-flat is called *spacelike*, *timelike*, or *null* in case $d^{ij}A_iA_j < 0$, $d^{ij}A_iA_j > 0$, or $d^{ij}A_iA_j = 0$ respectively.

We can classify 2-flats according to the value of

$$\Delta \equiv (a_i b^i)^2 - (a_i a^i)(b_j b^j),$$

with $a_i b^i \equiv d_{ij}a^i b^j$.

(i) In the case $\Delta > 0$, the 2-flat σ_2 is called *timelike*.
(ii) In the case $\Delta < 0$, the 2-flat σ_2 is called *spacelike*.
(iii) In the case $\Delta = 0$, the 2-flat σ_2 is called *null*.

Theorem (2.3.2): *Let the null-cone corresponding to the points* $N_0 \equiv \{x\colon d_{ij}x^i x^j = 0\}$ *intersect with a 2-flat corresponding to* Σ_2*, which passes through the origin.*

(i) *For a timelike 2-flat, the intersection contains two null-lines.*
(ii) *For a spacelike 2-flat, the intersection is at one event corresponding to the origin.*
(iii) *For a null 2-flat, the intersection is a single null-line.*

Proof: The parametric equations for Σ_2 are $x^i = a^i u + b^i v$, as in equation (2.3.9). In the case $a_i a^i \neq 0$, we shall retain this parametrization. In the case $a_i a^i = 0$, $b_i b^i \neq 0$, we shall interchange parameters and coefficients by putting $\bar{u} = v$, $\bar{v} = u$, $\bar{a}^i = b^i$, $\bar{b}^i = a^i$ and dropping the bars subsequently. In the case $a_i a^i = b_i b^i = 0$ (so that a^i and b^i are components of two null vectors not M-orthogonal) we shall change to new parameters and coefficients by putting $\bar{u} = u$, $\bar{v} = v - u$, $\bar{a}^i = a^i + b^i$, $\bar{b}^i = b^i$ ($\bar{a}_i \bar{a}^i = 2a_i b^i \neq 0$) and dropping bars subsequently. So, by proper parametrization we can always represent Σ_2 by equation (2.3.9), with $a_i a^i \neq 0$. Now, for the intersection with the null-cone, we must have

$$\begin{aligned} d_{ij} x^i x^j &= d_{ij}(a^i u + b^i v)(a^j u + b^j v) \\ &= (a_i a^i)u^2 + 2(a_i b^i)uv + (b_i b^i)v^2 \\ &= 0. \end{aligned} \tag{2.3.10}$$

Solving for u we find

$$u = (a_k a^k)^{-1}[-(a_i b^i) \pm \sqrt{\Delta}]v.$$

(i) For a timelike 2-flat $\Delta > 0$ and u has two distinct solutions. Putting these solutions back into (2.3.9) we get

$$\begin{aligned} x^i &= \mathscr{X}^i(v) \equiv \{a^i(a_k a^k)^{-1}[-(a_j b^j) + \sqrt{\Delta}] + b^i\}v \equiv n^i v, \\ x^i &= \mathscr{X}'^i(v) \equiv -\{a^i(a_k a^k)^{-1}[(a_j b^j) + \sqrt{\Delta}] - b^i\}v \equiv n'^i v, \end{aligned} \tag{2.3.11}$$

where $v \in \mathbb{R}$. Clearly (2.3.11) yields two null-lines, since $n_i n^i = n'_i n'^i = 0$.

(ii) For a spacelike 2-flat $\Delta < 0$ and u has complex roots unless $v = 0$. In that case $u = v = 0$, $x^i = 0$, which denotes the origin.

(iii) For a null 2-flat $\Delta = 0$ and u has one real solution. Substituting this solution into (2.3.9) we obtain

$$x^i = \mathscr{X}^i_{(0)}(v) \equiv [b^i - a^i(a_k a^k)^{-1}(a_j b^j)]v \equiv n^i_{(0)} v, \tag{2.3.12}$$

where $v \in \mathbb{R}$. Obviously, (2.3.12) yields a single null-line ($n_{(0)i} n^i_{(0)} = 0$) passing through the origin. ■

Examples: (i) An example of a timelike 2-flat is given by

$$\Sigma_2 \equiv \{x: x^i = \delta^i{}_1 u + \delta^i{}_4 v, (u, v) \in \mathbb{R}^2\}.$$

It is clear that $\Delta = 1$. The two null-lines of intersection with N_0 as given by equation (2.3.11) are

$$x^1 = x^4 = v, \qquad x^2 = x^3 = 0,$$

$$x^1 = -x^4 = -v, \qquad x^2 = x^3 = 0.$$

(ii) An example of a spacelike 2-flat is given by

$$\Sigma_2 \equiv \{x: x^i = \delta^i{}_1 u + \delta^i{}_2 v, (u, v) \in \mathbb{R}^2\}.$$

It is clear that $\Delta = -1$. The intersection with N_0 is only at the origin $x = 0$.

(iii) An example of a null 2-flat is given by

$$\Sigma_2 \equiv \{x: x^i = \delta^i{}_1 u + (\delta^i{}_4 - \delta^i{}_2)v, (u, v) \in \mathbb{R}^2\}.$$

It is clear that $\Delta = 0$. The single null-line of intersection with N_0 corresponds to the equation

$$x^2 = -x^4 = -v, \qquad x^1 = x^3 = 0. \quad \square$$

Now, we shall state and prove a *counterintuitive* theorem about a null 2-flat.

Theorem (2.3.3) (Synge): *Let Σ_2 and Γ_0 correspond to a null 2-flat passing through the origin and the single null-line of intersection of Σ_2 with the null-cone N_0 respectively. Then every straight world-line in Σ_2 that intersects Γ_0 is M-orthogonal to Γ_0.*

Proof: Suppose $x \in \Sigma_2$ and (χ_0, Σ_2) is a chart for Σ_2 as in Figure 14. Then we can write

$$\begin{gathered} x^i = a^i u + b^i v, \\ a_i a^i \neq 0, \qquad a^i \neq \lambda b^i, \qquad (a_i b^i)^2 = (a_i a^i)(b_j b^j). \end{gathered} \tag{2.3.13}$$

For two points x_0, $x^\#$ on Γ_0 we have

$$\begin{aligned} 0 &= d_{ij}(x_0^i - x^{\#i})(x_0^j - x^{\#j}) \\ &= (a_i a^i)[(u_0 - u^\#) + (a_j b^j)(a_k a^k)^{-1}(v_0 - v^\#)]^2. \end{aligned} \tag{2.3.14}$$

To check M-orthogonality of intersection between a line Γ_1 in Σ_2 and the

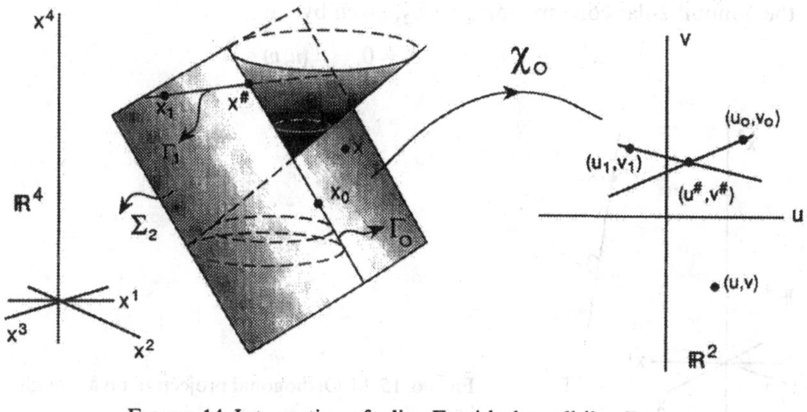

FIGURE 14. Intersection of a line Γ_1 with the null-line Γ_0.

single null-line Γ_0 at $x^{\#}$, we have to compute

$$\begin{aligned}
&d_{ij}(x_0^i - x^{\#i})(x_1^j - x^{\#j}) \\
&= [a_i(u_0 - u^{\#}) + b_i(v_0 - v^{\#})][a^i(u_1 - u^{\#}) + b^i(v_1 - v^{\#})] \\
&= (a_i a^i)(u_0 - u^{\#})(u_1 - u^{\#}) + (b_i b^i)(v_0 - v^{\#})(v_1 - v^{\#}) \\
&\quad + (a_i b^i)[(u_0 - u^{\#})(v_1 - v^{\#}) + (v_0 - v^{\#})(u_1 - u^{\#})] \\
&= (a_i a^i)(u_0 - u^{\#})(u_1 - u^{\#}) + \frac{(a_i b^i)^2 (v_0 - v^{\#})(v_1 - v^{\#})}{(a_j a^j)} \\
&\quad + (a_i b^i)[(u_0 - u^{\#})(v_1 - v^{\#}) + (v_0 - v^{\#})(u_1 - u^{\#})] \\
&= (a_i a^i)\left[(u_0 - u^{\#}) + \frac{(a_k b^k)(v_0 - v^{\#})}{(a_j a^j)}\right]\left[(u_1 - u^{\#}) + \frac{(a_l b^l)(v_1 - v^{\#})}{(a_m a^m)}\right].
\end{aligned}$$

By equation (2.3.14), the above expression vanishes. Thus Γ_1 and Γ_0 intersect M-orthogonally at $x^{\#}$. Furthermore, Γ_1 could be Γ_0 itself. ■

We shall now consider the *M-orthogonal projection* from an event to a straight world-line. Let Γ correspond to a nonnull straight world-line, given by the parametric equations

$$x^i = a^i t + b^i, \qquad a_i a^i \neq 0, \qquad t \in \mathbb{R}.$$

Let y correspond to an event from which an M-orthogonal projection is made onto the line Γ as in Figure 15. We want to obtain the point z on Γ corresponding to the projection. The M-orthogonality implies that $a_i(z^i - y^i) = 0$. Suppose that $z^i = a^i t_1 + b^i$ for some $t_1 \in \mathbb{R}$. The M-orthogonality yields $t_1 = a_i(a_k a^k)^{-1}(y^i - b^i)$. Substituting this value into the expression for z^i, we finally get

$$z^i = a^i a_j (a_k a^k)^{-1}(y^j - b^j) + b^i. \tag{2.3.15}$$

Now, we shall obtain the M-orthogonal projection z from the point y to the nonnull 2-flat corresponding to Σ_2, given by

$$x^i = a^i u + b^i v, \quad \Delta \neq 0, \quad (u, v) \in \mathbb{R}^2.$$

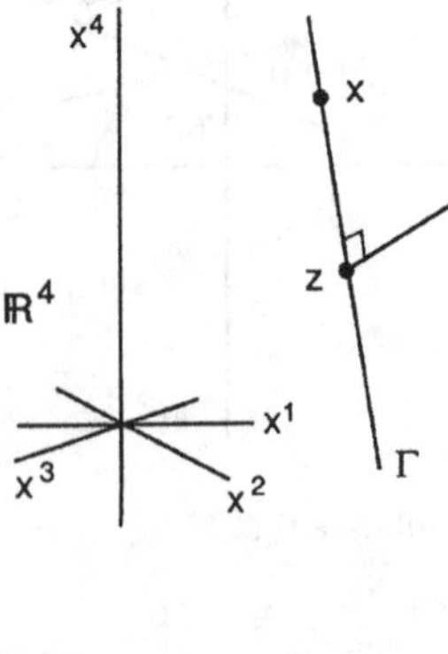

FIGURE 15. M-Orthogonal projection on a straight line.

From the M-orthogonality, we have $a_i(z^i - y^i) = b_i(z^i - y^i) = 0$. Let $z^i = a^i u_1 + b^i v_1$ for some $(u_1, v_1) \in \mathbb{R}^2$. The M-orthogonality yields solutions

$$u_1 = -(a_i y^i b_l b^l - a_j b^j b_k y^k)/\Delta,$$

$$v_1 = -(a_i a^i b_l y^l - a_j y^j a_k b^k)/\Delta.$$

Substituting these values, we get the expression

$$z^r = -a^r(a_i y^i b_l b^l - a_j b^j b_k y^k)/\Delta - b^r(a_i a^i b_l y^l - a_j y^j a_k b^k)/\Delta. \quad (2.3.16)$$

In the case $a_i b^i = 0$, the equation (2.3.16) reduces to

$$z^r = (a^r a_j y^j / a_k a^k) + (b^r b_j y^j / b_k b^k). \quad (2.3.17)$$

EXERCISES 2.3

1. Suppose that a nonnull 2-flat

$$\Sigma_2 \equiv \{x: x^i = a^i u + b^i v, \delta^{ij} a_i a_j > 0, \delta^{ij} v_i b_j > 0, a_i \neq \lambda b_i, u, v \in \mathbb{R}\}.$$

The M-orthogonal 2-flat is defined by

$$\Sigma_2^\perp \equiv \{x: a_i x^i = 0, b_i x^i = 0\}.$$

(i) Prove that the intersection $\Sigma_2 \cap \Sigma_2^\perp = \{0\}$.

(ii) Prove that every straight line in $\Sigma_2^\perp$ passing through the origin is M-orthogonal to every straight line in Σ_2 passing through the origin.

(iii) Prove that if Σ_2 corresponds to a null 2-flat then $\Sigma_2^\perp$ corresponds to another null 2-flat.

2. Suppose that y and $\Sigma_3 \equiv \{x: A_i x^i = 0, A_i A^i \neq 0\}$ correspond to an event and a 3-flat respectively ($y \notin \Sigma_3$). Obtain explicitly the point z that corresponds to the M-orthogonal projection of y onto Σ_3.

2.4. Minkowski Tensor Fields on $\mathbf{M}_4$

Consider two charts (χ, U) and $(\hat{\chi}, \hat{U})$ in $\mathbf{M}_4$ that are $\mathscr{C}^3$-related and the transformations are given by

$$\hat{x}^k = \hat{X}^k(x) \quad \text{for } x \in D_s,$$

$$x^k = X^k(\hat{x}) \quad \text{for } \hat{x} \in \hat{D}_s,$$

$$\frac{\partial(\hat{x}^1, \hat{x}^2, \hat{x}^3, \hat{x}^4)}{\partial(x^1, x^2, x^3, x^4)} \neq 0 \quad \text{for } x \in D_s.$$

Let us consider two continuous vector fields $\mathbf{u}(p)$, $\mathbf{v}(p)$ for all $p \in U \cap \hat{U}$. If both charts are Minkowski charts then the inner product is given in terms of the components by

$$\mathbf{u}(p) \cdot \mathbf{v}(p) = \mathbf{u}[\chi^{-1}(x)] \cdot \mathbf{v}[\chi^{-1}(x)] = d_{ij} u^i(x) v^j(x)$$

$$= \mathbf{u}[\hat{\chi}^{-1}(\hat{x})] \cdot \mathbf{v}[\hat{\chi}^{-1}(\hat{x})] = d_{ij} \hat{u}^i(\hat{x}) \hat{v}^j(\hat{x}).$$

By the transformation rules in equation (2.1.9), we obtain

$$\left[d_{kl}\frac{\partial \hat{X}^k(x)}{\partial x^i}\frac{\partial \hat{X}^l(x)}{\partial x^j} - d_{ij}\right]u^i(x)v^j(x) = 0.$$

Since the above equation is valid for every pair of continuous vector fields in $U \cap \hat{U}$, we must have

$$d_{kl}\frac{\partial \hat{X}^k(x)}{\partial x^i}\frac{\partial \hat{X}^l(x)}{\partial x^j} = d_{ij}. \tag{2.4.1}$$

An inhomogeneous linear transformation

$$\hat{x}^i = \hat{X}^i(x) \equiv a^i + l^i{}_j x^j, \tag{2.4.2}$$

where $a^i \in \mathbb{R}$ and $l^i{}_j$ are the entries of a Lorentz matrix, is called a *Poincaré transformation.*

Theorem (2.4.1): *The coordinate transformation from one Minkowski chart to another is a Poincaré transformation* [*equation* (2.4.2)].

Proof: We start with the system of first-order nonlinear partial differential equations (2.4.1). We denote partial derivatives by

$$\hat{X}^k_i \equiv \frac{\partial \hat{X}^k(x)}{\partial x^i}, \qquad \hat{X}^k_{is} \equiv \frac{\partial^2 \hat{X}^k(x)}{\partial x^s \partial x^i},$$

which are all continuous functions. Since $\hat{X}^k$ are $\mathscr{C}^3$-functions,

$$\hat{X}^k_{is} \equiv \frac{\partial^2 \hat{X}^k(x)}{\partial x^s \partial x^i} = \frac{\partial^2 \hat{X}^k(x)}{\partial x^i \partial x^s} = \hat{X}^k_{si}.$$

Differentiating equation (2.4.1) with respect to x^s we get

$$d_{kl}(\hat{X}^k_{is}\hat{X}^l_j + \hat{X}^k_i\hat{X}^l_{js}) = 0. \tag{2.4.3}$$

Cyclically permuting the indices yields

$$d_{kl}(\hat{X}^k_{ji}\hat{X}^l_s + \hat{X}^k_j\hat{X}^l_{si}) = 0. \tag{2.4.4}$$

Subtracting (2.4.3) from (2.4.4) and using $\hat{X}^k_{is} = \hat{X}^k_{si}$, we have

$$d_{kl}(\hat{X}^k_{ji}\hat{X}^l_s - \hat{X}^k_i\hat{X}^l_{js}) = 0. \tag{2.4.5}$$

Interchanging s and j in the last equation we get

$$d_{kl}(\hat{X}^k_{si}\hat{X}^l_j - \hat{X}^k_i\hat{X}^l_{sj}) = 0. \tag{2.4.6}$$

Adding equations (2.4.3) and (2.4.6) we have $2d_{kl}\hat{X}^k_{si}\hat{X}^l_j = 0$. Multiplying the above equation with $\frac{1}{2}(\partial X^j(\hat{x})/\partial \hat{x}^a)$ and recalling that $(\partial \hat{X}^l(x)/\partial x^j)\cdot(\partial X^j(\hat{x})/\partial \hat{x}^a) = \delta^l{}_a$, we get

$$d_{ka}\hat{X}^k_{si} = 0. \tag{2.4.7}$$

Multiplying the above equation by d^{ja} we obtain

$$\frac{\partial^2 \hat{X}^j(x)}{\partial x^i \partial x^s} = 0. \tag{2.4.8}$$

Integrating the above linear partial differential equations we have $\partial\hat{X}^i(x)/\partial x^j = l^i{}_j$, $\hat{x}^i = \hat{X}^i(x) = a^i + l^i{}_j x^j$, where the a^i and the $l^i{}_j$ are constants of integration. Since the equation (2.4.1) has to be satisfied, the matrix of the constants $L = [l^i{}_j]$ must satisfy $L^{\mathrm{T}}DL = D$. ■

The Lorentz matrix L, which satisfies the condition (1.2.8), has six independent constants $l^i{}_j$. In a Poincaré transformation (2.4.2), the four constants a^i are independent. Thus the Poincaré transformations involve ten independent parameters. In the case where $a^i \equiv 0$, the transformation reduces to

$$\hat{x}^i = l^i{}_j x^j, \qquad L^{\mathrm{T}}DL = D. \tag{2.4.9}$$

The above transformation, which involves six parameters, is called a *Lorentz transformation* of coordinates. Note that a Lorentz matrix is involved in *both* the Lorentz mapping of $\mathbf{V}_4$ and the Lorentz transformation of coordinates in $\mathbf{M}_4$.

Under a Poincaré transformation (2.4.2), the components $u^i(x)$ of a vector field will undergo the transformation

$$\hat{u}^k(\hat{x}) = \frac{\partial \hat{X}^k(x)}{\partial x^j} u^j(x) = l^k{}_j u^j(x). \tag{2.4.10}$$

The above equation reminds us of the equation (1.3.1). So we shall define the transformation rules for the components of an $(r + s)$-order (or rank) *Minkowski tensor field* under a Poincaré transformation by analogy to (1.3.8). Thus we assume

$$\begin{gathered} \hat{\tau}^{i_1 \cdots i_r}_{j_1 \cdots j_s}(\hat{x}) = l^{i_1}{}_{k_1} \cdots l^{i_r}{}_{k_r} a^{m_1}{}_{j_1} \cdots a^{m_s}{}_{j_s} \tau^{k_1 \cdots k_r}_{m_1 \cdots m_s}(x), \\ l^i{}_j a^j{}_k = a^i{}_j l^j{}_k = \delta^i{}_k. \end{gathered} \tag{2.4.11}$$

The *apparent switching* of the $l^i{}_k$, $a^m{}_j$ in (1.3.8) and (2.4.11) is due to similar switching in equations (1.3.1) and (2.4.10).

Example: Consider a scalar field $\tau(x)$ [a $(0 + 0)$-order tensor field]. The transformation of the scalar field by equation (2.4.11) is

$$\hat{\tau}(\hat{x}) = \tau(x). \tag{2.4.12}$$

Let this scalar field be defined by

$$\tau(x) \equiv (x^1)^2/2 + 3(x^4)^2$$

for $(x^1, x^2, x^3, x^4) \in \mathbb{R}^4$. In the hatted Minkowski coordinates, the scalar field is given by (see equation (2.4.2))

$$\hat{\tau}(\hat{x}) = [a^1{}_k(-a^k + \hat{x}^k)]^2/2 + 3[a^4{}_k(-a^k + \hat{x}^k)]^2. \quad \square$$

We shall generalize the theorems (1.3.2), (1.3.3), and (1.3.4) to create new Minkowski tensor fields out of known Minkowski tensor fields.

Theorem (2.4.2): *Let the components of scalar and Minkowski tensor fields*

$$\lambda(x),\ \mu(x),\ \tau^{i_1\cdots i_r}_{j_1\cdots j_s}(x),\ \sigma^{i_1\cdots i_r}_{j_1\cdots j_s}(x),\ \rho^{k_1\cdots k_p}_{m_1\cdots m_q}(x)$$

be all defined in $D \subset \mathbb{R}^4$. Then

(i) *the linear combination*

$$\lambda(x)\tau^{i_1\cdots i_r}_{j_1\cdots j_s}(x) + \mu(x)\sigma^{i_1\cdots i_r}_{j_1\cdots j_s}(x)$$

produces the components of an $(r + s)$-order Minkowski tensor field in D;

(ii) *the tensor product*

$$\tau^{i_1\cdots i_r}_{j_1\cdots j_s}(x)\rho^{k_1\cdots k_p}_{m_1\cdots m_q}(x)$$

gives the components of an $[(r + p) + (s + q)]$-order Minkowski tensor field in D:

(iii) *the single contraction*

$$\tau^{i_1\cdots i_{k-1}c\,i_{k+1}\cdots i_r}_{j_1\cdots j_{h-1}c\,j_{h+1}\cdots j_s}(x)$$

yields the components of an $[(r - 1) + (s - 1)]$-order Minkowski tensor field in D.

The proof of this theorem is entirely parallel to that of theorems (1.3.2), (1.3.3), and (1.3.4).

The *raising* and *lowering* of indices in a component of a Minkowski tensor field is defined exactly as in equation (1.3.10).

The *partial differentiation* of tensor field components generates a new tensor field. Let us denote a partial differentiation by a comma, that is,

$$\tau_{,a} \equiv \frac{\partial \tau(x)}{\partial x^a}, \qquad \tau^{i_1\cdots i_r}_{j_1\cdots j_s,q} \equiv \frac{\partial}{\partial x^q}\tau^{i_1\cdots i_r}_{j_1\cdots j_s}(x).$$

Theorem (2.4.3): *The partial derivatives $\tau^{i_1\cdots i_r}_{j_1\cdots j_s,a}$ transform as the components of a $[r + (s + 1)]$-order Minkowski tensor field.*

Proof: Using equations (2.4.2), (2.4.11) and the chain rule we have

$$\begin{aligned}
\frac{\partial}{\partial \hat{x}^b}\hat{t}^{k_1\cdots k_r}_{m_1\cdots m_s} &= \frac{\partial X^a}{\partial \hat{x}^b}\frac{\partial}{\partial x^a}[l^{k_1}{}_{i_1}\cdots l^{k_r}{}_{i_r}a^{j_1}{}_{m_1}\cdots a^{j_s}{}_{m_s}\tau^{i_1\cdots i_r}_{j_1\cdots j_s}] \\
&= \left\{\frac{\partial}{\partial \hat{x}^b}[a^a{}_c(-a^c + \hat{x}^c)]\right\}[l^{k_1}{}_{i_1}\cdots l^{k_r}{}_{i_r}a^{j_1}{}_{m_1}\cdots a^{j_s}{}_{m_s}\tau^{i_1\cdots i_r}_{j_1\cdots j_s,a}] \\
&= l^{k_1}{}_{i_1}\cdots l^{k_r}{}_{i_r}a^{j_1}{}_{m_1}\cdots a^{j_s}{}_{m_s}a^a{}_b\tau^{i_1\cdots i_r}_{j_1\cdots j_s,a}. \quad \blacksquare
\end{aligned}$$

Example: Consider a Poincaré transformation

$$\begin{aligned}\hat{x}^1 &= \beta(x^1 - vx^4),\\ \hat{x}^2 &= x^2 - 1\\ \hat{x}^3 &= x^3 + 1,\\ \hat{x}^4 &= \beta(-vx^1 + x^4),\end{aligned} \tag{2.4.13}$$

where $\beta \equiv (1 - v^2)^{-1/2}$ and $0 < |v| < 1$.

Let a contravariant Minkowski vector field be defined by

$$\tau^1(x) \equiv 0, \qquad \tau^2(x) \equiv 0, \qquad \tau^3(x) \equiv 0, \qquad \tau^4(x) \equiv (1/3)(x^4)^3$$

for $(x^1, x^2, x^3, x^4) \in \mathbb{R}^4$. Then the transformed component $\partial\hat{\tau}^1(\hat{x})/\partial\hat{x}^1$ under (2.4.13) is

$$\hat{\tau}^1{}_{,1} = (-\beta v)(\beta v)\tau^4{}_{,4} = -\beta^2 v^2 (x^4)^2 = -\beta^4 v^2 (v\hat{x}^1 + \hat{x}^4)^2. \quad \square$$

We will now briefly discuss integration on the flat space–time manifold $\mathbf{M}_4$. Although elegant treatments involving differential forms and chains can be used, we shall be content here with the usual multiple *Riemann integral*. Suppose that $(\chi, \mathbf{M}_4)$ is a Minkowski chart for $\mathbf{M}_4$. Let $\bar{U}$ be a closed region in $\mathbf{M}_4$ such that $\chi(\bar{U}) = D \cup \partial D$, where ∂D is the boundary of $D \subset \mathbb{R}^4$. Let $f\colon \bar{U} \subset \mathbb{R}^4 \to \mathbb{R}$ be a continuous function. Then the *volume* integral

$$\int_{\bar{U}} f(p)\, d_4 v \equiv \int_{\bar{D}} F(x)\, dx^1\, dx^2\, dx^3\, dx^4, \tag{2.4.14}$$

where $F(x) \equiv f[\chi^{-1}(x)] = f(p)$ and the integral on the right is a multiple Riemann-integral.

Example: Let $F(x) \equiv \exp[-(|x^1| + |x^2| + |x^3| + |x^4|)]$.

$$\begin{aligned}\int_{\mathbf{M}_4} f(p)\, d_4 v &= \int_{\mathbb{R}_4} \exp[-(|x^1| + |x^2| + |x^3| + |x^4|)]\, dx^1\, dx^2\, dx^3\, dx^4\\ &= \int_{\mathbb{R}_3} \left[\int_{-\infty}^{\infty} \exp[-(|x^1| + |x^2| + |x^3| + |x^4|)]\, dx^4\right] dx^1\, dx^2\, dx^3\\ &= 2\int_{\mathbb{R}_2} \left[\int_{-\infty}^{\infty} \exp[-(|x^1| + |x^2| + |x^3|)]\, dx^3\right] dx^1\, dx^2\\ &= 16. \quad \square\end{aligned}$$

To evaluate the above integral, we have used Fubini's Theorem and the definition of an improper Riemann integral of the first kind.

We now *state a very useful theorem* involving integration. It is called *Gauss's Theorem* or *Green's Theorem* on a four-dimensional manifold.

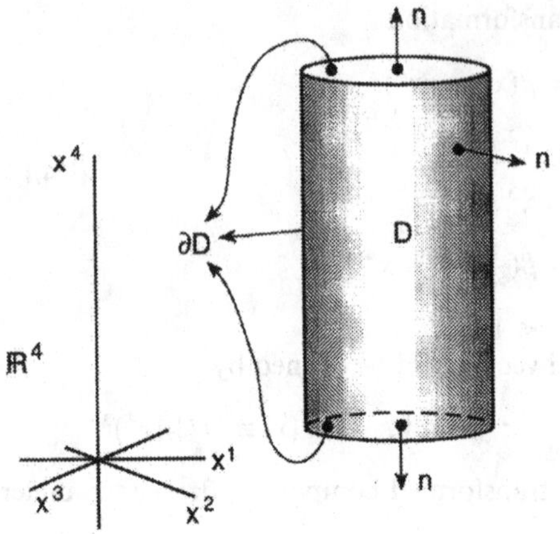

FIGURE 16. The region $D \cup \partial D$ of a four-dimensional integration.

Theorem (2.4.5): *Let $\bar{D} = D \cup \partial D$ be a four-dimensional region of $\mathbb{R}^4$ such that the boundary ∂D is continuous, piecewise differentiable and orientable. Let $n^i(x)$ be the components of a unit ($|d_{ij}n^i n^j| = 1$), outward normal vector to ∂D (as in Fig. 16). Let $\tau^i(x)$ be a differentiable field in $\bar{D}$. Then*

$$\int_{\bar{D}} \tau^i{}_{,i}\, dx^1\, dx^2\, dx^3\, dx^4 = \int_{\partial D} d_{ij}\tau^i(x) n^j(x)\, d_3\sigma, \qquad (2.4.15)$$

where $d_3\sigma$ denotes the invariant, three-dimensional hypersurface element of ∂D.

The proof of this important theorem can be found in references at the end of the chapter.

Example: See Figure 16. Consider the domain

$$D \equiv \{x: (x^1)^2 + (x^2)^2 + (x^3)^2 < 1, t_1 < x^4 < t_2\}$$

with boundary

$$\partial D = \partial D_1 \cup \partial D_2 \cup \partial D_3,$$

$$\partial D_1 = \{x: (x^1)^2 + (x^2)^2 + (x^3)^2 = 1, t_1 \le x^4 \le t_2\},$$

$$\partial D_2 = \{x: (x^1)^2 + (x^2)^2 + (x^3)^2 < 1, x^4 = t_1\},$$

$$\partial D_3 = \{x: (x^1)^2 + (x^2)^2 + (x^3)^2 < 1, x^4 = t_2\}.$$

On ∂D_2 and ∂D_3 the invariant hypersurface element is $d_3\sigma = dx^1\, dx^2\, dx^3$, whereas on ∂D_1 the hypersurface elements is

$$d_3\sigma = dx^1\, dx^2\, dx^4/\sqrt{1 - (x^1)^2 - (x^2)^2}. \quad \square$$

EXERCISES 2.4

1. Prove that, if the 4^{r+s} components $\hat{t}^{i_1 \cdots i_r}_{j_1 \cdots j_s}(\hat{x})$ of a Minkowski tensor field are zero everywhere, then in another coordinate system related by a Poincaré transformation $\tau^{i_1 \cdots i_r}_{j_1 \cdots j_s}(x) \equiv 0$.

2. Suppose that a twice differentiable scalar field $\phi(x)$ satisfies the *wave equation* $\Box\phi(x) \equiv d^{ij}\phi_{,ij} = 0$. Prove that in another Minkowski coordinate system $\hat{\Box}\hat{\phi}(\hat{x}) \equiv d^{ab}\hat{\phi}_{,ab} = 0$.

3. Suppose that $A_{ji}(x) = -A_{ij}(x)$ is an antisymmetric, twice differentiable second-order covariant Minkowski tensor field. Prove that $A_{ij,k} + A_{jk,i} + A_{ki,j}$ transforms as a totally antisymmetric $(0 + 3)$-order Minkowski tensor field.

4. Prove that the four-dimensional divergence equation $j^i{}_{,i} = 0$ can be solved by the expression $j^i = A^{ij}{}_{,j}$, where $A^{ij}(x)$ is an arbitrary antisymmetric twice differentiable $(2 + 0)$-order Minkowski tensor field.

References

1. N. J. Hicks, *Notes on differential geometry*, Van Nostrand, London, 1971. [pp. 27, 30]
2. D. Lovelock and H. Rund, *Tensors, differential forms, and variational principles*, John Wiley and Sons, New York, 1975. [p. 61]
3. W. Noll, Notes on Tensor Analysis prepared by C. C. Wang, The Johns Hopkins University, Mathematics Department, 1963. [p. 61]
4. J. L. Synge, *Relativity: The special theory*, North-Holland, Amsterdam, 1964. [pp. 35, 37, 38, 40, 41, 44, 45, 48, 50, 53, 55, 59]
5. J. L. Synge and A. Schild, *Tensor calculus*, University of Toronto Press, Toronto, 1966. [p. 61]

3
The Lorentz Transformation

3.1. Applications of the Lorentz Transformation

We shall derive the Lorentz transformation by physical arguments. Let us pretend for a short while that we do not know about Minkowski space–time and Minkowski coordinates. Instead, we are aware of space and time and inertial frames of reference. An *inertial observer*, an idealized point observer subject to no forces, is assumed to follow a straight line in Euclidean space E_3. In a suitably chosen Cartesian coordinate system, the straight line of the inertial observer can be given by

$$x^\alpha = \mathscr{X}^\alpha(x^4) \equiv 0,$$
$$x^4 = \mathscr{X}^4(x^4) \equiv x^4 \in \mathbb{R}.$$

The above equation may also be interpreted in an inertial frame of reference as the equation of a straight world-line in space and time. Similarly, a second inertial observer, in relative motion to the first one, can be characterized in another Cartesian coordinate and time system as

$$\hat{x}^\alpha = \hat{\mathscr{X}}^\alpha(\hat{x}^4) \equiv 0,$$
$$\hat{x}^4 = \hat{\mathscr{X}}^4(\hat{x}^4) \equiv \hat{x}^4 \in \mathbb{R}.$$

The transformation between these two coordinate systems in space and time can be given by

$$\begin{aligned} \hat{x}^\alpha &= \hat{X}^\alpha(x^1, x^2, x^3, x^4), \\ \hat{x}^4 &= \hat{X}^4(x^1, x^2, x^3, x^4), \end{aligned} \qquad (3.1.1)$$
$$\frac{\partial(\hat{x}^1, \hat{x}^2, \hat{x}^3, \hat{x}^4)}{\partial(x^1, x^2, x^3, x^4)} \neq 0.$$

At this stage the functions $\hat{X}^k$ are unknown. Now let both inertial observers locate another massive free particle traversing on a straight line in space or a straight world-line in space and time. The parametric equations for this line, in two coordinate systems, are given by

$$x^\alpha = \xi^\alpha(x^4) \equiv v^\alpha x^4 + x_0^\alpha, \qquad \hat{x}^\alpha = \hat{\xi}^\alpha(\hat{x}^4) \equiv \hat{v}^\alpha \hat{x}^4 + \hat{x}_0^\alpha, \tag{3.1.2}$$

where v^α, x_0^α, $\hat{v}^\alpha$, $\hat{x}_0^\alpha$ are constants. In the case (3.1.1) is given by a linear transformation (not necessarily homogeneous)

$$\hat{x}^k = \hat{\mathscr{X}}^k(x) \equiv a^k + \lambda^k{}_j x^j, \qquad \det[\lambda^i{}_j] \neq 0, \tag{3.1.3}$$

the two equations in (3.1.2) are mutually consistent. The converse is also true.

For the sake of simplicity, assume that two inertial observers meet at some initial time at the origin of space. This assumption simplifies (3.1.3) to

$$\hat{x}^k = \hat{\mathscr{X}}^k(x) \equiv \lambda^k{}_j x^j. \tag{3.1.4}$$

Finally we have to incorporate the results of the Michelson–Morley experiment, which implies the constancy of the speed of light relative to inertial observers. Thus we have

$$\begin{aligned} \text{speed of light} &= 1 \\ &= \frac{\sqrt{(x^1)^2 + (x^2)^2 + (x^3)^2}}{|x^4|} = \frac{\sqrt{(\hat{x}^1)^2 + (\hat{x}^2)^2 + (\hat{x}^3)^2}}{|\hat{x}^4|}, \\ \text{or} \quad & d_{ij} x^i x^j = d_{ij} \hat{x}^i \hat{x}^j = 0, \end{aligned} \tag{3.1.5}$$

for *all events* on the *null-cone* corresponding to $\mathscr{N}_0$. The equations (3.1.4) together with (3.1.5) *do not* yield the Lorentz transformation (see problem (3.1.#1)). However, it is usually *assumed* that

$$d_{ij} x^i x^j = d_{ij} \hat{x}^i \hat{x}^j \tag{3.1.6}$$

for all events in space–time. In that case, using (3.1.4) and the fact that $\{x^i x^j: 1 \le i \le j \le 4\}$ is a basis for the second-degree homogeneous polynomials in x^1, x^2, x^3, x^4, it follows that

$$\lambda^k{}_j = l^k{}_j, \tag{3.1.7}$$

where $[l^k{}_j]$ is a Lorentz matrix. Historically, the Lorentz transformation (2.4.9), (2.4.10) was derived from such a sequence of arguments.

Now we shall derive a particular Lorentz transformation and interpret it in physical terms. For that purpose we take inertial observers in a 2-flat $\Sigma_2 \equiv \{x: x^2 = x^3 = 0\}$, as Figure 17.

We restrict the Lorentz transformation so that $\hat{x}^2 = x^2$, $\hat{x}^3 = x^3$. Therefore, the resulting Lorentz transformation and the Lorentz matrix can be written as

$$\begin{aligned} \hat{x}^1 &= l^1{}_1 x^1 + l^1{}_4 x^4, \\ \hat{x}^2 &= x^2, \\ \hat{x}^3 &= x^3, \\ \hat{x}^4 &= l^4{}_1 x^1 + l^4{}_4 x^4, \end{aligned} \tag{3.1.8}$$

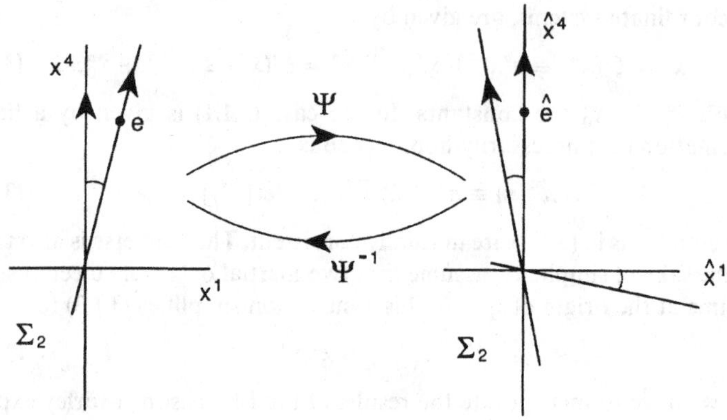

FIGURE 17. The special Lorentz transformation.

$$L = \begin{bmatrix} l^1{}_1 & 0 & 0 & l^1{}_4 \\ 0 & 1 & 0 & 0 \\ 0 & 0 & 1 & 0 \\ l^4{}_1 & 0 & 0 & l^4{}_4 \end{bmatrix}$$

From the condition $L^T DL = D$, we obtain three independent nonlinear coupled algebraic equations:

$$\begin{aligned} (l^1{}_1)^2 - (l^4{}_1)^2 &= 1, \\ (l^4{}_4)^2 - (l^1{}_4)^2 &= 1, \\ l^1{}_4 l^1{}_1 - l^4{}_4 l^4{}_1 &= 0. \end{aligned} \tag{3.1.9}$$

The first two of the above equations can be solved by putting (3.1.10)

$$\begin{aligned} (l^1{}_1)^2 &= \cosh^2\alpha', \\ (l^4{}_1)^2 &= \sinh^2\alpha', \\ (l^4{}_4)^2 &= \cosh^2\alpha, \\ (l^1{}_4)^2 &= \sinh^2\alpha \end{aligned} \tag{3.1.10}$$

for some real parameters α, α'. Now in Newtonian mechanics, which should be a good approximation for low relative velocities, the *Galilean transformation* for the two inertial systems is given by

$$\begin{aligned} \hat{x}^1 &= x^1 - vx^4, \\ \hat{x}^2 &= x^2, \\ \hat{x}^3 &= x^3, \\ \hat{x}^4 &= x^4. \end{aligned} \tag{3.1.11}$$

Comparing the relative signs of the coefficients in (3.1.8), and (3.1.11) we choose $l^1{}_1 > 0$, $l^4{}_4 > 0$. These inequalities yield from (3.1.10), the relations

$$\begin{aligned} l^1{}_1 &= \cosh\alpha', \\ l^4{}_1 &= \pm\sinh\alpha', \\ l^4{}_4 &= \cosh\alpha, \\ l^1{}_4 &= \pm\sinh\alpha. \end{aligned} \tag{3.1.12}$$

The third condition in (3.1.10) implies that $\pm\sinh(\alpha' \pm \alpha) = 0$, so $\alpha' = \pm\alpha$. Thus we can conclude that

$$\begin{aligned} l^1{}_1 &= \cosh\alpha, \\ l^4{}_1 &= \pm\sinh\alpha, \\ l^4{}_4 &= \cosh\alpha, \\ l^1{}_4 &= \pm\sinh\alpha. \end{aligned} \tag{3.1.13}$$

Consider an event corresponding to $\hat{\mathbf{e}} = (0, 0, 0, \hat{x}^4)$ on the hatted observer's world-line (see Fig. 17). The coordinate $\hat{x}^4$ is the proper time of the event $\hat{\mathbf{e}}$ measured by the moving standard clock. For this event, the coordinates of the first observer are $e = (x^1, 0, 0, x^4)$ such that

$$\hat{x}^1 = x^1\cosh\alpha \pm x^4\sinh\alpha = 0,$$

which is equivalent to

$$\text{speed} = |v| = \text{distance/time} = |x^1/x^4| = |\tanh\alpha|. \tag{3.1.14}$$

Since $|\tanh\alpha| < 1$ for all $\alpha \in \mathbb{R}$, we have $|v| < 1$. This inequality yields a *profound physical consequence: The speed of any moving observer or any massive particle must be strictly less than the speed of light.* From (3.1.8), (3.1.12), (3.1.14) we have

$$\begin{aligned} l^1{}_1 &= \beta, \\ l^4{}_1 &= \pm v\beta, \\ l^4{}_4 &= \beta, \\ l^1{}_4 &= \pm v\beta, \end{aligned}$$

where $\beta \equiv 1/\sqrt{1 - v^2}$ and $|v| < 1$. The ambiguity in the sign can be removed by comparison with equation (3.1.11) for small v. Thus we have obtained a special Lorentz transformation (boost)

$$\begin{aligned} \hat{x}^1 &= \beta(x^1 - vx^4), \\ \hat{x}^2 &= x^2, \\ \hat{x}^3 &= x^3, \\ \hat{x}^4 &= \beta(-vx^1 + x^4), \end{aligned} \tag{3.1.15}$$

where $\beta \equiv 1/\sqrt{1-v^2}$ and $|v| < 1$. This is the relativistic generalization of the Galilean transformation (3.1.11).

In general the transformation from the Minkowski coordinates associated with an inertial observer to those associated with another is given by the Poincaré transformation (2.4.2)

$$\hat{x}^i = a^i + l^i{}_j x^j.$$

We shall derive some of the physical consequences of the special Lorentz transformation (3.1.15).

(i) First we shall investigate the *simultaneity of events.* Suppose that in a Minkowski coordinate system associated with the first inertial observer two distinct events $x_{(1)} = (x^1_{(1)}, 0, 0, x^4_{(1)})$, $x_{(2)} = (x^1_{(2)}, 0, 0, x^4_{(2)})$ are simultaneous. Thus we must have $x^4_{(1)} = x^4_{(2)}$, $x^1_{(1)} \neq x^1_{(2)}$. By the Lorentz transformation (3.1.15) we get

$$\begin{aligned}\hat{x}^4_{(2)} - \hat{x}^4_{(1)} &= v\beta(x^1_{(1)} - x^1_{(2)}) + \beta(x^4_{(2)} - x^4_{(1)})\\ &= v\beta(x^1_{(1)} - x^1_{(2)}) \neq 0\end{aligned}$$

for $v \neq 0$. Therefore, the moving observer *does not* view these events as simultaneous. The notion of simultaneity is thus seen to be an observer-dependent relationship between events.

(ii) Secondly, we consider the apparent contraction of a moving rod. The space–time history of a straight rod or string is a world-sheet as in Figure 18. Consider the inertial observer with respect to whom the rod is at rest. Let the

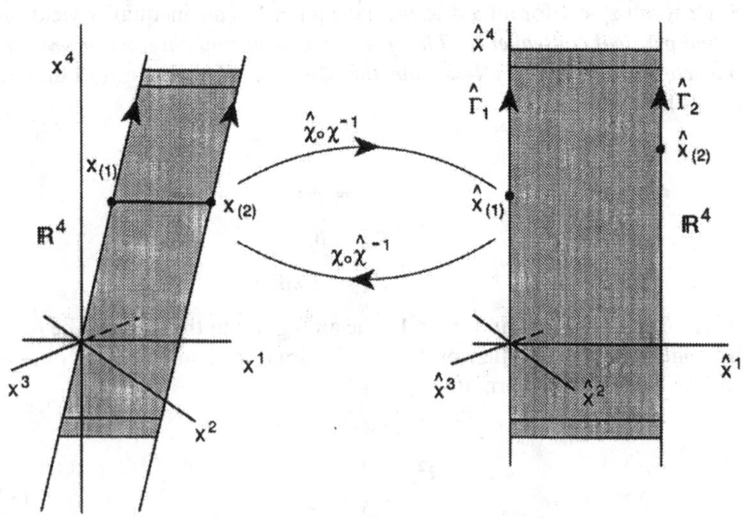

FIGURE 18. Lorentz-Fitzgerald contraction of length.

edges $\hat{\Gamma}_1, \hat{\Gamma}_2$ of the world-strip of the rod be given by

$$\hat{\Gamma}_1 = \{\hat{x}: \hat{x}^1 = \hat{x}^2 = \hat{x}^3 = 0, \hat{x}^4 = \hat{s}\},$$

$$\hat{\Gamma}_2 = \{\hat{x}: \hat{x}^1 = l_0, \hat{x}^2 = \hat{x}^3 = 0, \hat{x}^4 = \hat{s}\}.$$

The length of the rod, as measured by the inertial observer relative to whom the rod is moving with constant velocity v along the x^1-axis, must be

$$l = x^1_{(2)} - x^1_{(1)}, \qquad x_{(1)} = (x^1_{(1)}, 0, 0, x^4_{(1)}), \qquad x_{(2)} = (x^1_{(2)}, 0, 0, x^4_{(2)}),$$
$$x^4_{(2)} = x^4_{(1)}.$$

By the Lorentz transformation (3.1.15) we have

$$\begin{aligned} l_0 &= \hat{x}^1_{(2)} - \hat{x}^1_{(1)} = \beta[x^1_{(2)} - x^1_{(1)} - v(x^4_{(2)} - x^4_{(1)})] \\ &= \beta(x^1_{(2)} - x^1_{(1)}) = \beta l, \\ l &= l_0\sqrt{1 - v^2}. \end{aligned} \tag{3.1.16}$$

This effect is known as the *Lorentz–Fitzgerald contraction* of the moving rod, since, for $v \neq 0$, $l < l_0$. As the speed $|v|$ approaches 1, the speed of light, the moving length l tends to zero!

(iii) We now discuss the *time-retardation* or *time-dilation* of a moving clock. We consider two inertial observers in relative motion with constant velocity v in the direction of the x^1-axis as in Figure 19.

We note two distinct events $\hat{x}_{(2)}, \hat{x}_{(1)}$ on the standard clock of the relatively moving observer. The proper time elapsed between these events is

$$T_0 = \hat{x}^4_{(2)} - \hat{x}^4_{(1)}, \qquad \hat{x}_{(2)} = (0, 0, 0, \hat{x}^4_{(2)}), \qquad \hat{x}_{(1)} = (0, 0, 0, \hat{x}^4_{(1)}).$$

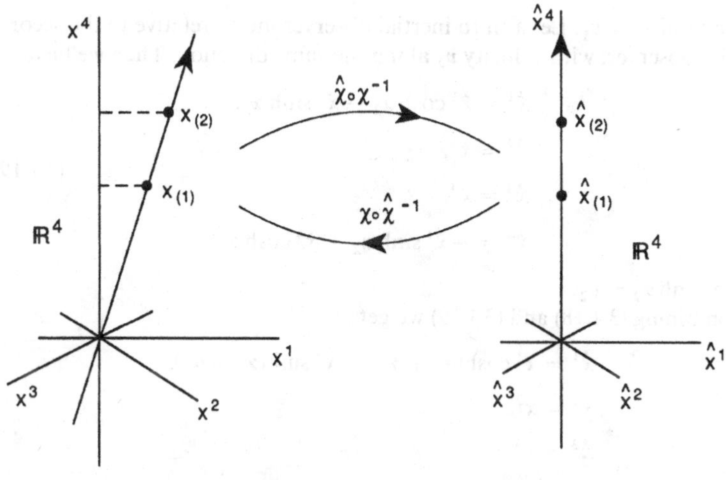

FIGURE 19. Time-Retardation of a moving clock.

The difference in the readings of time for these two events for the stationary observer is $T = x^4_{(2)} - x^4_{(1)}$. The inverse Lorentz transformation of (3.1.15) is given by

$$\begin{aligned}
x^1 &= \beta(\hat{x}^1 + v\hat{x}^4),\\
x^2 &= \hat{x}^2,\\
x^3 &= \hat{x}^3,\\
x^4 &= \beta(v\hat{x}^1 + \hat{x}^4),
\end{aligned}$$

where $\beta \equiv 1/\sqrt{1-v^2}$ and $|v| < 1$. So we can obtain

$$\begin{aligned}
T &= \beta[\hat{x}^4_{(2)} - \hat{x}^4_{(1)} + v(\hat{x}^1_{(2)} - \hat{x}^1_{(1)})]\\
&= \beta(\hat{x}^4_{(2)} - \hat{x}^4_{(1)}) = \beta T_0\\
&= T_0/\sqrt{1-v^2}.
\end{aligned} \tag{3.1.17}$$

Thus, for $v \neq 0$, $T_0 < T$, and time goes at a slower rate on the moving clock according to the judgement of the stationary observer. As the speed $|v|$ approaches the speed of light, the moving clock tends to stop!

(iv) Now we discuss the composition of relative velocities. Suppose that, relative to the first inertial observer, the second one moves with velocity v_1 along the x^1-axis, so that from (3.1.15) we get

$$\begin{aligned}
\hat{x}^1 &= x^1 \cosh\alpha_1 - x^4 \sinh\alpha_1,\\
\hat{x}^2 &= x^2,\\
\hat{x}^3 &= x^3,\\
\hat{x}^4 &= -x^1 \sinh\alpha_1 + x^4 \cosh\alpha_1,
\end{aligned} \tag{3.1.18}$$

where $\tanh\alpha_1 = v_1$. Let a third inertial observer move, relative to the second inertial observer, with velocity v_2 along the same direction. Then we have

$$\begin{aligned}
\hat{\hat{x}}^1 &= \hat{x}^1 \cosh\alpha_2 - \hat{x}^4 \sinh\alpha_2,\\
\hat{\hat{x}}^2 &= \hat{x}^2,\\
\hat{\hat{x}}^3 &= \hat{x}^3,\\
\hat{\hat{x}}^4 &= -\hat{x}^1 \sinh\alpha_2 + \hat{x}^4 \cosh\alpha_2,
\end{aligned} \tag{3.1.19}$$

where $\tanh\alpha_2 = v_2$.

Combining (3.1.18) and (3.1.19) we get

$$\begin{aligned}
\hat{\hat{x}}^1 &= x^1 \cosh(\alpha_1 + \alpha_2) - x^4 \sinh(\alpha_1 + \alpha_2),\\
\hat{\hat{x}}^2 &= x^2,\\
\hat{\hat{x}}^3 &= x^3,\\
\hat{\hat{x}}^4 &= -x^1 \sinh(\alpha_1 + \alpha_2) + x^4 \cosh(\alpha_1 + \alpha_2).
\end{aligned}$$

The above transformation is a Lorentz transformation, and the corresponding relative velocity v must be given by

$$v = \tanh \alpha \equiv \tanh(\alpha_1 + \alpha_2)$$
$$= \frac{\tanh \alpha_1 + \tanh \alpha_2}{1 + \tanh \alpha_1 \tanh \alpha_2} = \frac{v_1 + v_2}{1 + v_1 v_2}. \tag{3.1.20}$$

The above equation gives the velocity v, which is the *composition of velocities* v_1 and v_2. Suppose that $0 < |v_1| < 1$ and v_2 tends to the velocity of light. Then the limiting composite velocity is

$$v = \lim_{v_2 \to 1_-} \frac{v_1 + v_2}{1 + v_1 v_2} = 1.$$

Therefore, the *velocity of light cannot be exceeded* even by combining many relative velocities. Many more physical problems will be discussed in Chapter 5 on relativistic mechanics.

EXERCISES 3.1

1. Let $\hat{x}^i = \lambda^i{}_j x^j$, $\det[\lambda^i{}_j] \neq 0$. Furthermore, assume that $d_{ij}\hat{x}^i\hat{x}^j = 0$ if and only if $d_{ij}x^i x^j = 0$. Then prove that

$$\lambda^i{}_j = \mu l^i{}_j,$$

where $\mu \neq 0$ and $[l^i{}_j]$ is a Lorentz matrix.

2. Consider the equation of a null curve

$$d_{ij}\frac{d\mathscr{X}^i(t)}{dt}\frac{d\mathscr{X}^j(t)}{dt} = 0.$$

Prove that this equation remains invariant under the transformation of coordinates $\hat{x}^i = x^i/(d_{kl}x^k x^l)$, where $x \notin N_0$.

3. Assuming $|v_1| < 1$, $|v_2| < 1$, $|v_3| < 1$, prove that

$$\left|\frac{v_1 + v_2 + v_3 + v_1 v_2 v_3}{1 + v_1 v_2 + v_2 v_3 + v_3 v_1}\right| < 1.$$

Interpret the above inequality in physical terms.

4. Derive the Lorentz transformation from one inertial observer to another such that the components of the relative three-velocity are v_1, v_2, v_3.

3.2. The Lorentz Group $\mathscr{L}_4$

Let us start with the definition of a group. A *group* G is a set of elements with a composition rule which satisfies the following axioms:

$$\begin{aligned}
&\text{G.1.} \quad g_1 \circ g_2 \in G \text{ whenever } g_1, g_2 \in G \quad \text{(closure).}\\
&\text{G.2.} \quad g_1 \circ (g_2 \circ g_3) = (g_1 \circ g_2) \circ g_3 \quad \text{(associative).}\\
&\text{G.3.} \quad \text{There exist an element } e \in G \text{ such that}\\
&\qquad\quad e \circ g = g \circ e = g \text{ for all } g \in G \quad \text{(identity).}\\
&\text{G.4.} \quad \text{Given a } g \in G \text{ there exists } g^{-1} \in G \text{ such}\\
&\qquad\quad \text{that } g \circ g^{-1} = g^{-1} \circ g = e \quad \text{(inverse).}
\end{aligned} \tag{3.2.1}$$

A nonempty subset $H \subseteq G$ is called a *subgroup* of G provided that $h_1 \circ h_2^{-1} \in H$ whenever $h_1, h_2 \in H$. In case commutativity holds, i.e., $g_1 \circ g_2 = g_2 \circ g_1$ for all $g_1, g_2 \in G$, then the group G is called *abelian*. We shall give some simple examples of groups.

Example 1: The real number system $\mathbb{R}$ is an abelian group under the composition rule of addition. The number zero is to be identified with the identity element, and the negative of a number is to be identified with its inverse element. The number of elements in $\mathbb{R}$ is infinite. The subset $\mathbb{Z} \subset \mathbb{R}$ forms a subgroup, which contain a denumerable infinity of elements. □

Example 2: The set of permutations of integers $(1, 2, \ldots, n)$ forms a nonabelian group of permutations S_n called the *symmetric group* on n elements when $n > 2$. It contains $n!$ elements. The subset of even permutations is a subgroup. □

It has been proved in Theorem (2.4.1) that the transformation from one Minkowski coordinate system to another is given by a Poincaré transformation $P_1: \mathbb{R}^4 \to \mathbb{R}^4$, where

$$\hat{x} = P_1(x), \qquad \hat{x}^i = a^i_{(1)} + l_{(1)}{}^i{}_j x^j, \tag{3.2.2}$$

where $L_1 = [l_{(1)}{}^i{}_j]$ is a Lorentz matrix. A successive Poincaré transformation P_2 can be written as

$$\hat{\hat{x}} = P_2(\hat{x}), \qquad \hat{\hat{x}}^i = a^i_{(2)} + l_{(2)}{}^i{}_j \hat{x}^j, \tag{3.2.3}$$

where $L_2 = [l_{(2)}{}^i{}_j]$ is a Lorentz matrix. The composite transformation $P_2 \circ P_1$ is given by

$$\hat{\hat{x}} = (P_2 \circ P_1)(x), \qquad \hat{\hat{x}}^i = a^i_{(2)} + l_{(2)}{}^i{}_j a^j_{(1)} + l_{(2)}{}^i{}_j l_{(1)}{}^j{}_k x^k. \tag{3.2.4}$$

Theorem (3.2.1): *Under the rule of composition $P_2 \circ P_1$, as given in* (3.2.4), *the set of all Poincaré transformations $\mathscr{P}_4 \equiv \mathscr{I}\mathscr{O}(3, 1)$ forms a group.*

Proof: The equation (3.2.4) can be rewritten as

$$\begin{gathered}
\hat{\hat{x}}^i = a^i_{(3)} + \lambda^i{}_k x^k, \\
a^i_{(3)} \equiv a^i_{(2)} + l_{(2)}{}^i{}_j a^j_{(1)}, \qquad \lambda^i{}_k \equiv l_{(2)}{}^i{}_j l_{(1)}{}^j{}_k.
\end{gathered} \tag{3.2.5}$$

Defining the matrix $\Lambda \equiv [\lambda^i{}_k]$, we have

$$\begin{aligned}\Lambda &= L_2 L_1,\\ \Lambda^{\mathrm{T}} D \Lambda &= (L_2 L_1)^{\mathrm{T}} D (L_2 L_1) = L_1{}^{\mathrm{T}} (L_2{}^{\mathrm{T}} D L_2) L_1 \\ &= L_1{}^{\mathrm{T}} D L_1 = D.\end{aligned} \tag{3.2.6}$$

Thus Λ is a Lorentz matrix and the transformation in (3.2.5) is a Poincaré transformation. This proves the closure axiom G.1 in (3.2.1).

The equations (3.2.2)–(3.2.4) can be expressed as matrix equations:

$$\begin{aligned}\hat{x} &= a_{(1)} + L_1 x,\\ \hat{\hat{x}} &= a_{(2)} + L_2 \hat{x} = a_{(2)} + L_2 a_{(1)} + L_2 L_1 x.\end{aligned} \tag{3.2.7}$$

Here we have viewed x, $\hat{x}$, $\hat{\hat{x}}$ as column vectors.

We introduce a third Poincaré transformation

$$x^{\#} = P_3(\hat{\hat{x}}), \qquad x^{\#} = a_3 + L_3 \hat{\hat{x}}.$$

Computing $x^{\#} = P_3(\hat{\hat{x}})$ and $x^{\#} = (P_3 \circ P_2)(\hat{x})$ we get

$$\begin{aligned}x^{\#} &= a_{(3)} + L_3 a_{(2)} + L_3[L_2 a_{(1)} + L_2 L_1 x],\\ x^{\#} &= a_{(3)} + L_3 a_{(2)} + L_3 L_2 [a_{(1)} + L_1 x],\end{aligned} \tag{3.2.8}$$

which are identical due to the associativity of matrix multiplication. Thus $P_3 \circ (P_2 \circ P_1) = (P_3 \circ P_2) \circ P_1$ and the axiom G.2 of (3.2.1) holds.

The identity matrix I is a Lorentz matrix since $\mathrm{I}^{\mathrm{T}} D \mathrm{I} = D$. We choose the identity Poincaré transformation $P_1 = \mathrm{I}$ by putting $a^i_{(1)} \equiv 0$ and $l_{(1)}{}^i{}_j = \delta^i{}_j$. Then equations (3.2.2), (3.2.3) reduce to

$$\begin{aligned}\hat{x}^i &= x^i,\\ \hat{\hat{x}}^i &= a^i_{(2)} + l_{(2)}{}^i{}_j \hat{x}^j = a^i_{(2)} + l_{(2)}{}^i{}_j x^j.\end{aligned}$$

The above proves $P_2 \circ \mathrm{I} = P_2$ and similarly $\mathrm{I} \circ P_2 = P_2$. Thus the axiom G.3 of (3.2.1) holds.

In equation (1.2.11) it was proved that $\det L = \pm 1$ for any Lorentz matrix. Thus every Lorentz matrix has an inverse; furthermore, L^{-1} is a Lorentz matrix (see problem 2 of Exercise 1.2). Recalling the Poincaré transformations in (3.2.2)–(3.2.4), we define the inverse Poincaré transformation $P_2 \equiv P_1^{-1}$ by putting $L_2 \equiv L_1^{-1}$, and $a_2 \equiv -L_1^{-1} a_1$.

Thus equation (3.2.4) reduces to $\hat{\hat{x}}^i = x^i$. Therefore, $P_1^{-1} \circ P_1 = \mathrm{I}$ and similarly $P_1 \circ P_1^{-1} = \mathrm{I}$. This shows the validity of G.4 in (3.2.1). ■

The group $\mathscr{P}_4$ of Poincaré transformations has many subgroups. The translation subgroup associated with the set of Poincaré transformations is characterized by $l_{(1)}{}^i{}_j = \delta^i{}_j$. For such a transformation

$$\hat{x}^i = a^i_{(1)} + \delta^i{}_j x^j = a^i_{(1)} + x^i. \tag{3.2.9}$$

For a second successive translation,

$$\hat{\hat{x}}^i = a^i_{(2)} + \hat{x}^i,$$

and the composite transformation is

$$\hat{\hat{x}}^i = a^i_{(1)} + a^i_{(2)} + x^i.$$

Since $a^i_{(1)} + a^i_{(2)} = a^i_{(2)} + a^i_{(1)}$, the two translations commute. Thus the subgroup of translations is an abelian subgroup of $\mathscr{P}_4$. In the case that the translation parameters $a^i \equiv 0$, the Poincaré transformation reduces to the Lorentz transformation $\hat{x}^i = l^i{}_j x^j$, where $L = [l^i{}_j]$ is a Lorentz matrix. It is easy to prove that the set of all Lorentz transformations forms a subgroup of the Poincaré group $\mathscr{P}_4$. This subgroup is called the *general Lorentz group* $\mathscr{L}_4 \equiv O(3,1)$. The general Lorentz group $\mathscr{L}_4$ has two important subsets. The *orthochronous subset* of $\mathscr{L}_4$ (characterized by $l^4{}_4 \geq 1$) is denoted by $\mathscr{L}_4^+$. The *proper subset* of $\mathscr{L}_4$ with $\det[l^i{}_j] = 1$ is denoted $\mathscr{L}_{4+}$. Furthermore the *proper orthochronous subset* of $\mathscr{L}_4$ with $l^4{}_4 \geq 1$ and $\det[l^i{}_j] = 1$ is denoted by $\mathscr{L}_{4+}^+$.

Theorem (3.2.2): (i) *The subsets $\mathscr{L}_4^+$, $\mathscr{L}_{4+}$ are subgroups of $\mathscr{L}_4$.*
(ii) *$\mathscr{L}_{4+}^+$ is a subgroup of both $\mathscr{L}_4^+$ and $\mathscr{L}_{4+}$.*

Proof: (i) From the (4,4)-entries of the matrix equations

$$L_1 L_2^{-1} \equiv L_1 A_2 = L,$$
$$L_1 D L_1^T = D = A_2^T D A_2,$$

we obtain

$$l_{(1)}{}^4{}_1 a_{(2)}{}^1{}_4 + l_{(1)}{}^4{}_2 a_{(2)}{}^2{}_4 + l_{(1)}{}^4{}_3 a_{(2)}{}^3{}_4 + l_{(1)}{}^4{}_4 a_{(2)}{}^4{}_4 = l^4{}_4,$$
$$(l_{(1)}{}^4{}_1)^2 + (l_{(1)}{}^4{}_2)^2 + (l_{(1)}{}^4{}_3)^2 - (l_{(1)}{}^4{}_4)^2 = -1,$$
$$(a_{(2)}{}^1{}_4)^2 + (a_{(2)}{}^2{}_4)^2 + (a_{(2)}{}^3{}_4)^2 - (a_{(2)}{}^4{}_4)^2 = -1.$$

Using the Schwarz inequality (1.1.13) and $l_{(1)}{}^4{}_4 \geq 1$, $a_{(2)}{}^4{}_4 \geq 1$ (see problem 4 of Exercise 1.2), we get

$$-\sqrt{[(l_{(1)}{}^4{}_1)^2 + (l_{(1)}{}^4{}_2)^2 + (l_{(1)}{}^4{}_3)^2][(a_{(2)}{}^1{}_4)^2 + (a_{(2)}{}^2{}_4)^2 + (a_{(2)}{}^3{}_4)^2]}$$
$$\leq l_{(1)}{}^4{}_1 a_{(2)}{}^1{}_4 + l_{(1)}{}^4{}_2 a_{(2)}{}^2{}_4 + l_{(1)}{}^4{}_3 a_{(2)}{}^3{}_4,$$

or equivalently

$$0 \leq |l_{(1)}{}^4{}_4||a_{(2)}{}^4{}_4|[1 - \sqrt{(1 - |l_{(1)}{}^4{}_4|^{-2})(1 - |a_{(2)}{}^4{}_4|^{-2})}]$$
$$\leq l_{(1)}{}^4{}_1 a_{(2)}{}^1{}_4 + l_{(1)}{}^4{}_2 a_{(2)}{}^2{}_4 + l_{(1)}{}^4{}_3 a_{(2)}{}^3{}_4 + l_{(1)}{}^4{}_4 a_{(2)}{}^4{}_4 = l^4{}_4.$$

Therefore, $l^4{}_4 \geq 1$ and $L_1 \circ L_2^{-1} \in \mathscr{L}_4^+$. Thus $\mathscr{L}_4^+$ is a subgroup. The subset $\mathscr{L}_{4+} \subset \mathscr{L}_4$ is characterized by $\det L = 1$. Suppose that $L_1, L_2 \in L_{4+}$. Then $\det L_1 L_2^{-1} = \det L_1/\det L_2 = 1$. Thus $L_1 \circ L_2^{-1} \in \mathscr{L}_{4+}$ and $\mathscr{L}_{4+}$ is a subgroup of $\mathscr{L}_4$.

(ii) The straightforward proof is ommited. ■

The subset of *improper* Lorentz transformations is characterized by $\det L = -1$. Examples of the improper Lorentz transformation are the space reflection P (1.2.14) and the time reversal T (1.2.15). The subset of *improper* Lorentz transformations is *not a subgroup.*

EXERCISES 3.2

1. Prove that the subset of special Lorentz transformations in (3.1.15), form an abelian subgroup of $\mathscr{L}^+_{4+}$.

2. Consider the subset of Lorentz transformations for which $l^4{}_4 \leq -1$. Determine whether or not this subset is a subgroup.

3. Let $L^+ \in \mathscr{L}^+_4$. Prove that $L^+ \in \mathscr{L}^+_{4+}$, or $L^+ = P \circ L^+_+$, where P denotes the space reflection (1.2.14) and L^+_+ is an element of $\mathscr{L}^+_{4+}$.

3.3. Real Representations of the Lorentz Group $\mathscr{L}_4$

In this section we shall define real matrix representations of a group G and apply this theory to the Lorentz group $\mathscr{L}_4$. Suppose that the set of $n \times n$ matrices with real entries is denoted by M_{nn} (note that there is no sum on n). A mapping $\rho: G \to M_{nn}$ is called a *real matrix representation* of G if

$$\begin{aligned} \rho(e) &= \mathrm{I}, \\ \rho(g_1 \circ g_2) &= \rho(g_1)\rho(g_2), \end{aligned} \tag{3.3.1}$$

for all $g_1, g_2 \in G$. The real vector space of column vectors $V_n \equiv M_{n1}$ is called the *n-dimensional space of representations* ρ of G. The integer n is called the *degree of the representation.*

We shall be interested in real representations of the Lorentz group $\mathscr{L}_4$. The simplest representation is the one-dimensional scalar representation. The next simplest is the four-dimensional self-representation $\mathscr{I}: \mathscr{L}_4 \to M_{44}$, where $\mathscr{I}(L) \equiv L$. For more general representations we have to introduce the concept of the *Kronecker product* (also called the direct product) of two matrices A, B of sizes $m_1 \times n_1$, $m_2 \times n_2$ respectively. The Kronecker product $C = A \times B$, which is of size $m_1 m_2 \times n_1 n_2$, is defined to be

$$C \equiv \begin{bmatrix} a_{11}B & a_{12}B & \cdots & a_{1n_1}B \\ a_{21}B & a_{22}B & \cdots & a_{2n_1}B \\ \vdots & \vdots & & \vdots \\ a_{m_11}B & a_{m_12}B & \cdots & a_{m_1n_1}B \end{bmatrix}$$

$$= \begin{bmatrix} a_{11}b_{11} & a_{11}b_{12} & \cdots & a_{1n_1}b_{1n_2} \\ a_{11}b_{21} & a_{11}b_{22} & \cdots & a_{1n_1}b_{2n_2} \\ \vdots & \vdots & & \vdots \\ a_{m_11}b_{m_21} & a_{m_11}b_{m_22} & \cdots & a_{m_1n_1}b_{m_2n_2} \end{bmatrix} \tag{3.3.2}$$

We will show some elementary examples now.

Example 1:

$$\begin{bmatrix} 1 & -1 \\ 2 & \pi \end{bmatrix} \times [4 - 5] = \begin{bmatrix} 4 & -5 & -4 & 5 \\ 8 & -10 & 4\pi & -5\pi \end{bmatrix}. \quad \square$$

Example 2:

$$\begin{bmatrix} a_1 \\ a_2 \end{bmatrix} \times \begin{bmatrix} b_1 \\ b_2 \\ b_3 \end{bmatrix} = \begin{bmatrix} a_1 b_1 \\ a_1 b_2 \\ a_1 b_3 \\ a_2 b_1 \\ a_2 b_2 \\ a_2 b_3 \end{bmatrix}. \quad \square$$

The main theorem regarding the computations of Kronecker products will be stated below.

Theorem (3.3.1): *Let A, B, C be matrices of sizes $m_1 \times n_1$, $m_2 \times n_2$, $m_3 \times n_3$ respectively. Then the following equations must hold*:

$$\begin{aligned} &\text{(i)} \quad A \times (B \times C) = (A \times B) \times C, \\ &\text{(ii)} \quad (A \times B)^{\mathsf{T}} = A^{\mathsf{T}} \times B^{\mathsf{T}}, \\ &\text{(iii)} \quad \text{If } \hat{A}, \hat{B} \text{ have sizes } n_1 \times m_1, n_2 \times m_2 \text{ respectively then} \\ &\qquad (A \times B)(\hat{A} \times \hat{B}) = (A\hat{A}) \times (B\hat{B}). \end{aligned} \tag{3.3.3}$$

Proof: The proof is left as problem 1 of Exercises 3.3. ■

Suppose that we have a Minkowski tensor field

$$\tau^{k_1 \cdots k_r}(x) \equiv v^{k_1}_{(1)}(x) v^{k_2}_{(2)}(x) \cdots v^{k_r}_{(r)}(x),$$

where $v^{k_i}_{(i)}(x)$ are vector fields. The transformation rule (2.4.11) of this tensor at a special point x_0 is

$$\hat{v}^{i_1}_{(1)} \hat{v}^{i_2}_{(2)} \cdots \hat{v}^{i_r}_{(r)} = l^{i_1}{}_{k_1} \cdots l^{i_r}{}_{k_r} v^{k_1}_{(1)} v^{k_2}_{(2)} \cdots v^{k_r}_{(r)}, \tag{3.3.4}$$

where the $l^i{}_k$ are the entries of a Lorentz matrix L, the hatted quantities are evaluated at $\hat{x}_0$, and the unhatted quantities are evaluated at x_0. Defining a 4×1 column matrix

$$\mathbf{v}_{(i)} \equiv \begin{bmatrix} v^1_{(i)}(x_0) \\ v^2_{(i)}(x_0) \\ v^3_{(i)}(x_0) \\ v^4_{(i)}(x_0) \end{bmatrix},$$

and using equations (3.3.2), (3.3.3) we can rewrite (3.3.4) in matrix language as:

$$\hat{\mathbf{v}}_{(1)} \times \cdots \times \hat{\mathbf{v}}_{(r)} = (L \times \cdots \times L)(\mathbf{v}_{(1)} \times \cdots \times \mathbf{v}_{(r)}). \tag{3.3.5}$$

We define $\Lambda \equiv L \times \cdots \times L$, and it is a square matrix of size $4^r \times 4^r$. We also define $\mathbf{W} \equiv \mathbf{v}_{(1)} \times \cdots \times \mathbf{v}_{(r)}$, and it is a column matrix of size $4^r \times 1$. So the equation (3.3.5) can be expressed as

$$\hat{\mathbf{W}} = \Lambda \mathbf{W}. \tag{3.3.6}$$

Theorem (3.3.2): *The mapping $\rho: \mathscr{L}_4 \to M_{4^r 4^r}$, defined by $\Lambda \equiv \rho(L)$, is a matrix representation of degree 4^r of the general Lorentz group $\mathscr{L}_4$.*

Proof: Let the identity matrix of size $4^r \times 4^r$ be denoted by $\mathrm{I}_{(r)}$ so that $\mathrm{I}_{(r)} = \mathrm{I} \times \cdots \times \mathrm{I}$ (r factors), where I is the 4×4 identity matrix. The matrix I is the self-representation of the identity $\mathrm{I} \in \mathscr{L}_4$. We define $\mathrm{I}_{(r)} \equiv \rho(\mathrm{I})$, $\Lambda \equiv \rho(L)$. Using Theorem (3.3.1), we obtain

$$\begin{aligned}\rho(L_1)\rho(L_2) = \Lambda_1\Lambda_2 &= (L_1 \times \cdots \times L_1)(L_2 \times \cdots \times L_2)\\ &= (L_1L_2) \times \cdots \times (L_1L_2) = \rho(L_1L_2).\end{aligned}$$

Thus by (3.3.1), ρ is a matrix representation of $\mathscr{L}_4$ and the corresponding degree is 4^r. ■

We can conclude that the set of $(r + 0)$-order Minkowski tensor fields at an event is isomorphic to the 4^r-dimensional space of representation of the general Lorentz group $\mathscr{L}_4$.

Let us go back to the matrix equation (3.3.6), viz. $\hat{\mathbf{W}} = \Lambda\mathbf{W}$. Let the set of all column vectors of the size $4^r \times 1$ be denoted $\mathscr{W}$. It may so happen that, for a subspace $\mathscr{W}' \subset \mathscr{W}$, the column vector $\Lambda \mathrm{w}' \in \mathscr{W}'$ for every $\mathrm{w}' \in \mathscr{W}'$. Such a subspace $\mathscr{W}'$ is called an *invariant subspace* of the particular linear mapping corresponding to Λ. In the case $\mathscr{W}'$ is invariant under every Λ, we call $\mathscr{W}'$ an invariant subspace of the matrix representation of the group $\mathscr{L}_4$. The matrix representation is called *irreducible* in case it does not have any invariant subspace other that $\mathscr{W}$ or $\{\mathbf{0}\}$. If the matrix representation has two nontrivial invariant subspaces $\mathscr{W}'$, $\mathscr{W}''$ such that $\mathscr{W}' \cup \mathscr{W}'' = \mathscr{W}$, $\mathscr{W}' \cap \mathscr{W}'' = \{\mathbf{0}\}$, and $\dim \mathscr{W}' > 0$, $\dim \mathscr{W}'' > 0$, the representation is called *completely reducible*. For a completely reducible representation every matrix $\Lambda = \rho(L)$ assumes a block-diagonal form:

$$\Lambda = \left[\begin{array}{c|c} \Lambda' & 0 \\ \hline 0 & \Lambda'' \end{array}\right]. \tag{3.3.7}$$

Example: Consider a $(2 + 0)$-order Minkowski tensor field $\tau^{ab}(x)$ and its transformation properties (2.4.11) at an event x_0. Thus

$$\hat{\tau}^{ij}(\hat{x}_0) = l^i{}_m l^j{}_n \tau^{mn}(x_0). \tag{3.3.8}$$

The matrix representation Λ inherent in the above equation is sixteen-dimensional. But

$$\begin{aligned} \tau^{mn}(x_0) &\equiv s^{mn} + a^{mn}, \\ s^{mn} &\equiv [\tau^{mn}(x_0) + \tau^{nm}(x_0)]/2 = s^{nm}, \\ a^{mn} &\equiv [\tau^{mn}(x_0) - \tau^{nm}(x_0)]/2 = -a^{nm}, \\ \hat{s}^{ij} &= l^i{}_m l^j{}_n s^{mn}, \qquad \hat{a}^{ij} = l^i{}_m l^j{}_n a^{mn}. \end{aligned} \tag{3.3.9}$$

It can easily be proved that $\hat{s}^{ij} = \hat{s}^{ji}$, $\hat{a}^{ij} = -\hat{a}^{ji}$. Therefore, the symmetric tensors are transformed into symmetric tensors and antisymmetric tensors are transformed into antisymmetric tensors. The only tensor that is both symmetric and antisymmetric is the zero tensor. Furthermore, there are ten and six independent components of s^{mn} and a^{mn} respectively. Thus the dimension of the vector space associated with the symmetric tensors is ten and the dimension of the vector space associated with the antisymmetric tensors is six. So the 16×16 matrix representation Λ is completely reducible. To obtain this reduction explicitly we label the tensor components as

$$\begin{gathered} w_1' \equiv s^{11}, \qquad w_2' \equiv s^{12}, \qquad w_3' \equiv s^{13}, \\ w_4' \equiv s^{14}, \qquad w_5' \equiv s^{22}, \qquad w_6' \equiv s^{23}, \\ w_7' \equiv s^{24}, \qquad w_8' \equiv s^{33}, \qquad w_9' \equiv s^{34}, \qquad w_{10}' \equiv s^{44}, \\ w_1'' \equiv a^{12}, \qquad w_2'' \equiv a^{13}, \qquad w_3'' \equiv a^{14}, \\ w_4'' \equiv a^{23}, \qquad w_5'' \equiv a^{24}, \qquad w_6'' \equiv a^{34}. \end{gathered} \tag{3.3.10}$$

The transformation equation (3.3.8) is equivalent to the matrix equation

$$\begin{bmatrix} \hat{\mathbf{w}}' \\ \hline \hat{\mathbf{w}}'' \end{bmatrix} = \left[\begin{array}{c|c} \Lambda' & 0 \\ \hline 0 & \Lambda'' \end{array}\right] \begin{bmatrix} \mathbf{w}' \\ \hline \mathbf{w}'' \end{bmatrix}, \tag{3.3.11}$$

where $\hat{\mathbf{w}}'$ and $\mathbf{w}'$ are 10×1 column vectors, $\hat{\mathbf{w}}''$ and $\mathbf{w}''$ are 6×1 column vectors, and Λ' and Λ'' are given by large matrices, which have been left as an exercise.

The matrix representation Λ'' of the Lorentz group is irreducible, whereas the representation Λ' is completely reducible. The reason behind the reducibility of Λ' is that for a symmetric tensor s^{ij} one has

$$\begin{aligned} s^{ij} &= s_0^{ij} + d^{ij}(s/4), \\ s_0^{ij} &\equiv s^{ij} - d^{ij}(s/4) = s_0^{ji}, \\ s &\equiv d_{ij} s^{ij}, \qquad d_{ij} s_0^{ij} \equiv 0. \end{aligned} \tag{3.3.12}$$

Under a Lorentz transformation, $\hat{s} = s$, $\hat{s}_0^{ij} = l^i{}_a l^j{}_b s_0^{ab} = \hat{s}_0^{ji}$, $d_{ij}\hat{s}_0^{ij} \equiv 0$. Therefore, the one-dimensional subspace of the scalar s and the nine-dimensional subspace of the trace-free symmetric tensor s_0^{ij} are invariant subspaces of the matrix representation of the Lorentz group $\mathscr{L}_4$. Thus

$$\Lambda' = \left[\begin{array}{c|c} \Lambda_1' & 0 \\ \hline 0 & \Lambda_2' \end{array}\right] = \left[\begin{array}{c|c} \Lambda_1' & 0 \\ \hline 0 & 1 \end{array}\right],$$

where Λ' is a 10×10 matrix, Λ_1' is a 9×9 matrix, and Λ_1', Λ_2' are irreducible representations. □

Higher-order tensor fields may exhibit *intermediate* symmetries called *Young symmetries*, which are between total symmetry and total antisymmetry. These intermediate symmetries also produce irreducible representations.

EXERCISES 3.3

1. (i) Prove the Theorem (3.3.1).
(ii) Let A and B be two square matrices of sizes $m \times m$, $n \times n$ respectively. Then prove that $\operatorname{trace}(A \times B) = \operatorname{trace} A \cdot \operatorname{trace} B$.

2. Let $\Lambda \equiv L \times \cdots \times L$, $\mathscr{D} \equiv D \times \cdots \times D$, where there are r factors in each of the definitions and L is a Lorentz matrix. Show that $\Lambda^{\mathrm{T}}\mathscr{D}\Lambda = \mathscr{D}$.

3. Let $T^{ijk}(x)$ be an arbitrary Minkowski $(3 + 0)$-order tensor field with sixty-four components. One can write the identity at any point x_0

$$\begin{aligned}
T^{ijk} &\equiv S^{ijk} + A^{ijk} + Y^{ijk} + Y'^{ijk}, \\
S^{ijk} &\equiv [T^{ijk} + T^{jki} + T^{kij} + T^{ikj} + T^{jik} + T^{kji}]/6, \\
A^{ijk} &\equiv [T^{ijk} + T^{jki} + T^{kij} - T^{ikj} - T^{jik} - T^{kji}]/6, \\
Y^{ijk} &\equiv [T^{ijk} - T^{jki} + T^{jik} - T^{kji}]/3, \\
Y'^{ijk} &\equiv [T^{ijk} + T^{kji} - T^{jik} - T^{kij}]/3.
\end{aligned}$$

From the above identities prove that the sixty-four-dimensional representation of the Lorentz group decomposes into the sum of three twenty-dimensional representations and one four-dimensional representation of the Lorentz group.

4. Explicitly find the matrices Λ' and Λ'' in the block decomposition (3.1.11).

3.4. The Lie Group $\mathscr{L}_{4+}^{+}$

Let us obtain the Lorentz matrix of an arbitrary proper orthochronous Lorentz mapping in terms of the six canonical parameters. In case of a three-dimensional rotation, the rotation matrix can be conveniently expressed in terms of three Eulerian angles. In this scheme one orthonormal triad is rotated into another in three steps, each step being a rotation about an axis.

In expressing a proper orthochronous Lorentz matrix, we shall carry out a scheme closely analogous to the Eulerian scheme, using six steps instead of three. We first choose an orthonormal tetrad $\{\mathbf{e}_1, \mathbf{e}_2, \mathbf{e}_3, \mathbf{e}_4\}$, satisfying equation (1.1.7). Changing the notation, we shall write

$$\begin{gathered}\mathbf{i} \equiv \mathbf{e}_1, \quad \mathbf{j} \equiv \mathbf{e}_2, \quad \mathbf{k} \equiv \mathbf{e}_3, \quad \mathbf{l} \equiv \mathbf{e}_4, \\ \mathbf{i}\cdot\mathbf{i} = \mathbf{j}\cdot\mathbf{j} = \mathbf{k}\cdot\mathbf{k} = -\mathbf{l}\cdot\mathbf{l} = 1, \\ \mathbf{i}\cdot\mathbf{j} = \mathbf{i}\cdot\mathbf{k} = \mathbf{i}\cdot\mathbf{l} = \mathbf{j}\cdot\mathbf{k} = \mathbf{j}\cdot\mathbf{l} = \mathbf{k}\cdot\mathbf{l} = 0.\end{gathered} \tag{3.4.1}$$

The equation (1.2.16) can be expressed as

$$\begin{aligned}\hat{\mathbf{i}} &= \mathbf{i}\cos\theta + \mathbf{j}\sin\theta \\ \hat{\mathbf{j}} &= -\mathbf{i}\sin\theta + \mathbf{j}\cos\theta, \\ \hat{\mathbf{k}} &= \mathbf{k}, \\ \hat{\mathbf{l}} &= \mathbf{l},\end{aligned} \tag{3.4.2}$$

where $-\pi < \theta < \pi$.

The above proper orthochronous Lorentz mapping can be called a spacelike rotation by an angle θ of the 2-flat spanned by $\{\mathbf{i}, \mathbf{j}\}$ about the fixed 2-flat spanned by $\{\mathbf{k}, \mathbf{l}\}$. Another proper orthochronous Lorentz mapping can be given as

$$\begin{aligned}\hat{\mathbf{i}} &= \mathbf{i}, \\ \hat{\mathbf{j}} &= \mathbf{j}, \\ \hat{\mathbf{k}} &= \mathbf{k}\cosh\alpha + \mathbf{l}\sinh\alpha, \\ \hat{\mathbf{l}} &= \mathbf{k}\sinh\alpha + \mathbf{l}\cosh\alpha,\end{aligned} \tag{3.4.3}$$

where $\alpha \in \mathbb{R}$.

The above Lorentz mapping can be called a timelike pseudorotation by the pseudoangle α of the 2-flat spanned by $\{\mathbf{k}, \mathbf{l}\}$ about the fixed 2-flat spanned by $\{\mathbf{i}, \mathbf{j}\}$.

The six steps of Lorentz mapping from the basis $\{\mathbf{i}, \mathbf{j}, \mathbf{k}, \mathbf{l}\}$ to $\{\hat{\mathbf{i}}, \hat{\mathbf{j}}, \hat{\mathbf{k}}, \hat{\mathbf{l}}\}$ can be schematically exhibited by

$$\begin{aligned}\{\mathbf{i}, \mathbf{j}, \mathbf{k}, \mathbf{l}\} &\xrightarrow{\alpha_1} \{\mathbf{i}_1, \mathbf{j}_1, \mathbf{k}_1, \mathbf{l}_1\} \xrightarrow{\alpha_2} \{\mathbf{i}_2, \mathbf{j}_2, \mathbf{k}_2, \mathbf{l}_2\} \xrightarrow{\alpha_3} \{\mathbf{i}_3, \mathbf{j}_3, \mathbf{k}_3, \mathbf{l}_3\} \\ &\xrightarrow{\theta_1} \{\mathbf{i}_4, \mathbf{j}_4, \mathbf{k}_4, \mathbf{l}_4\} \xrightarrow{\theta_2} \{\mathbf{i}_5, \mathbf{j}_5, \mathbf{k}_5, \mathbf{l}_5\} \xrightarrow{\theta_3} \{\hat{\mathbf{i}}, \hat{\mathbf{j}}, \hat{\mathbf{k}}, \hat{\mathbf{l}}\}.\end{aligned}$$

We shall employ the following abbreviations in the sequel:

$$C_\rho \equiv \cosh\alpha_\rho, \qquad S_\rho \equiv \sinh\alpha_\rho, \qquad c_\rho \equiv \cos\theta_\rho, \qquad s_\rho \equiv \sin\theta_\rho,$$

for $\rho = 1, 2, 3$. The six mappings can be expressed as

$$\begin{array}{lll}\mathbf{i}_1 = C_1\mathbf{i} + S_1\mathbf{l}, & \mathbf{i}_2 = \mathbf{i}_1, & \mathbf{i}_3 = \mathbf{i}_2, \\ \mathbf{j}_1 = \mathbf{j}, & \mathbf{j}_2 = C_2\mathbf{j}_1 + S_2\mathbf{l}_1, & \mathbf{j}_3 = \mathbf{j}_2, \\ \mathbf{k}_1 = \mathbf{k}, & \mathbf{k}_2 = \mathbf{k}_1, & \mathbf{k}_3 = C_3\mathbf{k}_2 + S_3\mathbf{l}_2, \\ \mathbf{l}_1 = S_1\mathbf{i} + C_1\mathbf{l}, & \mathbf{l}_2 = S_2\mathbf{j}_1 + C_2\mathbf{l}_1, & \mathbf{l}_3 = S_3\mathbf{k}_2 + C_3\mathbf{l}_2,\end{array}$$

$$\begin{aligned}
&\mathbf{i}_4 = c_1\mathbf{i}_3 + s_1\mathbf{j}_3, &&\mathbf{i}_5 = c_2\mathbf{i}_4 - s_2\mathbf{k}_4, &&\hat{\mathbf{i}} = c_3\mathbf{i}_5 + s_3\mathbf{j}_5,\\
&\mathbf{j}_4 = -s_1\mathbf{i}_3 + c_1\mathbf{j}_3, &&\mathbf{j}_5 = \mathbf{j}_4, &&\hat{\mathbf{j}} = -s_3\mathbf{i}_5 + c_3\mathbf{j}_5,\\
&\mathbf{k}_4 = \mathbf{k}_3, &&\mathbf{k}_5 = s_2\mathbf{i}_4 + c_2\mathbf{k}_4, &&\hat{\mathbf{k}} = \mathbf{k}_5,\\
&\mathbf{l}_4 = \mathbf{l}_3, &&\mathbf{l}_5 = \mathbf{l}_4, &&\hat{\mathbf{l}} = \mathbf{l}_5.
\end{aligned} \tag{3.4.4}$$

The above six mappings can be combined into

$$\begin{aligned}
\hat{\mathbf{i}} &= [(c_1c_2c_3 - s_1s_3)C_1 + (s_1c_2c_3 + c_1s_3)S_1S_2 - s_2c_3S_1C_2S_3]\mathbf{i}\\
&\quad + [(s_1c_2c_3 + c_1s_3)C_2 - s_2c_3S_2S_3]\mathbf{j} - s_2c_3C_3\mathbf{k}\\
&\quad + [(c_1c_2c_3 - s_1s_3)S_1 + (s_1c_2c_3 + c_1s_3)C_1S_2 - s_2c_3C_1C_2S_3]\mathbf{l},\\
\hat{\mathbf{j}} &= [-(c_1c_2s_3 + s_1c_3)C_1 + (c_1c_3 - s_1c_2s_3)S_1S_2 + s_2s_3S_1C_2S_3]\mathbf{i}\\
&\quad + [(c_1c_3 - s_1c_2s_3)C_2 + s_2s_3S_2S_3]\mathbf{j} + s_2s_3C_3\mathbf{k}\\
&\quad + [-(c_1c_2s_3 + s_1c_3)S_1 + (c_1c_3 - s_1c_2s_3)C_1S_2 + s_2s_3C_1C_2S_3]\mathbf{l},\\
\hat{\mathbf{k}} &= [c_1s_2C_1 + s_1s_2S_1S_2 + c_2S_1C_2S_3]\mathbf{i} + [s_1s_2C_2 + c_2S_2S_3]\mathbf{j}\\
&\quad + c_2C_3\mathbf{k} + [c_1s_2S_1 + s_1s_2C_1S_2 + c_2C_1C_2S_3]\mathbf{l},\\
\hat{\mathbf{l}} &= S_1C_2C_3\mathbf{i} + S_2C_3\mathbf{j} + S_3\mathbf{k} + C_1C_2C_3\mathbf{l}.
\end{aligned} \tag{3.4.5}$$

Using equations (1.2.2) and (3.4.5) one can obtain the entries of the corresponding Lorentz matrix $L = [l^i{}_j]$ as

$$\begin{aligned}
l^1{}_1 &= (c_1c_2c_3 - s_1s_3)C_1 + (s_1c_2c_3 + c_1s_3)S_1S_2 - s_2c_3S_1C_2S_3,\\
l^1{}_2 &= -(c_1c_2s_3 + s_1c_3)C_1 + (c_1c_3 - s_1c_2s_3)S_1S_2 + s_2s_3S_1C_2S_3,\\
l^1{}_3 &= c_1s_2C_1 + s_1s_2S_1S_2 + c_2S_1C_2S_3,\\
l^1{}_4 &= S_1C_2C_3,\\
l^2{}_1 &= (s_1c_2c_3 + c_1s_3)C_2 - s_2c_3S_2S_3,\\
l^2{}_2 &= (c_1c_3 - s_1c_2s_3)C_2 + s_2s_3S_2S_3,\\
l^2{}_3 &= s_1s_2C_2 + c_2S_2S_3,\\
l^2{}_4 &= S_2C_3,\\
l^3{}_1 &= -s_2c_3C_3,\\
l^3{}_2 &= s_2s_3C_3,\\
l^3{}_3 &= c_2C_3,\\
l^3{}_4 &= S_3,\\
l^4{}_1 &= (c_1c_2c_3 - s_1s_3)S_1 + (s_1c_2c_3 + c_1s_3)C_1S_2 - s_2c_3C_1C_2S_3,\\
l^4{}_2 &= -(c_1c_2s_3 + s_1c_3)S_1 + (c_1c_3 - s_1c_2s_3)C_1S_2 + s_2s_3C_1C_2S_3,\\
l^4{}_3 &= c_1s_2S_1 + s_1s_2C_1S_2 + c_2C_1C_2S_3,\\
l^4{}_4 &= C_1C_2C_3;
\end{aligned} \tag{3.4.6}$$

$$\alpha_\rho \in \mathbb{R},\ -\pi < \theta_1 < \pi,\ 0 < \theta_2 < \pi,\ -\pi < \theta_3 < \pi.$$

The above equations provide a convenient parametrization of an arbitrary proper orthochronous Lorentz matrix, in terms of six parameters α_ρ, θ_ρ.

Let us consider the domain of the six parameters

$$D^{\#} \equiv \{(\alpha_1, \alpha_2, \alpha_3, \theta_1, \theta_2, \theta_3):$$
$$\alpha_\rho \in \mathbb{R}^3,\ -\pi < \theta_1 < \pi, 0 < \theta_2 < \pi,\ -\pi < \theta_3 < \pi\}. \quad (3.4.7)$$

The domain $D^{\#} \subset \mathbb{R}^6$ may be considered as the image $\chi^{\#}(U^{\#})$, $U^{\#} \subset \mathcal{M}_6$, a six-dimensional differentiable manifold. But all possible proper orthochronous Lorentz matrices do not correspond to the points contained in $D^{\#}$. To accomplish that we first note that $D^{\#} = \mathbb{R}^3 \times D_0^{\#}$, where $D_0^{\#} \equiv \{(\theta_1, \theta_2, \theta_3): -\pi < \theta_1 < \pi, 0 < \theta_2 < \pi, -\pi < \theta_3 < \pi\}$. The domain $D_0^{\#}$ is shown in Figure 20.

We add boundary points corresponding to $\theta_2 = 0$ and $\theta_2 = \pi$ to $D_0^{\#}$. Then we add boundary points $\theta_1 = \pi$, $\theta_1 = -\pi$ and topologically identify these sets of points. Similarly we add points $\theta_3 = \pi$, $\theta_3 = -\pi$ and identify. The resulting region, denoted by $\bar{D}_0^{\#}$, is homeomorphic to the projective space $\mathbb{P}_3$. The differentiable manifold $\mathcal{M}_6 = \chi^{-1}[\mathbb{R}^3 \times \bar{D}_0^{\#}]$ is called the *group manifold* for $\mathcal{L}_{4+}^{+}$. Such a group, which is associated with a group manifold, gives rise to a *Lie group*. In the present context, we shall define a Lie group of coordinate transformations. Consider a differentiable coordinate transformation $f: D \times D^{\#} \subset \mathbb{R}^n \times \mathbb{R}^r \to \mathbb{R}^n$ given by:

(i)

$$\hat{x}^i = [\pi^i \circ f](x; \alpha) = f^i(x; \alpha) = f(x^1, \ldots, x^n; \alpha_1, \ldots, \alpha_r),$$
$$\frac{\partial(\hat{x}^1, \ldots, \hat{x}^n)}{\partial(x^1, \ldots, x^n)} \neq 0. \quad (3.4.8)$$

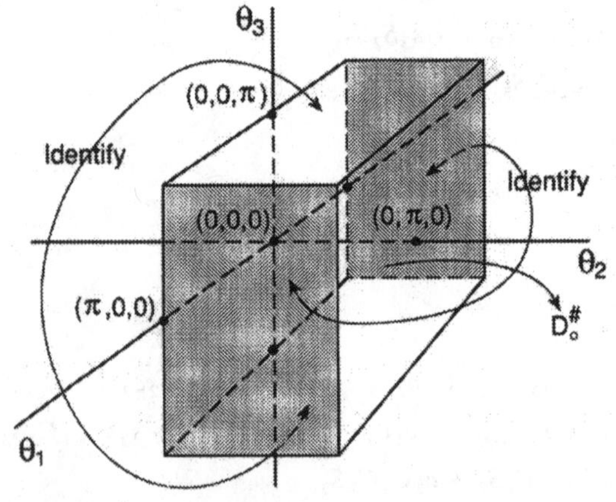

FIGURE 20. The three-dimensional domain $D_0^{\#}$.

Here $x = (x^1, \ldots, x^n) \in D \subset \mathbb{R}^n$, $\alpha = (\alpha^1, \ldots, \alpha^r) \in D^{\#} \subset \mathbb{R}^r$, such that the functions f^i are *real analytic functions* of the variables $\alpha_1, \ldots, \alpha_r$. Recall that a real analytic function F allows a power series expansion of $F(\alpha_1, \ldots, \alpha_r)$ in a convex domain $D^{\#}$.

(ii) The Composition rule of two successive transformations is

$$\begin{aligned} \hat{\hat{x}}^i &= f^i(\hat{x}^1, \ldots, \hat{x}^n; \beta_1, \ldots, \beta_r) \\ &= f^i[f^1(x;\alpha), \ldots, f^n(x;\alpha); \beta_1, \ldots, \beta_r] \\ &= f^i(x^1, \ldots, x^n; \gamma_1, \ldots, \gamma_r), \end{aligned}$$

where real analytic functions ϕ_μ must exist such that

$$\gamma_\mu = \phi_\mu(\alpha_1, \ldots, \alpha_r; \beta_1, \ldots, \beta_r) \equiv \phi_\mu(\alpha; \beta). \tag{3.4.9}$$

(iii) The associativity of three successive transformations characterized by parameters α, β, γ imply that

$$\phi_\mu(\phi(\alpha;\beta);\gamma) = \phi_\mu(\alpha;\phi(\beta;\gamma)). \tag{3.4.10}$$

(iv) For the identity transformation there must exist parameters $(\alpha_1^0, \ldots, \alpha_r^0) \in D^{\#}$ such that

$$\hat{x}^i = f^i(x;\alpha^0) = x^i. \tag{3.4.11}$$

(v) For the inverse transformation there must exist parameters $(\alpha_1^-, \ldots, \alpha_r^-) \in D^{\#}$ such that

$$\phi_\mu(\alpha;\alpha^-) = \phi_\mu(\alpha^-;\alpha) = \alpha_\mu^0. \tag{3.4.12}$$

Example: Consider a rotation in $\mathbb{R}^2$ given by the coordinate transformation

$$\begin{aligned} \hat{x}^1 &= f^1(x;\alpha) \equiv x^1 \cos\alpha - x^2 \sin\alpha, \\ \hat{x}^2 &= f^2(x;\alpha) \equiv x^1 \sin\alpha + x^2 \cos\alpha, \end{aligned}$$

where $\alpha \in D^{\#} \equiv \{\alpha: -\pi < \alpha < \pi\}$ and $x \in D \equiv \mathbb{R}^2$.

To obtain every transformation of the rotation group, the end points $\alpha = \pm\pi$ have to be added to $D^{\#}$ and then the end points are identified. In that case $\bar{D}^{\#}$ and $\mathscr{M}_1$ are homeomorphic to the unit circle S^1. The Jacobian of the transformation is

$$\partial(\hat{x}^1, \hat{x}^2)/\partial(x^1, x^2) = 1.$$

Since $\cos\alpha$ and $\sin\alpha$ are real analytic functions of $\alpha \in D^{\#}$, it can be concluded that $f^i(x;\alpha)$ are real analytic functions of α. The composition of two transformations is

$$\begin{aligned} \hat{\hat{x}}^1 &= \hat{x}^1 \cos\beta - \hat{x}^2 \sin\beta = x^1 \cos(\beta + \alpha) - x^2 \sin(\beta + \alpha), \\ \hat{\hat{x}}^2 &= \hat{x}^1 \sin\beta + \hat{x}^2 \cos\beta = x^1 \sin(\beta + \alpha) + x^2 \cos(\beta + \alpha); \end{aligned}$$

thus, $\phi(\alpha;\beta) \equiv \alpha + \beta$. More precisely $\phi(\alpha;\beta) \equiv (\beta + \alpha) \bmod 2\pi - \pi$ so that $-\pi \leq \phi(\alpha;\beta) \leq \pi$. Note that $\phi(\alpha;\beta)$ is a real analytic function of α, β in a neighborhood of the origin. The associativity can be checked by noticing that

$$\begin{aligned}\phi(\phi(\alpha;\beta);\gamma) &= (\gamma + (\beta + \alpha)) \bmod 2\pi - \pi \\ &= ((\gamma + \beta) + \alpha) \bmod 2\pi - \pi \\ &= \phi(\alpha;\phi(\beta;\gamma)).\end{aligned}$$

The identity transformation corresponds to $\alpha^0 \equiv 0$. The inverse transformation corresponds to $\alpha^- \equiv -\alpha$, so that

$$\phi(\alpha;\alpha^-) = -\alpha + \alpha = \alpha + (-\alpha) = \phi(\alpha^-;\alpha). \quad \square$$

We are ready to state a theorem on the structure of the Lie group $\mathscr{L}_{4+}^{+}$.

Theorem (3.4.1): *The set $\mathscr{L}_{4+}^{+}$ of all proper orthochronous Lorentz transformations of coordinates on $\mathbb{R}^4$ given by the matrix* (3.4.6) *constitute a six-parameter Lie group.*

Proof: We shall give a *partial proof* of this theorem. Consider the proper orthochronous Lorentz transformation

$$\hat{x}^i = l^i{}_j x^j,$$

where the entries are explicitly given in equation (3.4.6). Clearly we have

$$\frac{\partial(\hat{x}^1,\hat{x}^2,\hat{x}^3,\hat{x}^4)}{\partial(x^1,x^2,x^3,x^4)} = \det L = 1$$

for all $x = (x^1,x^2,x^3,x^4) \in \mathbb{R}^4$. The entries $l^i{}_j$ are the values of the functions $g^i{}_j$ of the parameters in (3.4.6) so that

$$l^i{}_j = g^i{}_j(\alpha_1,\alpha_2,\alpha_3,\theta_1,\theta_2,\theta_3),$$

where $(\alpha_1,\alpha_2,\alpha_3,\theta_1,\theta_2,\theta_3) \in D^{\#}$. The domain $D^{\#}$ and the region $\bar{D}^{\#}$ have been determined in equation (3.4.7) and the subsequent discussions.

Consider, for example, the entry

$$l^3{}_3 = g^3{}_3(\alpha_1,\alpha_2,\alpha_3,\theta_1,\theta_2,\theta_3) \equiv \cos\theta_2 \cosh\alpha_3.$$

Since the expansions

$$\cos\theta_2 = 1 - \frac{(\theta_2)^2}{2!} + \frac{(\theta_2)^4}{4!} - \cdots,$$

$$\cosh\alpha_3 = 1 + \frac{(\alpha_3)^2}{2!} + \frac{(\alpha_3)^4}{4!} + \cdots$$

are absolutely convergent in $D^{\#}$, we can write the product

$$\cos\theta_2\cosh\alpha_3 = 1 - \frac{(\theta_2)^2}{2!} + \frac{(\alpha_3)^2}{2!} + \frac{(\theta_2)^4}{4!} + \frac{(\alpha_3)^4}{4!} - \frac{(\theta_2\alpha_3)^2}{(2!)^2} - \cdots$$

as an absolutely convergent series in $D^{\#}$. Therefore, the function $g^3{}_3$ is a real analytic function of the six parameters. Similarly all of the $g^i{}_j$ are real analytic functions in $D^{\#}$.

The composition of two Lorentz transformations can be written as

$$\begin{aligned}\hat{\hat{x}}^i &= l'^i{}_j \hat{x}^j = l'^i{}_j l^j{}_k x^k \\ &= g^i{}_k(\alpha_1', \alpha_2', \alpha_3', \theta_1', \theta_2', \theta_3') g^k{}_j(\alpha_1, \alpha_2, \alpha_3, \theta_1, \theta_2, \theta_3) x^j \\ &= g^i{}_j(\gamma_1, \gamma_2, \gamma_3, \psi_1, \psi_2, \psi_3) x^j. \end{aligned} \tag{3.4.13}$$

Considering a special case of (3.4.13) and using (3.4.6), we can obtain

$$\begin{aligned}\hat{\hat{x}}^3 &= [\cdots]x^1 + [\cdots]x^2 + [\cdots]x^3 \\ &\quad + [-s_2' c_3' C_3' S_1 C_2 C_3 + s_2' s_3' C_3' S_2 C_3 + c_2' C_3' S_3 + S_3' C_1 C_2 C_3] x^4 \\ &= -(\sin\psi_2 \cos\psi_3 \cosh\gamma_3) x^1 + (\sin\psi_2 \sin\psi_3 \cosh\gamma_3) x^2 \\ &\quad + (\cos\psi_2 \cosh\gamma_3) x^3 + (\sinh\gamma_3) x^4. \end{aligned}$$

Comparing the coefficients of x^4 in the above equation we have

$$\sinh\gamma_3 = -s_2' c_3' C_3' S_1 C_2 C_3 + s_2' s_3' C_3' S_2 C_3 + c_2' C_3' S_3 + S_3' C_1 C_2 C_3.$$

So we find by inversion of sinh that

$$\begin{aligned}\gamma_3 &= \phi_3(\alpha_1, \alpha_2, \alpha_3, \theta_1, \theta_2, \theta_3; \alpha_1', \alpha_2', \alpha_3', \theta_1', \theta_2', \theta_3') \\ &\equiv \operatorname{Argsinh}[-\sin\theta_2' \cos\theta_3' \cosh\alpha_3' \sinh\alpha_1 \cosh\alpha_2 \cosh\alpha_3 \\ &\qquad + \sin\theta_2' \sin\theta_3' \cosh\alpha_3' \sinh\alpha_2 \cosh\alpha_3 \\ &\qquad + \cos\theta_2' \cosh\alpha_3' \sinh\alpha_3 \\ &\qquad + \sinh\alpha_3' \cosh\alpha_1 \cosh\alpha_2 \cosh\alpha_3]. \end{aligned}$$

The above function ϕ_3 is real analytic in the twelve variables. Similarly the other five real analytic functions

$$\begin{aligned}\gamma_1 &= \phi_1(\alpha_1, \alpha_2, \alpha_3, \theta_1, \theta_2, \theta_3; \alpha_1', \alpha_2', \alpha_3', \theta_1', \theta_2', \theta_3'), \\ \gamma_2 &= \phi_2(\alpha_1, \alpha_2, \alpha_3, \theta_1, \theta_2, \theta_3; \alpha_1', \alpha_2', \alpha_3', \theta_1', \theta_2', \theta_3'), \\ \psi_1 &= \phi_4(\alpha_1, \alpha_2, \alpha_3, \theta_1, \theta_2, \theta_3; \alpha_1', \alpha_2', \alpha_3', \theta_1', \theta_2', \theta_3'), \\ \psi_2 &= \phi_5(\alpha_1, \alpha_2, \alpha_3, \theta_1, \theta_2, \theta_3; \alpha_1', \alpha_2', \alpha_3', \theta_1', \theta_2', \theta_3'), \\ \psi_3 &= \phi_6(\alpha_1, \alpha_2, \alpha_3, \theta_1, \theta_2, \theta_3; \alpha_1', \alpha_2', \alpha_3', \theta_1', \theta_2', \theta_3') \end{aligned}$$

can be found. These functions must satisfy the associative property (3.4.10) by Theorem (3.2.2).

The identity transformation corresponds to parameters

$$(\alpha_1^0, \alpha_2^0, \alpha_3^0, \theta_1^0, \theta_2^0, \alpha_3^0) = (0, 0, 0, 0, 0, 0).$$

The inverse of a transformation corresponds to parameters that are the solutions of six equations. One of these six equations is

$$\phi_3(\alpha_1, \alpha_2, \alpha_3, \theta_1, \theta_2, \theta_3; \alpha_1^-, \alpha_2^-, \alpha_3^-, \theta_1^-, \alpha_2^-, \theta_3^-) = 0,$$

or equivalently

$$\begin{aligned} &-\sin\theta_2' \cos\theta_3' \cosh\alpha_3' \sinh\alpha_1 \cosh\alpha_2 \cosh\alpha_3 \\ &\quad + \sin\theta_2' \sin\theta_3' \cosh\alpha_3' \sinh\alpha_2 \cosh\alpha_3 \\ &\quad + \cos\theta_2' \cosh\alpha_3' \sinh\alpha_3 + \sinh\alpha_3' \cosh\alpha_1 \cosh\alpha_2 \cosh\alpha_3 = 0. \quad \blacksquare \end{aligned}$$

EXERCISES 3.4

1. Consider the set of projective transformations given by

$$\hat{x} = \frac{ax + b}{cx + d},$$

where $ad - bc \neq 0$ and $x \in \mathbb{R}$. Prove that this set of transformations constitutes a Lie group.

2. Consider a subset of $\mathscr{L}_{4+}^{+}$, characterized by

$$\begin{aligned} \hat{x}^1 &= x^1 \cos\theta - x^2 \sin\theta, \\ \hat{x}^2 &= x^1 \sin\theta + x^2 \cos\theta, \\ \hat{x}^3 &= x^3 \cosh\alpha - x^4 \sinh\alpha, \\ \hat{x}^4 &= -x^3 \sinh\alpha + x^4 \cosh\alpha. \end{aligned}$$

Prove explicitly that this subset is an abelian Lie subgroup and its group submanifold is homeomorphic to $S^1 \times \mathbb{R}$.

3. Consider the one-dimensional wave equation

$$\frac{\partial^2 \phi}{\partial x^2}(x, t) - \frac{\partial^2 \phi}{\partial t^2}(x, t) = 0.$$

The characteristic or null coordinates (or light-cone coordinates) are given by

$$u = x - t, \qquad v = x + t.$$

(i) Show that the wave equation reduces to the form

$$\frac{\partial^2 \Phi}{\partial u \partial v}(u, v) = 0,$$

where $\Phi(u, v) \equiv \phi(x, t)$.

(ii) Prove that the wave equation remains invariant under a coordinate transformation

$$\hat{u} = e^{(\alpha_1)^3} u, \qquad \hat{v} = e^{-(\alpha_2)^3} v,$$

where $(\alpha_1, \alpha_2) \in \mathbb{R}^2$.

(iii) Prove that the set of transformations in part (ii) constitutes a Lie group.

References

1. Y. Choquet-Bruhat, C. De Witt-Morette, and M. Dillard-Bleick, *Analysis, manifolds and, physics*, North-Holland, Amsterdam, 1977. [p. 96]
2. A. Einstein, Ann. Physik 17 (1905), 891–921. [p. 69]
3. I. M. Gelfand, R. A. Minlos, and Z. Y. Shapiro, *Representations of the rotation and Lorentz groups, and their applications*, The MacMillan Co., New York, 1963. [p. 81]
4. M. Hammermesh, *Group theory*, Addison-Wesley, MA, 1962. [pp. 84, 97, 89]
5. J. L. Synge, *Relativity: The special theory*, North-Holland, Amsterdam, 1964. [pp. 91, 93]

4
Pauli Matrices, Spinors, Dirac Matrices, and Dirac Bispinors

4.1. Pauli Matrices, Rotations, and Lorentz Transformations

A rotation in the Euclidean plane can be characterized by the following matrix equation:

$$\begin{bmatrix} \hat{x}^1 \\ \hat{x}^2 \end{bmatrix} = \begin{bmatrix} \cos\phi & \sin\phi \\ -\sin\phi & \cos\phi \end{bmatrix} \begin{bmatrix} x^1 \\ x^2 \end{bmatrix},$$

where $\phi \in (-\pi, \pi)$ and x^1, x^2 are Cartesian coordinates. Introducing complex variables $z \equiv x^1 + ix^2$, $\hat{z} \equiv \hat{x}^1 + i\hat{x}^2$, the above transformation becomes the conformal transformation in the complex z-plane,

$$\hat{z} = uz, \tag{4.1.1}$$

where $u \equiv e^{-i\phi}$, so that $u\bar{u} = 1$. Thus u is *unimodular*. Here we have used a bar to denote *complex conjugation.*

A general rotation in three-dimensional Euclidean space can be characterized by two systems of Cartesian coordinates (x^1, x^2, x^3), (x'^1, x'^2, x'^3), and Euler angles θ, ϕ, ψ. This rotation can be obtained as a composition of three rotations: the rotation ϕ about the x^3-axis, the rotation θ about the new $\hat{x}^1$-axis, and the rotation ψ about the new $\hat{\hat{x}}^3$-axis. Thus the composite rotation is given by the matrix equation

$$\begin{bmatrix} x'^1 \\ x'^2 \\ x'^3 \end{bmatrix} = \begin{bmatrix} \cos\psi & \sin\psi & 0 \\ -\sin\psi & \cos\psi & 0 \\ 0 & 0 & 1 \end{bmatrix} \begin{bmatrix} 1 & 0 & 0 \\ 0 & \cos\theta & \sin\theta \\ 0 & -\sin\theta & \cos\theta \end{bmatrix} \begin{bmatrix} \cos\phi & \sin\phi & 0 \\ -\sin\phi & \cos\phi & 0 \\ 0 & 0 & 1 \end{bmatrix} \begin{bmatrix} x^1 \\ x^2 \\ x^3 \end{bmatrix}. \tag{4.1.2a}$$

We can explicitly carry out the matrix multiplication to get the composite matrix equation

$$\begin{bmatrix} x'^1 \\ x'^2 \\ x'^3 \end{bmatrix} = \begin{bmatrix} \cos\psi\cos\phi - \cos\theta\sin\phi\sin\psi, & \cos\psi\sin\phi + \cos\theta\cos\phi\sin\psi & \sin\psi\sin\theta \\ -\sin\psi\cos\phi - \cos\theta\sin\phi\cos\psi, & -\sin\psi\sin\phi + \cos\theta\cos\phi\cos\psi & \cos\psi\sin\theta \\ \sin\theta\sin\phi & -\sin\theta\cos\phi & \cos\theta \end{bmatrix} \begin{bmatrix} x^1 \\ x^2 \\ x^3 \end{bmatrix}. \tag{4.1.2b}$$

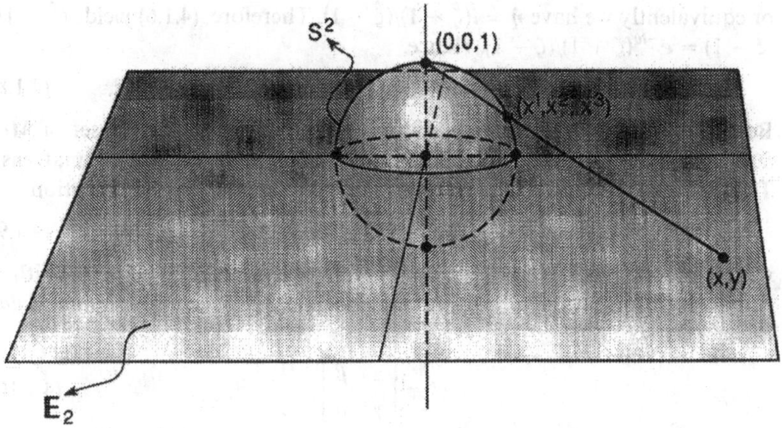

FIGURE 21. Stereographic projection of S^2 onto $\mathbb{E}_2$.

A three-dimensional rotation can correspond to a complex conformal transformation by the following device. Project S^2, the surface of the unit sphere given by $(x^1)^2 + (x^2)^2 + (x^3)^2 = 1$, from the point $(0, 0, 1)$ onto the Euclidean plane $\mathbb{E}_2$ given by $x^3 = 0$. This projection is called a *stereographic projection.* Let (x, y) be the projection of (x^1, x^2, x^3) as in Figure 21.

By the similarity of triangles, it can be shown that

$$x^1/x = x^2/y = (1 - x^3)/1. \tag{4.1.3}$$

One can introduce a complex variable

$$\zeta \equiv (x^1 + ix^2)/(1 - x^3) = (1 + x^3)/(x^1 - ix^2). \tag{4.1.4}$$

A rotation of angle ϕ about the x^3-axis can be expressed as the complex conformal transformation

$$\begin{aligned}\hat{\zeta} &= (\hat{x}^1 + i\hat{x}^2)/(1 - \hat{x}^3) = (\hat{x}^1 + i\hat{x}^2)/(1 - x^3)\\ &= e^{-i\phi}(x^1 + ix^2)/(1 - x^3)\\ &= e^{-i\phi}\zeta = (e^{-i\phi/2}\zeta + 0)/(0 + e^{i\phi/2}).\end{aligned} \tag{4.1.5}$$

To treat a rotation by an angle θ about the x^1-axis, we can introduce the complex transformation

$$\begin{aligned}\hat{\eta} &\equiv (\hat{x}^2 + i\hat{x}^3)/(1 - \hat{x}^1)\\ &= e^{-i\theta}(x^2 + ix^3)/(1 - x^1)\\ &= e^{-i\theta}\eta.\end{aligned} \tag{4.1.6}$$

We see now that

$$\begin{aligned}(\hat{\eta} + i)/(\hat{\eta} - i) &= [-(\hat{x}^1 + i\hat{x}^2) + (1 + \hat{x}^3)]/[(\hat{x}^1 - i\hat{x}^2) - (1 - \hat{x}^3)]\\ &= [-\hat{\zeta}(1 - \hat{x}^3) + (1 + \hat{x}^3)]/[(\hat{\zeta})^{-1}(1 + \hat{x}^3) - (1 - \hat{x}^3)]\\ &= \hat{\zeta},\end{aligned} \tag{4.1.7}$$

or equivalently we have $\hat{\eta} = i(\hat{\zeta} + 1)/(\hat{\zeta} - 1)$. Therefore, (4.1.6) yields $(\hat{\zeta} + 1)/(\hat{\zeta} - 1) = e^{-i\theta}(\zeta + 1)/(\zeta - 1)$. Hence,

$$\hat{\zeta} = [\cos(\theta/2)\zeta - i\sin(\theta/2)]/[-i\sin(\theta/2)\zeta + \cos(\theta/2)]. \tag{4.1.8}$$

Both of the transformations (4.1.5) and (4.1.8) belong to the subset of *Mobius transformations* (also called fractional linear conformal transformations). These special Möbius transformations are characterized by the equation

$$\hat{\zeta} = (\alpha\zeta + \beta)/(\gamma\zeta + \delta), \tag{4.1.9}$$

where $\alpha\delta - \beta\gamma = 1$, $|\alpha|^2 + |\gamma|^2 = |\beta|^2 + |\delta|^2 = 1$, and $\bar{\alpha}\beta + \bar{\gamma}\delta = 0$. With a Möbius transformation (4.1.9), one can associate a 2×2 complex *unimodular unitary matrix*

$$U \equiv \begin{bmatrix} \alpha & \beta \\ \gamma & \delta \end{bmatrix}, \tag{4.1.10}$$

such that $\det U = 1$, and $UU^\dagger = U^\dagger U = \mathrm{I}$ where the dagger denotes the *hermitian conjugation*.

The two-dimensional complex vector space V_2, for which the matrix U represents a linear mapping, is called the *spinor space*. We shall now introduce three remarkable 2×2 matrices called the *Pauli matrices*. The Pauli matrices are defined by

$$\sigma_1 \equiv \begin{bmatrix} 0 & 1 \\ 1 & 0 \end{bmatrix}, \quad \sigma_2 \equiv \begin{bmatrix} 0 & -i \\ i & 0 \end{bmatrix}, \quad \sigma_3 \equiv \begin{bmatrix} 1 & 0 \\ 0 & -1 \end{bmatrix}. \tag{4.1.11}$$

Theorem (4.1.1): *The three Pauli matrices σ_α satisfy*

$$\begin{aligned}
&\text{(i)} \quad \sigma_\alpha^\dagger = \sigma_\alpha, \\
&\text{(ii)} \quad \det \sigma_\alpha = -1, \\
&\text{(iii)} \quad \operatorname{Trace} \sigma_\alpha = 0, \\
&\text{(iv)} \quad \sigma_\alpha\sigma_\beta = \delta_{\alpha\beta}\mathrm{I} + i\eta_{\alpha\beta\gamma}\sigma_\gamma, \\
&\text{(v)} \quad \sigma_\alpha\sigma_\beta + \sigma_\beta\sigma_\alpha = 2\delta_{\alpha\beta}\mathrm{I}, \\
&\text{(vi)} \quad \sigma_\alpha\sigma_\beta - \sigma_\beta\sigma_\alpha = 2i\eta_{\alpha\beta\gamma}\sigma_\gamma,
\end{aligned} \tag{4.1.12}$$

where we let the Greek indices take values $\{1, 2, 3\}$ and the totally antisymmetric pseudotensor $\eta_{\alpha\beta\gamma}$ in Cartesian coordinates is defined analogously to (1.3.13*b*).

Proof: The proof of this theorem follows from direct computations, so we will just prove (iv). (Note that parts (v) and (vi) follow directly from part (iv).) To prove part (iv) we notice that

$$\sigma_1\sigma_1 = \mathrm{I} = \mathrm{I} + 0 = \delta_{11}\mathrm{I} + i\eta_{11\gamma}\sigma_\gamma,$$

$$\sigma_1\sigma_2 = \begin{bmatrix} i & 0 \\ 0 & -i \end{bmatrix} = i\sigma_3 = \delta_{12}\mathrm{I} + i\eta_{12\gamma}\sigma_\gamma.$$

The other equations can be proved similarly. ■

The next theorem about the Pauli matrices is instructive.

Theorem (4.1.2): *The set of Pauli matrices* $\{\sigma_1, \sigma_2, \sigma_3\}$ *is a basis for the vector subspace of* 2×2 *complex matrices with zero trace.*

Proof: Let $M = \begin{bmatrix} \lambda & \mu \\ \nu & \tau \end{bmatrix}$ such that Trace $M = 0$. Then $\lambda + \tau = 0$, so $M = \begin{bmatrix} \lambda & \mu \\ \nu & -\lambda \end{bmatrix} = [(\mu + \nu)/2]\sigma_1 + [i(\mu - \nu)/2]\sigma_2 + \lambda\sigma_3$. This equation proves that the set $\{\sigma_1, \sigma_2, \sigma_3\}$ spans the vector space of complex traceless 2×2 matrices. To investigate linear independence, we examine the matrix equation

$$\alpha_1\sigma_1 + \alpha_2\sigma_2 + \alpha_3\sigma_3 = \begin{bmatrix} \alpha_3 & \alpha_1 - i\alpha_2 \\ \alpha_1 + i\alpha_2 & -\alpha_3 \end{bmatrix} = \begin{bmatrix} 0 & 0 \\ 0 & 0 \end{bmatrix}.$$

This equation has only $\alpha_1 = \alpha_2 = \alpha_3 = 0$ as solution. Thus the set $\{\sigma_1, \sigma_2, \sigma_3\}$ is linearly independent and, hence, is a basis. ■

Every 2×2 hermitian matrix with zero trace is a linear combination of σ_α with *real* coefficients. Thus a point (x^1, x^2, x^3) in three-dimensional Euclidean space can be associated with a 2×2 hermitian matrix P with zero trace in the following manner:

$$P \equiv \begin{bmatrix} x^3 & x^1 - ix^2 \\ x^1 + ix^2 & -x^3 \end{bmatrix} = x^1\sigma_1 + x^2\sigma_2 + x^3\sigma_3 \equiv x^\alpha\sigma_\alpha. \quad (4.1.13)$$

Consider a 2×2 unitary matrix U and a similarity transformation

$$\hat{P} = UPU^{-1} = UPU^\dagger. \quad (4.1.14)$$

Since Trace $\hat{P}$ = Trace UPU^{-1} = Trace $UU^{-1}P$ = Trace $P = 0$ and also $\hat{P}^\dagger = (UPU^\dagger)^\dagger = UPU^\dagger = \hat{P}$, we can write

$$\hat{P} = \begin{bmatrix} \hat{x}^3 & \hat{x}^1 - i\hat{x}^2 \\ \hat{x}^1 + i\hat{x}^2 & -\hat{x}^3 \end{bmatrix} \quad (4.1.15)$$

for some real numbers $\hat{x}^\alpha$. Thus $\det \hat{P} = \det U \det P / \det U = \det P$. Therefore, we must have $(\hat{x}^1)^2 + (\hat{x}^2)^2 + (\hat{x}^3)^2 = (x^1)^2 + (x^2)^2 + (x^3)^2$ under the transformation (4.1.14). So we see that the transformation (4.1.14) induces an orthogonal transformation (which can be proved to be a rotation) in the three-dimensional Euclidean space.

Example 1: Let us choose the unitary unimodular matrix

$$U_1 \equiv \begin{bmatrix} e^{i\phi/2} & 0 \\ 0 & e^{-i\phi/2} \end{bmatrix},$$

where $\phi \in (-\pi, \pi)$. The transformation (4.1.14) yields $\hat{P} = U_1 P U_1^\dagger$, so

$$\begin{bmatrix} \hat{x}^3 & \hat{x}^1 - i\hat{x}^2 \\ \hat{x}^1 + i\hat{x}^2 & -\hat{x}^3 \end{bmatrix} = \begin{bmatrix} e^{i\phi/2} & 0 \\ 0 & e^{-i\phi/2} \end{bmatrix} \begin{bmatrix} x^3 & x^1 - ix^2 \\ x^1 + ix^2 & -x^3 \end{bmatrix} \begin{bmatrix} e^{-i\phi/2} & 0 \\ 0 & e^{i\phi/2} \end{bmatrix}$$
$$= \begin{bmatrix} x^3 & e^{i\phi}(x^1 - ix^2) \\ e^{-i\phi}(x^1 + ix^2) & -x^3 \end{bmatrix}. \tag{4.1.16}$$

The above equation implies that

$$\hat{x}^1 = x^1 \cos\phi + x^2 \sin\phi,$$
$$\hat{x}^2 = -x^1 \sin\phi + x^2 \cos\phi,$$
$$\hat{x}^3 = x^3.$$

This is a rotation by an angle ϕ about the x^3-axis. The unitary matrix U_1 corresponds to the transformation (4.1.5). □

Example 2: Consider the unitary unimodular matrix

$$U_2 \equiv \begin{bmatrix} \cos(\theta/2) & i\sin(\theta/2) \\ i\sin(\theta/2) & \cos(\theta/2) \end{bmatrix},$$

where $\theta \in (0, \pi)$. This is the matrix in equation (4.1.10) corresponding to the transformation (4.1.8). The matrix equation $\hat{\hat{P}} = U_2 \hat{P} U_2^\dagger$ implies that

$$\hat{\hat{x}}^1 = \hat{x}^1,$$
$$\hat{\hat{x}}^2 = \hat{x}^2 \cos\theta + \hat{x}^3 \sin\theta,$$
$$\hat{\hat{x}}^3 = -\hat{x}^2 \sin\theta + \hat{x}^3 \cos\theta.$$

The above equation yields a rotation about the $\hat{x}^1$-axis. □

Three successive rotations by Euler angles ϕ, θ, ψ as in (4.1.2) can be accomplished by the following unitary unimodular transformations

$$\hat{P} = U_1 P U_1^\dagger, \qquad \hat{\hat{P}} = U_2 \hat{P} U_2^\dagger, \qquad P' = U_3 \hat{\hat{P}} U_3^\dagger = U P U^\dagger, \tag{4.1.17a}$$

where U is given by

$$U \equiv U_3 U_2 U_1$$
$$= \begin{bmatrix} e^{i\psi/2} & 0 \\ 0 & e^{-i\psi/2} \end{bmatrix} \begin{bmatrix} \cos(\theta/2) & i\sin(\theta/2) \\ i\sin(\theta/2) & \cos(\theta/2) \end{bmatrix} \begin{bmatrix} e^{i\phi/2} & 0 \\ 0 & e^{-i\phi/2} \end{bmatrix}$$
$$= \begin{bmatrix} e^{i(\psi+\phi)/2}\cos(\theta/2) & ie^{-i(\phi-\psi)/2}\sin(\theta/2) \\ ie^{i(\psi-\phi)/2}\sin(\theta/2) & e^{-i(\psi+\phi)/2}\cos(\theta/2) \end{bmatrix}. \tag{4.1.17b}$$

Note that another unitary unimodular matrix $U' \equiv -U$ induces exactly the same rotation. In general, the same rotation in three dimensions is induced by *two* unitary unimodular matrices $\pm U$, which represent transformations in spinor space.

We shall generalize these techniques for the Lorentz transformation by adding to the set of three Pauli matrices in (4.1.11) a fourth matrix, namely,

$$\sigma^4 \equiv \begin{bmatrix} 1 & 0 \\ 0 & 1 \end{bmatrix} = \mathrm{I} = -\sigma_4. \tag{4.1.18}$$

Theorem (4.1.3): *The set of four matrices $\{\sigma^1, \sigma^2, \sigma^3, \sigma^4\}$ is a basis for the vector space of complex 2×2 matrices. Furthermore every 2×2 hermitian matrix is a linear combination of $\{\sigma^1, \sigma^2, \sigma^3, \sigma^4\}$ with real coefficients.*

Proof: For an arbitrary 2×2 complex matrix

$$\begin{bmatrix} \lambda & \mu \\ \nu & \tau \end{bmatrix} = \frac{\sigma^1(\mu + \nu)}{2} + \frac{i\sigma^2(\mu - \nu)}{2} + \frac{\sigma^3(\lambda - \tau)}{2} + \frac{\sigma^4(\lambda + \tau)}{2}.$$

This equation shows that $\{\sigma^1, \sigma^2, \sigma^3, \sigma^4\}$ spans the space of 2×2 complex matrices. For linear independence of $\{\sigma^1, \sigma^2, \sigma^3, \sigma^4\}$ we investigate the matrix equation

$$0 = \alpha_1\sigma^1 + \alpha_2\sigma^2 + \alpha_3\sigma^3 + \alpha_4\sigma^4 = \begin{bmatrix} \alpha_4 + \alpha_3 & \alpha_1 - i\alpha_2 \\ \alpha_1 + i\alpha_2 & \alpha_4 - \alpha_3 \end{bmatrix} = \begin{bmatrix} 0 & 0 \\ 0 & 0 \end{bmatrix}.$$

The only solution of the above equation is $\alpha_1 = \alpha_2 = \alpha_3 = \alpha_4 = 0$, which proves the linear independence of $\{\sigma^1, \sigma^2, \sigma^3, \sigma^4\}$. Thus $\{\sigma^1, \sigma^2, \sigma^3, \sigma^4\}$ is a basis for the vector space of 2×2 complex matrices.

For a hermitian matrix $H = H^\dagger$ we see the matrix of H has the form $H = \begin{bmatrix} a & \beta \\ \bar{\beta} & c \end{bmatrix}$, where $a, c \in \mathbb{R}$. From this we get

$$\begin{bmatrix} a & \beta \\ \bar{\beta} & c \end{bmatrix} \equiv \frac{\sigma^1(\beta + \bar{\beta})}{2} + \frac{i\sigma^2(\beta - \bar{\beta})}{2} + \frac{\sigma^3(a - c)}{2} + \frac{\sigma^4(a + c)}{2}.$$

The coefficients on the right-hand side are obviously real. ∎

The Minkowski coordinates (x^1, x^2, x^3, x^4) of an event in space–time can be associated with a 2×2 hermitian matrix X in the following way

$$X \equiv x^k\sigma_k = \begin{bmatrix} x^3 - x^4 & x^1 - ix^2 \\ x^1 + ix^2, & -(x^3 + x^4) \end{bmatrix}. \tag{4.1.19}$$

In analogy to equation (4.1.14), we define a transformation

$$\hat{X} = \Lambda X \Lambda^\dagger, \tag{4.1.20}$$

where $\det \Lambda = 1$. The matrix Λ is assumed to be unimodular but not necessarily unitary. Clearly $\det \hat{X} = \det \Lambda \det X \det \Lambda^\dagger = \det X$. Therefore by (4.1.19) we must have

$$(\hat{x}^1)^2 + (\hat{x}^2)^2 + (\hat{x}^3)^2 - (\hat{x}^4)^2 = (x^1)^2 + (x^2)^2 + (x^3)^2 - (x^4)^2.$$

The transformation (4.1.20) induces a Lorentz transformation (which is proper and orthochronous).

Example 1: Consider either of the two matrices

$$\Lambda \equiv \pm \begin{bmatrix} e^{\chi/2} & 0 \\ 0 & e^{-\chi/2} \end{bmatrix},$$

where $\chi \in \mathbb{R}$. Equation (4.1.20) yields

$$\begin{bmatrix} \hat{x}^3 - \hat{x}^4 & \hat{x}^1 - i\hat{x}^2 \\ \hat{x}^1 + i\hat{x}^2 & -(\hat{x}^3 + \hat{x}^4) \end{bmatrix} = \begin{bmatrix} e^{\chi/2} & 0 \\ 0 & e^{-\chi/2} \end{bmatrix} \begin{bmatrix} x^3 - x^4 & x^1 - ix^2 \\ x^1 + ix^2 & -(x^3 + x^4) \end{bmatrix} \begin{bmatrix} e^{\chi/2} & 0 \\ 0 & e^{-\chi/2} \end{bmatrix}$$

$$= \begin{bmatrix} e^{\chi}(x^3 - x^4) & x^1 - ix^2 \\ x^1 + ix^2 & -e^{-\chi}(x^3 + x^4) \end{bmatrix}.$$

The above equation implies that

$$\hat{x}^1 = x^1,$$

$$\hat{x}^2 = x^2,$$

$$\hat{x}^3 = x^3 \cosh \chi - x^4 \sinh \chi,$$

$$\hat{x}^4 = -x^3 \sinh \chi + x^4 \cosh \chi.$$

These equations define a well-known proper orthochronous Lorentz transformation. □

Example 2: Consider either of the two matrices

$$\Lambda \equiv \pm \begin{bmatrix} \sqrt{(\cosh \chi + 1)/2} & -\sqrt{(\cosh \chi - 1)/2} \\ -\sqrt{(\cosh \chi - 1)/2} & \sqrt{(\cosh \chi + 1)/2} \end{bmatrix}.$$

Equation (4.1.20) produces

$$\begin{bmatrix} \hat{x}^3 - \hat{x}^4 & \hat{x}^1 - i\hat{x}^2 \\ \hat{x}^1 + i\hat{x}^2 & -(\hat{x}^3 + \hat{x}^4) \end{bmatrix}$$

$$= \begin{bmatrix} -x^1 \sinh \chi + x^3 - x^4 \cosh \chi & x^1 \cosh \chi - ix^2 + x^4 \sinh \chi \\ x^1 \cosh \chi + ix^2 + x^4 \sinh \chi & -x^1 \sinh \chi - x^3 - x^4 \cosh \chi \end{bmatrix}.$$

The above matrix equation yields the proper orthochronous Lorentz transformation

$$\hat{x}^1 = x^1 \cosh \chi + x^4 \sinh \chi,$$

$$\hat{x}^2 = x^2,$$

$$\hat{x}^3 = x^3,$$

$$\hat{x}^4 = x^1 \sinh \chi + x^4 \cosh \chi. \quad \square$$

EXERCISES 4.1

1. Let $P \equiv x^\alpha \sigma_\alpha$, $\hat{P} \equiv \hat{x}^\alpha \sigma_\alpha$, where the σ_α are Pauli matrices. Consider a unitary unimodular matrix U and the matrix equation $\hat{P} = UPU^\dagger$. Prove that it is impossible to obtain the transformation $\hat{x}^\alpha = -x^\alpha$ from the above matrix equation.

2. Consider the matrix equations

$$X \equiv x^k \sigma_k, \qquad \hat{X} \equiv \hat{x}^k \sigma_k = \Lambda X \Lambda^\dagger,$$

where $\det \Lambda = 1$. Construct the Lorentz transformation inherent in the above equations in terms of the entries of the matrix Λ. Furthermore, prove that this Lorentz transformation has to be proper and orthochronous.

3. Consider the six-parameter Lorentz transformation given by (3.4.6). Obtain explicitly the two 2×2 complex matrices Λ that induce that Lorentz transformation.

4.2. Spinors and Spinor-Tensors

The spinor space V_2 has already been defined as a complex two-dimensional vector space. Let $\{\mathbf{e}_1, \mathbf{e}_2\}$ be a basis (dyad) for V_2, and let $\psi \in V_2$. Then $\psi = \sum_{A=1}^{2} \psi^A \mathbf{e}_A = \psi^A \mathbf{e}_A$ for some complex components ψ^A. An invertible linear mapping Σ: $V_2 \to V_2$ can be characterized by the equation

$$\hat{\mathbf{e}}_B = \Sigma(\mathbf{e}_B) = \Sigma^A{}_B \mathbf{e}_A, \tag{4.2.1}$$

where $\det \Sigma \neq 0$. In this case the components of $\psi = \psi^A \mathbf{e}_A = \hat{\psi}^A \hat{\mathbf{e}}_A$ undergo the transformation [recall Theorem (1.2.2)]

$$\hat{\psi}^A = \Lambda^A{}_B \psi^B, \qquad \Sigma^A{}_B \Lambda^B{}_C = \Lambda^A{}_B \Sigma^B{}_C = \delta^A{}_C. \tag{4.2.2}$$

We shall subsequently assume the *unimodular* conditions

$$\det \Sigma = \det \Lambda = 1. \tag{4.2.3}$$

The complex conjugate components transform under the rule:

$$\hat{\psi}^{\bar{A}} \equiv \overline{\hat{\psi}^A} = \overline{\Lambda^A{}_B \psi^B} \equiv \Lambda^{\bar{A}}{}_{\bar{B}} \psi^{\bar{B}}. \tag{4.2.4}$$

Here the repeated barred capital indices and the repeated capital indices are summed disjointly over the range $\{1, 2\}$. The components of a covariant complex vector (which is a spinor) and its complex conjugate components transform as follows:

$$\hat{\chi}_A = \Sigma^B{}_A \chi_B, \qquad \hat{\chi}_{\bar{A}} = \Sigma^{\bar{B}}{}_{\bar{A}} \chi_{\bar{B}}. \tag{4.2.5}$$

Higher-order spinor components transform by the rule

$$\begin{aligned} \hat{\psi}^{A_1 \cdots A_k \bar{B}_1 \cdots \bar{B}_l}{}_{C_1 \cdots C_m \bar{D}_1 \cdots \bar{D}_n} &= \Lambda^{A_1}{}_{E_1} \cdots \Lambda^{A_k}{}_{E_k} \Lambda^{\bar{B}_1}{}_{\bar{F}_1} \cdots \Lambda^{\bar{B}_l}{}_{\bar{F}_l} \Sigma^{G_1}{}_{C_1} \cdots \Sigma^{G_m}{}_{C_m} \\ &\quad \times \Sigma^{\bar{H}_1}{}_{\bar{D}_1} \cdots \Sigma^{\bar{H}_n}{}_{\bar{D}_n} \psi^{E_1 \cdots E_k \bar{F}_1 \cdots \bar{F}_l}{}_{G_1 \cdots G_m \bar{H}_1 \cdots \bar{H}_n}. \end{aligned} \tag{4.2.6}$$

The totally antisymmetric permutation symbol is defined by

$$\begin{aligned} \varepsilon_{12} &= -\varepsilon_{21} \equiv 1, \\ \varepsilon_{11} &= \varepsilon_{22} \equiv 0. \end{aligned} \tag{4.2.7a}$$

Thus in matrix form we have

$$[\varepsilon_{AB}] \equiv \begin{bmatrix} 0 & 1 \\ -1 & 0 \end{bmatrix} \equiv [\varepsilon^{AB}] \equiv [\varepsilon^{\bar{A}\bar{B}}] \equiv [\varepsilon_{\bar{A}\bar{B}}]. \tag{4.2.7b}$$

It follows that

$$\begin{aligned} \varepsilon_{AC}\varepsilon^{CB} &= -\delta^{B}{}_{A}, \\ \varepsilon_{\bar{A}\bar{C}}\varepsilon^{\bar{C}\bar{B}} &= -\delta^{\bar{B}}{}_{\bar{A}}, \\ \varepsilon_{AC}\varepsilon^{AC} &= \varepsilon_{\bar{A}\bar{C}}\varepsilon^{\bar{A}\bar{C}} = 2. \end{aligned} \tag{4.2.8}$$

The transformation properties of the permutation symbol (compare the equation (1.3.13a))

$$\begin{aligned} \hat{\varepsilon}^{AB} &= (\det \Lambda)\Lambda^{A}{}_{C}\Lambda^{B}{}_{D}\varepsilon^{CD} = \Lambda^{A}{}_{C}\Lambda^{B}{}_{D}\varepsilon^{CD}, \\ \hat{\varepsilon}^{AB} &= \varepsilon^{AB}, \\ \hat{\varepsilon}_{AB} &= \varepsilon_{AB}. \end{aligned} \tag{4.2.9}$$

Under a unimodular transformation ε_{AB}, ε^{AB} behave as *numerical spinors*, with unchanged values. The antisymmetric bilinear combination

$$\varepsilon_{AB}\psi^{A}\chi^{B} = \psi^{1}\chi^{2} - \psi^{2}\chi^{1} = \det\begin{bmatrix} \psi_1 & \chi_1 \\ \psi_2 & \chi_2 \end{bmatrix}$$

remains invariant under a unimodular transformation. This quantity is certainly *not* an inner product between two complex vectors ψ, χ, but it is formally the "complex area" between two complex vectors ψ, χ. Therefore, the spinor space V_2 is *not* an inner product space and ε_{AB} is *not* the metric spinor as it is often loosely called. Since V_2 is endowed with an antisymmetric bilinear form $\varepsilon(\psi, \chi) \equiv \varepsilon_{AB}\psi^{A}\chi^{B}$, it is a *symplectic* space. In such a space, a vector $\psi \in V_2$ has a *dual* covector naturally associated with it via the antisymmetric bilinear form ε [compare (1.3.15)]

$$\begin{aligned} \psi^{*}_{A} &\equiv \varepsilon_{AB}\psi^{B}, \\ \psi^{*}_{\bar{A}} &\equiv \varepsilon_{\bar{A}\bar{B}}\psi^{\bar{B}}. \end{aligned} \tag{4.2.10}$$

In the literature the star is usually dropped and the above equations appear as

$$\begin{aligned} \psi_{A} &\equiv \varepsilon_{AB}\psi^{B}, \\ \psi_{\bar{A}} &\equiv \varepsilon_{\bar{A}\bar{B}}\psi^{\bar{B}}. \end{aligned} \tag{4.2.11}$$

We shall follow the same notation as in the literature. Inverting the above equations via (4.2.8) we get

$$\begin{aligned} \psi^{C} &= \varepsilon^{AC}\psi_{A} = -\varepsilon^{CA}\psi_{A}, \\ \psi^{\bar{C}} &= \varepsilon^{\bar{A}\bar{C}}\psi_{\bar{A}} = -\varepsilon^{\bar{C}\bar{A}}\psi_{\bar{A}}. \end{aligned} \tag{4.2.12}$$

Equations (4.2.11), (4.2.12) indicate how to "raise or lower indices" for spinors. But one should be careful to note that

$$\chi_A\phi^A = \varepsilon_{AB}\chi^B\phi^A = -\varepsilon_{BA}\phi^A\chi^B = -\phi_B\chi^B = -\chi^A\phi_A,$$
$$\chi_A\chi^A \equiv 0, \qquad \phi_{\bar{A}}\phi^{\bar{A}} \equiv 0. \tag{4.2.13}$$

The $2^{k+l+m+n+2(r+s)}$ components of a mixed spinor-tensor transform, under the combined unimodular and Lorentz transformations, according to

$$\begin{aligned}\hat{\psi}^{i_1\cdots i_r A_1\cdots A_k \bar{B}_1\cdots \bar{B}_l}{}_{j_1\cdots j_s C_1\cdots C_m \bar{D}_1\cdots \bar{D}_n} &= a^{i_1}{}_{p_1}\cdots a^{i_r}{}_{p_r} l^{q_1}{}_{j_1}\cdots l^{q_s}{}_{j_s}\Lambda^{A_1}{}_{E_1}\cdots \Lambda^{A_k}{}_{E_k}\Lambda^{\bar{B}_1}{}_{\bar{F}_1}\cdots \Lambda^{\bar{B}_l}{}_{\bar{F}_l} \\ &\times \Sigma^{G_1}{}_{C_1}\cdots \Sigma^{G_m}{}_{C_m}\Sigma^{\bar{H}_1}{}_{\bar{D}_1}\cdots \Sigma^{\bar{H}_n}{}_{\bar{D}_n}\psi^{p_1\cdots p_r E_1\cdots E_k \bar{F}_1\cdots \bar{F}_l}{}_{q_1\cdots q_s G_1\cdots G_m \bar{H}_1\cdots \bar{H}_n}.\end{aligned} \tag{4.2.14}$$

We choose a special set of *numerical mixed tensor-spinors* by the following choice:

$$\begin{aligned}
[\sigma^{1\bar{A}B}] &= \begin{bmatrix}\sigma^{1\bar{1}1} & \sigma^{1\bar{1}2}\\ \sigma^{1\bar{2}1} & \sigma^{1\bar{2}2}\end{bmatrix} \equiv \begin{bmatrix}0 & 1\\ 1 & 0\end{bmatrix} = \sigma^1,\\
[\sigma^{2\bar{A}B}] &= \begin{bmatrix}\sigma^{2\bar{1}1} & \sigma^{2\bar{1}2}\\ \sigma^{2\bar{2}1} & \sigma^{2\bar{2}2}\end{bmatrix} \equiv \begin{bmatrix}0 & -i\\ i & 0\end{bmatrix} = \sigma^2,\\
[\sigma^{3\bar{A}B}] &= \begin{bmatrix}\sigma^{3\bar{1}1} & \sigma^{3\bar{1}2}\\ \sigma^{3\bar{2}1} & \sigma^{3\bar{2}2}\end{bmatrix} \equiv \begin{bmatrix}1 & 0\\ 0 & -1\end{bmatrix} = \sigma^3,\\
[\sigma^{4\bar{A}B}] &= \begin{bmatrix}\sigma^{4\bar{1}1} & \sigma^{4\bar{1}2}\\ \sigma^{4\bar{2}1} & \sigma^{4\bar{2}2}\end{bmatrix} \equiv \begin{bmatrix}1 & 0\\ 0 & 1\end{bmatrix} = \sigma^4.
\end{aligned} \tag{4.2.15}$$

Lowering the spinor indices by the rule [see (4.2.11)]

$$\sigma^k{}_{\bar{A}B} = -\varepsilon_{\bar{A}\bar{C}}\sigma^{k\bar{C}D}\varepsilon_{DB},$$

we obtain

$$\begin{aligned}
[\sigma^1{}_{\bar{A}B}] &= -\sigma^1,\\
[\sigma^2{}_{\bar{A}B}] &= \sigma^2,\\
[\sigma^3{}_{\bar{A}B}] &= -\sigma^3,\\
[\sigma^4{}_{\bar{A}B}] &= \sigma^4.
\end{aligned} \tag{4.2.16}$$

Recall Theorem (4.1.1), which is applicable to some of the matrices in (4.2.15). The appropriate generalization of this theorem is stated next.

Theorem (4.2.1): *The components $\sigma^{k\bar{A}B}$ of the numerical spinor-tensor satisfy*

$$\begin{aligned}
&\text{(i)}\quad \sigma^{k\bar{B}A} = \overline{\sigma^{k\bar{A}B}},\\
&\text{(ii)}\quad \sigma_a{}^{\bar{B}A}\sigma_{b\bar{B}C} = -d_{ab}\delta^A{}_C + (i/2)\eta_{abkl}\sigma^{k\bar{B}A}\sigma^l{}_{\bar{B}C},
\end{aligned} \tag{4.2.17}$$

where η_{abcd} is the totally antisymmetric pseudotensor [equation (1.3.13b)].

Proof: (i) By the hermitian properties $\sigma^{k\dagger} = \sigma^k$ the equations $\sigma^{kB\bar{A}} = \sigma^{k\bar{A}B}$ follow.

(ii) The proof emerges by direct computations. For example,

$$\begin{aligned}
\sigma_1{}^{\bar{B}1}\sigma_{1\bar{B}1} &= \sigma^{1\bar{B}1}\sigma^1{}_{\bar{B}1} = \sigma^{1\bar{1}1}\sigma^1{}_{\bar{1}1} + \sigma^{1\bar{2}1}\sigma^1{}_{\bar{2}1} \\
&= 0 + (-1) = (-1) + 0 \\
&= -d_{11}\delta^1{}_1 + (i/2)\eta_{11kl}\sigma^{k\bar{B}1}\sigma^l{}_{\bar{B}1}, \\
\sigma_1{}^{\bar{B}1}\sigma_{2\bar{B}1} &= \sigma^{1\bar{B}1}\sigma^2{}_{\bar{B}1} = \sigma^{1\bar{1}1}\sigma^2{}_{\bar{1}1} + \sigma^{1\bar{2}1}\sigma^2{}_{\bar{2}1} \\
&= 0 + (i) = 0 + (i/2)[(1)(1) - (1)(-1)] \\
&= 0 + (i/2)[\sigma^{3\bar{B}1}\sigma_{4\bar{B}1} - \sigma^{4\bar{B}1}\sigma_{3\bar{B}1}] \\
&= -d_{12}\delta^1{}_1 + (i/2)\eta_{12kl}\sigma^{k\bar{B}1}\sigma^l{}_{\bar{B}1}.
\end{aligned}$$

Similarly all the other equations follow. ■

From (4.2.17) the following three other useful equations emerge:

$$\begin{aligned}
\sigma_a{}^{\bar{B}A}\sigma_{b\bar{B}C} + \sigma_b{}^{\bar{B}A}\sigma_{a\bar{B}C} &= -2d_{ab}\delta^A{}_C, \\
\sigma_a{}^{\bar{B}A}\sigma_{b\bar{B}C} - \sigma_b{}^{\bar{B}A}\sigma_{a\bar{B}C} &= i\eta_{abkl}\sigma^{k\bar{B}A}\sigma^l{}_{\bar{B}C}, \qquad (4.2.18)\\
\sigma_a{}^{\bar{B}A}\sigma_{b\bar{B}A} &= -2d_{ab}.
\end{aligned}$$

Now we shall discuss the transformation property of a second-order spinor $T^{\bar{A}B}$.

Theorem (4.2.2): *If $T^{\bar{A}B}$ are the components of a second-order spinor then $\det[T^{\bar{A}B}]$ is invariant under a spinor transformation* (4.2.6).

Proof: Under (4.2.6) one has $\hat{T}^{\bar{A}B} = \Lambda^{\bar{A}}{}_{\bar{C}}\Lambda^B{}_D T^{\bar{C}D}$. Thus taking the determinant of both sides one obtains

$$\begin{aligned}
\det[\hat{T}^{\bar{A}B}] &= \hat{T}^{\bar{1}1}\hat{T}^{\bar{2}2} - \hat{T}^{\bar{1}2}\hat{T}^{\bar{2}1} = (1/2)\varepsilon_{\bar{C}\bar{A}}\varepsilon_{DB}\hat{T}^{\bar{A}B}\hat{T}^{\bar{C}D} \\
&= (1/2)\hat{T}^{\bar{C}D}\hat{T}_{\bar{C}D} = (1/2)T^{\bar{C}D}T_{\bar{C}D} = \det[T^{\bar{A}B}]. \quad ■
\end{aligned}$$

In the special case where the matrix $[X^{\bar{A}B}]$ is hermitian, then by Theorem (4.1.3) we can write

$$X^{\bar{A}B} = x^k\sigma_k{}^{\bar{A}B} \qquad (4.2.19)$$

for some real numbers x^k, and the above equation is equivalent to equation (4.1.19). Under a spinor transformation

$$\hat{X}^{\bar{A}B} = \Lambda^{\bar{A}}{}_{\bar{C}}\Lambda^B{}_D X^{\bar{C}D},$$

the matrix $\hat{X}^{\bar{A}B}$ remains hermitian. Thus we have the linear combination

$$\hat{X}^{\bar{A}B} = \hat{x}^k\sigma_k{}^{\bar{A}B} \qquad (4.2.20)$$

for some real numbers $\hat{x}^k$. By equation (4.1.19) and Theorem (4.2.2), we have under a spinor transformation

$$-d_{kl}\hat{x}^k\hat{x}^l = \det[\hat{X}^{\bar{A}B}] = \det[X^{\bar{A}B}] = -d_{kl}x^k x^l. \tag{4.2.21}$$

Since the above equation is also a consequence of a Lorentz transformation, we can link a Lorentz transformation with a spinor transformation.

Theorem (4.2.3): *Let a spinor transformation be given by the unimodular matrix* $[\Lambda^A{}_B]$. *There exists a corresponding proper orthochronous Lorentz matrix* $[l^a{}_b]$ *given by*

$$l^a{}_b = (-1/2)\Lambda^{\bar{A}}{}_{\bar{C}}\Lambda^B{}_D\sigma^a{}_{\bar{A}B}\sigma_b{}^{\bar{C}D}. \tag{4.2.22a}$$

Proof: Recall that in (4.2.18) we had $\sigma_k{}^{\bar{A}B}\sigma^b{}_{\bar{A}B} = -2\delta^b{}_k$. Using this we get

$$\sigma^b{}_{\bar{A}B}X^{\bar{A}B} = (x^k\sigma_k{}^{\bar{A}B})\sigma^b{}_{\bar{A}B} = -2x^b;$$

thus, $x^b = (-1/2)\sigma^b{}_{\bar{A}B}X^{\bar{A}B}$. Similarly we have

$$\begin{aligned}\hat{x}^a &= (-1/2)\sigma^a{}_{\bar{A}B}\hat{X}^{\bar{A}B} = (-1/2)\sigma^a{}_{\bar{A}B}(\Lambda^{\bar{A}}{}_{\bar{C}}\Lambda^B{}_D X^{\bar{C}D}) \\ &= (-1/2)\Lambda^{\bar{A}}{}_{\bar{C}}\Lambda^B{}_D\sigma^a{}_{\bar{A}B}(\sigma_b{}^{\bar{C}D}x^b).\end{aligned}$$

But under a Lorentz transformation we have $\hat{x}^a = l^a{}_b x^b$. Thus we see

$$[l^a{}_b + (1/2)\Lambda^{\bar{A}}{}_{\bar{C}}\Lambda^B{}_D\sigma^a{}_{\bar{A}B}\sigma_b{}^{\bar{C}D}]x^b \equiv 0,$$

where x^k can take *arbitrary* real values. Differentiating with respect to x^b, the above identity yields the equation (4.2.22*a*). It can be proved that the Lorentz transformation in (4.2.22*a*) is proper and orthochronous (see Exercises 4.2, problem 3). ■

The equation

$$l^a{}_b\sigma^b{}_{\bar{A}B} = \Lambda^{\bar{C}}{}_{\bar{A}}\Lambda^D{}_B\sigma^a{}_{\bar{C}D} \tag{4.2.22b}$$

is equivalent to (4.2.22*a*).

Example: Consider the spinor transformation given by the unimodular matrix

$$[\Lambda^A{}_B] = \begin{bmatrix} e^{\chi/2} & 0 \\ 0 & e^{-\chi/2}\end{bmatrix} = [\Lambda^{\bar{A}}{}_{\bar{B}}], \qquad \chi \in \mathbb{R}.$$

By (4.2.22*a*)

$$\begin{aligned} l^3{}_4 &= -(1/2)\Lambda^{\bar{A}}{}_{\bar{C}}\Lambda^B{}_D\sigma^3{}_{\bar{A}B}\sigma_4{}^{\bar{C}D} \\ &= -(1/2)[\Lambda^{\bar{1}}{}_{\bar{1}}\Lambda^1{}_1\sigma^3{}_{\bar{1}1}\sigma_4{}^{\bar{1}1} + \Lambda^{\bar{2}}{}_{\bar{2}}\Lambda^2{}_2\sigma^3{}_{\bar{2}2}\sigma_4{}^{\bar{2}2}] \\ &= -(1/2)[(e^{\chi/2})^2(-1)(-1) + (e^{-\chi/2})^2(1)(-1)] \\ &= -\sinh\chi.\end{aligned}$$

This result agrees with the previous example worked out after equation (4.1.20). □

Consider now a higher-order spinor $\psi^{A_1 \cdots A_j \bar{B}_1 \cdots \bar{B}_k}$ that is totally symmetric with respect to the indices $A_1, \ldots, A_j$ and the indices $\bar{B}_1, \ldots, \bar{B}_k$. The components of such a spinor correspond to a vector in a complex vector space of dimension $(2j + 1)(2k + 1)$. Precisely such vector spaces constitute all irreducible representations for the proper orthochronous Lorentz group $\mathscr{L}^{\uparrow}_{4+}$. The proof of this important statement is beyond the scope of this book. (See the book by Gelfand et al in the references after this chapter.)

It has been mentioned before that improper Lorentz transformations are *not* induced by spinor transformations. However, when we consider the *conjugate linear mappings* of the spinor space V_2, the improper Lorentz transformations can be accommodated. Recall that for a unimodular conjugate linear mapping Σ of V_2 we have

$$\Sigma(\lambda_1 \psi_1 + \lambda_2 \psi_2) = \bar{\lambda}_1 \Sigma(\psi_1) + \bar{\lambda}_2 \Sigma(\psi_2).$$

In terms of the components, we can express a conjugate linear mapping as

$$\begin{gathered}
\hat{\psi}^{\bar{A}} = \Lambda^{\bar{A}}{}_B \psi^B, \qquad \hat{\psi}^A = \Lambda^A{}_{\bar{B}} \psi^{\bar{B}}, \\
\Sigma^{\bar{A}}{}_B \Lambda^B{}_{\bar{C}} = \Lambda^{\bar{A}}{}_B \Sigma^B{}_{\bar{C}} = \delta^{\bar{A}}{}_{\bar{C}}, \qquad \Sigma^A{}_{\bar{B}} \Lambda^{\bar{B}}{}_C = \Lambda^A{}_{\bar{B}} \Sigma^{\bar{B}}{}_C = \delta^A{}_C, \qquad (4.2.23) \\
\det[\Lambda^{\bar{A}}{}_B] = \det[\Lambda^A{}_{\bar{B}}] = \det[\Sigma^{\bar{A}}{}_B] = \det[\Sigma^A{}_{\bar{B}}] = 1.
\end{gathered}$$

Consider again the hermitian matrix $X^{\bar{A}B}$ and the equation (4.2.20). Under an improper Lorentz transformation we have

$$\begin{aligned}
l^a{}_b x^b = \hat{x}^a &= (-1/2)\sigma^a{}_{\bar{A}B} \hat{X}^{\bar{A}B} \\
&= (-1/2)\sigma^a{}_{\bar{A}B} \Lambda^{\bar{A}}{}_C \Lambda^B{}_{\bar{D}} X^{C\bar{D}} \\
&= (-1/2)\sigma^a{}_{\bar{A}B} \Lambda^{\bar{A}}{}_C \Lambda^B{}_{\bar{D}} X^{\bar{D}C} \\
&= -[(1/2)\Lambda^{\bar{A}}{}_C \Lambda^B{}_{\bar{D}} \sigma^a{}_{\bar{A}B} \sigma_b{}^{\bar{D}C}] x^b.
\end{aligned}$$

From the above equation we can derive

$$l^a{}_b = (-1/2)\Lambda^{\bar{A}}{}_C \Lambda^B{}_{\bar{D}} \sigma^a{}_{\bar{A}B} \sigma_b{}^{\bar{D}C}. \qquad (4.2.24)$$

Example: As a special case consider the unimodular matrix $[\Lambda^{\bar{A}}{}_B] = \begin{bmatrix} -1 & 0 \\ 0 & -1 \end{bmatrix}$. According to (4.2.24), (4.2.20), (4.2.16) we have

$$\begin{aligned}
\hat{x}^1 &= (-1/2)[\Lambda^{\bar{1}}{}_1 \Lambda^2{}_{\bar{2}} \sigma^1{}_{\bar{1}2} X^{\bar{2}1} + \Lambda^{\bar{2}}{}_2 \Lambda^1{}_{\bar{1}} \sigma^1{}_{\bar{2}1} X^{\bar{1}2}] \\
&= (-1/2)[(-1)(x^1 + ix^2) + (-1)(x^1 - ix^2)] = x^1, \\
\hat{x}^2 &= (-1/2)[\sigma^2{}_{\bar{1}2} X^{\bar{2}1} + \sigma^2{}_{\bar{2}1} X^{\bar{1}2}] = -x^2, \\
\hat{x}^3 &= (-1/2)[\sigma^3{}_{\bar{1}1} X^{\bar{1}1} + \sigma^3{}_{\bar{2}2} X^{\bar{2}2}] = x^3, \\
\hat{x}^4 &= (-1/2)[\sigma^4{}_{\bar{1}1} X^{\bar{1}1} + \sigma^4{}_{\bar{2}2} X^{\bar{2}2}] = x^4.
\end{aligned}$$

The above set of equations denotes a special improper Lorentz transformation. □

EXERCISES 4.2

1. (i) Prove that for an arbitrary symmetric spinor s_{AB} one has $s_A{}^A \equiv 0$.
(ii) Prove that for an arbitrary spinor α_{AB} one must have $\alpha_{AB} - \alpha_{BA} - \varepsilon_{AB}\alpha^C{}_C \equiv 0$.

2. Prove that $\sigma^a{}_{\bar{B}C}\sigma_{a\bar{D}F} = -2\varepsilon_{\bar{B}\bar{D}}\varepsilon_{CF}$.

3. Prove explicitly that the Lorentz transformation

$$l^a{}_b = (-1/2)\Lambda^{\bar{A}}{}_{\bar{C}}\Lambda^B{}_D\sigma^a{}_{\bar{A}B}\sigma_b{}^{\bar{C}D},$$

where $\Lambda^D{}_B$ is a spinor transformation, has to be proper and orthochronous.

4. Obtain the conjugate linear unimodular transformation $\pm\Lambda^{\bar{A}}{}_B$, which induces the improper Lorentz transformation of space reflection **P**.

4.3. Dirac Matrices and Dirac Bispinors

We shall define a *Dirac bispinor* as a vector in a four-dimensional *complex vector space* $V_4^\#$. Note that in the literature a Dirac bispinor is commonly called a Dirac spinor or simply a spinor. We shall adopt the simple convention that a bispinor refers to a Dirac bispinor. Relative to a basis of $V_4^\#$, the four components of a bispinor are expressed as the column vector

$$\psi = [\psi^u] = \begin{bmatrix} \psi^1 \\ \psi^2 \\ \psi^3 \\ \psi^4 \end{bmatrix}. \tag{4.3.1}$$

The indices u, v, w, t will be used for bispinors, and each will take values in the set $\{1, 2, 3, 4\}$. Furthermore, the bispinor components are identified with the components of *two* spinors by the rule

$$\begin{bmatrix} \psi^1 \\ \psi^2 \\ \psi^3 \\ \psi^4 \end{bmatrix} \equiv \begin{bmatrix} \chi_1 \\ \chi_2 \\ \phi^{\bar{1}} \\ \phi^{\bar{2}} \end{bmatrix}. \tag{4.3.2}$$

The components of a bispinor transform under a unimodular transformation as

$$\begin{gathered} \hat{\psi}^u = \mathcal{T}^u{}_v\psi^v, \qquad \psi^v = \Omega^v{}_u\hat{\psi}^u, \\ \mathcal{T}^u{}_v\Omega^v{}_w = \Omega^u{}_v\mathcal{T}^v{}_w = \delta^u{}_w, \end{gathered} \tag{4.3.3}$$

where $\det[\mathcal{T}^u{}_v] = \det[\Omega^u{}_v] = 1$. Here the summation convention is adopted.

The components of a higher-order mixed bispinor-tensor transform under a combined Lorentz and bispinor transformation according to

$$\begin{aligned}\psi'^{i_1\cdots i_r u_1\cdots u_k}{}_{j_1\cdots j_s v_1\cdots v_l} &= a^{i_1}{}_{k_1}\cdots a^{i_r}{}_{k_r} l^{m_1}{}_{j_1}\cdots l^{m_s}{}_{j_s}\mathcal{T}^{u_1}{}_{w_1}\cdots\mathcal{T}^{u_k}{}_{w_k}\\ &\quad\times \Omega^{t_1}{}_{v_1}\cdots\Omega^{t_l}{}_{v_l}\psi^{k_1\cdots k_r w_1\cdots w_k}{}_{m_1\cdots m_s t_1\cdots t_l}.\end{aligned} \tag{4.3.4}$$

Example: Consider the transformation of the components $\alpha^{iu}{}_v$ of a mixed bispinor-tensor:

$$\hat{\alpha}^{iu}{}_v = a^i{}_k\mathcal{T}^u{}_w\Omega^t{}_v\alpha^{kw}{}_t. \quad \square \tag{4.3.5}$$

Let us consider a special *numerical mixed bispinor-tensor* with components relative to a suitable basis given by

$$\begin{aligned}\gamma^1 = [\gamma^{1u}{}_v] &\equiv \begin{bmatrix} 0 & 0 & 0 & 1\\ 0 & 0 & 1 & 0\\ 0 & 1 & 0 & 0\\ 1 & 0 & 0 & 0\end{bmatrix},\\ \gamma^2 = [\gamma^{2u}{}_v] &\equiv \begin{bmatrix} 0 & 0 & 0 & -i\\ 0 & 0 & i & 0\\ 0 & -i & 0 & 0\\ i & 0 & 0 & 0\end{bmatrix},\\ \gamma^3 = [\gamma^{3u}{}_v] &\equiv \begin{bmatrix} 0 & 0 & 1 & 0\\ 0 & 0 & 0 & -1\\ 1 & 0 & 0 & 0\\ 0 & -1 & 0 & 0\end{bmatrix},\\ \gamma^4 = [\gamma^{4u}{}_v] &\equiv \begin{bmatrix} 0 & 0 & -1 & 0\\ 0 & 0 & 0 & -1\\ 1 & 0 & 0 & 0\\ 0 & 1 & 0 & 0\end{bmatrix}.\end{aligned} \tag{4.3.6}$$

The matrices γ^a are known as *Dirac matrices.* In terms of the 2×2 sub-matrices σ^k, the Dirac matrices can be expressed as

$$\gamma^\alpha = \begin{bmatrix} 0 & \sigma_\alpha\\ \sigma_\alpha & 0\end{bmatrix}, \quad \gamma^4 = \begin{bmatrix} 0 & -\mathrm{I}\\ \mathrm{I} & 0\end{bmatrix}. \tag{4.3.7}$$

In the literature various authors make various choices for the Dirac matrices. All these choices, however, must satisfy the following equations:

$$\gamma^a\gamma^b + \gamma^b\gamma^a = 2d^{ab}\mathrm{I}, \tag{4.3.8a}$$

$$\text{Trace } \gamma^a = 0, \tag{4.3.8b}$$

$$\gamma^{\alpha\dagger} = \gamma^\alpha, \tag{4.3.8c}$$

$$\gamma^{4\dagger} = -\gamma^4. \tag{4.3.8d}$$

Equation (4.3.8*a*) follows from (4.3.7) by blockwise matrix multiplication, and equations (4.3.8*b*), (4.3.8*c*), (4.3.8*d*) follow directly from (4.3.6).

We can introduce another 4 × 4 matrix

$$\gamma^5 \equiv \eta_{abcd}\gamma^a\gamma^b\gamma^c\gamma^d/4! = \gamma^1\gamma^2\gamma^3\gamma^4 = i\begin{bmatrix} I & 0 \\ 0 & -I \end{bmatrix}. \tag{4.3.9}$$

The special properties of γ^5 are summarized in the following equations:

$$\begin{aligned} (\gamma^5)^2 &= -I, \\ \gamma^{5\dagger} &= -\gamma^5, \\ \gamma^5\gamma^a + \gamma^a\gamma^5 &= 0. \end{aligned} \tag{4.3.10}$$

The transformation properties of the γ^a matrices are given by the equation (4.3.5), i.e.,

$$\hat{\gamma}^{iu}{}_v = a^i{}_k\mathcal{T}^u{}_w\Omega^t{}_v\gamma^{kw}{}_t. \tag{4.3.11}$$

However, given a Lorentz transformation $a^i{}_j$, there exist unimodular matrices $\mathcal{T}^u{}_w$ and $\Omega^u{}_w$ representing bispinor transformations, such that the numerical components $\gamma^{iu}{}_v$ remain unchanged. For such a case

$$\begin{aligned} \gamma^{iu}{}_v &= a^i{}_k\mathcal{T}^u{}_w\Omega^t{}_v\gamma^{kw}{}_t, \\ l^a{}_b\gamma^{bu}{}_v &= \mathcal{T}^u{}_w\Omega^t{}_v\gamma^{aw}{}_t, \\ l^a{}_b\gamma^b &= \mathcal{T}\gamma^a\mathcal{T}^{-1}. \end{aligned} \tag{4.3.12}$$

Instead of proving the equation (4.3.12) in general, we shall give some examples. Note that the last equation above generalizes equation (4.2.22*a*) to the 4 × 4 matrix form.

Example: Let us choose

$$[l^a{}_b] = \begin{bmatrix} 1 & 0 & 0 & 0 \\ 0 & 1 & 0 & 0 \\ 0 & 0 & \cosh\chi & -\sinh\chi \\ 0 & 0 & -\sinh\chi & \cosh\chi \end{bmatrix}.$$

One of the equations in (4.3.12) yields

$$l^3{}_b\gamma^b = \gamma^3\cosh\chi - \gamma^4\sinh\chi = \begin{bmatrix} 0 & 0 & e^\chi & 0 \\ 0 & 0 & 0 & -e^{-\chi} \\ e^{-\chi} & 0 & 0 & 0 \\ 0 & -e^\chi & 0 & 0 \end{bmatrix}$$

$$
= \begin{bmatrix} e^{\chi/2} & 0 & 0 & 0 \\ 0 & e^{-\chi/2} & 0 & 0 \\ 0 & 0 & e^{-\chi/2} & 0 \\ 0 & 0 & 0 & e^{\chi/2} \end{bmatrix} \begin{bmatrix} 0 & 0 & 1 & 0 \\ 0 & 0 & 0 & -1 \\ 1 & 0 & 0 & 0 \\ 0 & -1 & 0 & 0 \end{bmatrix}
$$

$$
\times \begin{bmatrix} e^{-\chi/2} & 0 & 0 & 0 \\ 0 & e^{\chi/2} & 0 & 0 \\ 0 & 0 & e^{\chi/2} & 0 \\ 0 & 0 & 0 & e^{-\chi/2} \end{bmatrix}
$$

$$
= \mathscr{T} \gamma^3 \mathscr{T}^{-1}.
$$

All other equations in (4.3.12) for this case are also satisfied with the same choice for $\mathscr{T}$. □

Example 2: Consider the Lorentz matrix with entries $l^a{}_b = -\delta^a{}_b$. The corresponding Lorentz transformation is proper but not orthochronous. In this case we have

$$
l^a{}_b \gamma^b = -\gamma^a = -\gamma^5 \gamma^a \gamma^5 = \gamma^5 \gamma^a (\gamma^5)^{-1}.
$$

Therefore, in this case we obtain the unimodular 4×4 matrix $\mathscr{T} = \pm\gamma^5$. □

EXERCISES 4.3

1. Consider the sixteen 4×4 matrices I, γ^a, $\gamma^a\gamma^b$ $(a < b)$, γ^5, $\gamma^5\gamma^a$.
(i) Prove that except for I all other fifteen matrices have a zero trace.
(ii) Prove that these sixteen matrices form a basis for the vector space of 4×4 complex matrices.

2. Consider the (improper) time reversal Lorentz transformation, viz.,

$$
\hat{x}^\alpha = x^\alpha, \qquad \hat{x}^4 = -x^4.
$$

Obtain the two 4×4 unimodular matrices $\pm\mathscr{T}$, corresponding to bispinor transformations, which induce this time reversal.

References

1. W. L. Bade and H. Jehle, Rev. Mod. Phys. **25** (1953), 714–728. [pp. 104, 112, 115]
2. E. M. Corson, *Introduction to tensors, spinors, and relativistic wave-equations*, Blackie and Son, London, 1955. [pp. 112, 115, 121]
3. I. M. Gelfand, R. A. Milnos, and Z. Y. Shapiro, *Representations of the rotation and Lorentz groups, and their applications*, The MacMillan Co., New York, 1963. [pp. 102, 103, 235]

5
The Special Relativistic Mechanics

5.1. The Prerelativistic Particle Mechanics

We shall start with a short review of the prerelativistic mechanics of Newton, Lagrange, and Hamilton. For simplicity we restrict ourselves to systems of a single-point particle having (constant) mass $m > 0$. Let the parameterized motion curve be given by $x^\alpha = \mathscr{X}^\alpha(t)$, in the Euclidean space $\mathbb{E}_3$. Let the three components of the *force vector* be given by $f^\alpha(t, \mathbf{x}, \mathbf{v})$, which are functions of seven real variables. Here t stands for the time variable, $\mathbf{x} \equiv (x^1, x^2, x^3)$ are the spatial coordinates in a Cartesian coordinate system, and $\mathbf{v} \equiv (v^1, v^2, v^3)$ are the velocity variables. *Newton's equations of motion* for a single particle are given by

$$m \frac{d^2\mathscr{X}^\alpha(t)}{dt^2} = f^\alpha(t, \mathbf{x}, \mathbf{v})_{|x^\alpha = \mathscr{X}^\alpha(t),\, v^\alpha = d\mathscr{X}^\alpha(t)/dt}. \tag{5.1.1}$$

These equations imply that

$$\frac{d}{dt}\left[(1/2)m\delta_{\alpha\beta}\frac{d\mathscr{X}^\alpha(t)}{dt}\frac{d\mathscr{X}^\beta(t)}{dt}\right]$$
$$= \delta_{\alpha\beta}[v^\alpha f^\beta(t, \mathbf{x}, \mathbf{v})]_{|x^\alpha = \mathscr{X}^\alpha(t),\, v^\alpha = d\mathscr{X}^\alpha(t)/dt}. \tag{5.1.2}$$

The above equations can be interpreted physically as "*the rate of increase of kinetic energy of a particle is equal to the rate of work performed by the external force.*"

In the special case of a *monogenic* force, there exists a *work function* $U(t, \mathbf{x}, \mathbf{v})$ such that

$$\delta_{\alpha\beta} f^\beta(t, \mathbf{x}, \mathbf{v})_{|..} = \frac{\partial U(t, \mathbf{x}, \mathbf{v})}{\partial x^\alpha}_{|..} - \frac{d}{dt}\left\{\left[\frac{\partial U(t, \mathbf{x}, \mathbf{v})}{\partial v^\alpha}\right]_{|..}\right\}. \tag{5.1.3}$$

In such a case, the equations of motion (5.1.1) are the Euler-Lagrange equations derived from a variational principle. To explain such a statement we define a *Lagrangian* function $L\colon [t_1, t_2] \times D_6 \subset \mathbb{R}^7 \to \mathbb{R}$ such that $L \in \mathscr{C}^2([t_1, t_2] \times D_6)$, and in $[t_1, t_2] \times D_6$ we have $\det[\partial^2 L(t, \mathbf{x}, \mathbf{v})/\partial v^\alpha \partial v^\beta] > 0$.

Consider the twice differentiable parameterized curve γ in $\mathbb{E}_3$ given by $x^\alpha = \mathscr{X}^\alpha(t)$, $t \in [t_1, t_2]$. The set of all such curves will be denoted by $\mathscr{C}^2[t_1, t_2]$. The *action functional* is a mapping $J: \mathscr{C}^2[t_1, t_2] \to \mathbb{R}$ such that

$$J[\gamma] \equiv \int_{t_1}^{t_2} [L(t, \mathbf{x}, \mathbf{v})_{|x^\alpha = \mathscr{X}^\alpha(t),\, v^\alpha = d\mathscr{X}^\alpha(t)/dt}]\, dt. \tag{5.1.4}$$

A "slightly varied" curve $\hat{\gamma}$ in $\mathbb{E}_3$, having the same end points, is given by

$$x^\alpha = \mathscr{X}^\alpha(t) + \varepsilon h^\alpha(t),$$
$$h^\alpha(t_1) = h^\alpha(t_2) = 0,$$

where $h^\alpha \in \mathscr{C}^2[t_1, t_2]$ and $\varepsilon > 0$ is a small positive number.

The "slight variation" in the action integral in (5.1.4) is given by

$$\begin{aligned}
\delta J[\gamma] &\equiv J[\hat{\gamma}] - J[\gamma] \\
&= \int_{t_1}^{t_2} [L(t, \mathbf{x}, \mathbf{v})_{|x^\alpha = \mathscr{X}^\alpha(t) + \varepsilon h^\alpha(t),\, v^\alpha = d\mathscr{X}^\alpha(t)/dt + \varepsilon(dh^\alpha(t)/dt)} \\
&\quad - L(t, \mathbf{x}, \mathbf{v})_{|x^\alpha = \mathscr{X}^\alpha(t),\, v^\alpha = d\mathscr{X}^\alpha(t)/dt}]\, dt \\
&= \varepsilon \int_{t_1}^{t_2} \left\{ h^\alpha(t) \left[\frac{\partial L(t, \mathbf{x}, \mathbf{v})}{\partial x^\alpha} \right]_{|x^\alpha = \mathscr{X}^\alpha(t),\, v^\alpha = d\mathscr{X}^\alpha(t)/dt} \right. \\
&\quad \left. + \frac{dh^\alpha(t)}{dt} \left[\frac{\partial L(t, \mathbf{x}, \mathbf{v})}{\partial v^\alpha} \right]_{|x^\alpha = \mathscr{X}^\alpha(t),\, v^\alpha = d\mathscr{X}^\alpha(t)/dt} \right\} dt + \mathcal{O}(\varepsilon^2) \\
&= \varepsilon \int_{t_1}^{t_2} \left\{ h^\alpha(t) \left[\frac{\partial L(t, \mathbf{x}, \mathbf{v})}{\partial x^\alpha} \right]_{|..} \right\} dt + \varepsilon h^\alpha(t) \left[\frac{\partial L(t, \mathbf{x}, \mathbf{v})}{\partial v^\alpha} \right]_{|..} \Bigg|_{t_1}^{t_2} \\
&\quad - \varepsilon \int_{t_1}^{t_2} \left\{ h^\alpha(t) \frac{d}{dt} \left[\frac{\partial L(t, \mathbf{x}, \mathbf{v})}{\partial v^\alpha} \right]_{|..} \right\} dt + \mathcal{O}(\varepsilon^2) \\
&= \varepsilon \int_{t_1}^{t_2} h^\alpha(t) \left\{ \left[\frac{\partial L(t, \mathbf{x}, \mathbf{v})}{\partial x^\alpha} \right]_{|..} - \frac{d}{dt} \left[\frac{\partial L(t, \mathbf{x}, \mathbf{v})}{\partial v^\alpha} \right]_{|..} \right\} dt + \mathcal{O}(\varepsilon^2).
\end{aligned}$$

If we impose that *stationary condition* on the action integral of the curve γ, then we have $\lim_{\varepsilon \to 0+} (\delta J/\varepsilon) = 0$. In that case we obtain

$$\int_{t_1}^{t_2} h^\alpha(t) \left\{ \left[\frac{\partial L(t, \mathbf{x}, \mathbf{v})}{\partial x^\alpha} \right]_{|..} - \frac{d}{dt} \left[\frac{\partial L(t, \mathbf{x}, \mathbf{v})}{\partial v^\alpha} \right]_{|..} \right\} dt = 0.$$

Since the $h^\alpha(t)$ are arbitrary twice differentiable functions in $[t_1, t_2]$ for $\alpha = 1$, 2, 3, we obtain by the lemma of Dubois–Reymond (see the book by Gelfand and Fomin referred at the end) that

$$0 = \left[\frac{\partial L(t, \mathbf{x}, \mathbf{v})}{\partial x^\alpha} \right]_{|..} - \frac{d}{dt} \left\{ \left[\frac{\partial L(t, \mathbf{x}, \mathbf{v})}{\partial v^\alpha} \right]_{|..} \right\}. \tag{5.1.5}$$

The above equations are called the *Euler–Lagrange* equations.

If we can choose the special Lagrangian

$$L(t, \mathbf{x}, \mathbf{v}) = T(\mathbf{x}, \mathbf{v}) + U(t, \mathbf{x}, \mathbf{v}), \qquad T(\mathbf{x}, \mathbf{v}) \equiv (m/2)\delta_{\alpha\beta}v^\alpha v^\beta, \tag{5.1.6}$$

then the Euler-Lagrange equations (5.1.5) yield

$$\left[\frac{\partial U(t, \mathbf{x}, \mathbf{v})}{\partial x^\alpha}\right]_{|\ldots} - \frac{d}{dt}\left\{\left[\frac{\partial U(t, \mathbf{x}, \mathbf{v})}{\partial v^\alpha}\right]_{|\ldots}\right\} - \frac{d}{dt}\left[m\delta_{\alpha\beta}\frac{d\mathscr{X}^\beta(t)}{dt}\right] = 0.$$

The above equations are precisely the equations of motion (5.1.1) for the case of a monogenic force given by (5.1.3).

We define *generalized or canonical momentum variables by*

$$p_\alpha \equiv \frac{\partial L(t, \mathbf{x}, \mathbf{v})}{\partial v^\alpha}, \tag{5.1.7}$$

where $\det[\partial^2 L(t, \mathbf{x}, \mathbf{v})/\partial v^\alpha \partial v^\beta] = \det[\partial p_\alpha/\partial v^\beta] > 0$.

In the next step we introduce the *Hamiltonian function* by a *partial Legendre transformation*:

$$H^\#(t, \mathbf{x}, \mathbf{v}, \mathbf{p}) \equiv p_\alpha v^\alpha - L(t, \mathbf{x}, \mathbf{v}), \tag{5.1.8}$$

where $\mathbf{p} \equiv (p_1, p_2, p_3)$. But we notice that

$$\frac{\partial H^\#(t, \mathbf{x}, \mathbf{v}, \mathbf{p})}{\partial v^\alpha} = p_\alpha - \frac{\partial L(t, \mathbf{x}, \mathbf{v})}{\partial v^\alpha} \equiv 0.$$

Therefore, we can take the Hamiltonian function to be

$$H(t, \mathbf{x}, \mathbf{p}) \equiv p_\alpha v^\alpha - L(t, \mathbf{x}, \mathbf{v}). \tag{5.1.9}$$

It follows that

$$\begin{aligned} \frac{\partial H(t, \mathbf{x}, \mathbf{p})}{\partial p_\alpha} &= v^\alpha, \\ \frac{\partial H(t, \mathbf{x}, \mathbf{p})}{\partial x^\alpha} &= -\frac{\partial L(t, \mathbf{x}, \mathbf{v})}{\partial x^\alpha}, \\ \frac{\partial H(t, \mathbf{x}, \mathbf{p})}{\partial t} &= -\frac{\partial L(t, \mathbf{x}, \mathbf{v})}{\partial t}. \end{aligned} \tag{5.1.10}$$

The momentum components *along* the motion curve γ is given by (5.1.7) as

$$P_\alpha(t) = \frac{\partial L(t, \mathbf{x}, \mathbf{v})}{\partial v^\alpha}_{|x^\alpha = \mathscr{X}^\alpha(t),\, v^\alpha = d\mathscr{X}^\alpha(t)/dt}.$$

By Euler-Lagrange equations (5.1.5) and equation (5.1.10) we have along the corresponding motion curve $\gamma^\#$ in the *phase space* [six-dimensional manifold coordinatized by $(x^1, x^2, x^3, p_1, p_2, p_3)$] the equations:

$$\frac{d}{dt}P_\alpha(t) = \frac{\partial L(t,\mathbf{x},\mathbf{v})}{\partial x^\alpha}_{|x^\alpha = \mathscr{X}^\alpha(t),\, v^\alpha = d\mathscr{X}^\alpha(t)/dt},$$
$$= -\frac{\partial H(t,\mathbf{x},\mathbf{p})}{\partial x^\alpha}_{|x^\alpha = \mathscr{X}^\alpha(t),\, p_\alpha = P_\alpha(t)}, \tag{5.1.11}$$
$$\frac{d}{dt}\mathscr{X}^\alpha(t) = \frac{\partial H(t,\mathbf{x},\mathbf{p})}{\partial p_\alpha}_{|x^\alpha = \mathscr{X}^\alpha(t),\, p_\alpha = P_\alpha(t)}.$$

The above six equations are called *Hamilton's canonical equations* (for a single particle). The total time derivative of the Hamiltonian function [using (5.1.11)] is

$$\frac{d}{dt}H(t,\mathbf{x},\mathbf{p})_{|..} = \frac{\partial H(t,\mathbf{x},\mathbf{p})}{\partial t}_{|..} + \frac{d\mathscr{X}^\alpha(t)}{dt}\frac{\partial H(t,\mathbf{x},\mathbf{p})}{\partial x^\alpha}_{|..} + \frac{dP_\alpha(t)}{dt}\frac{\partial H(t,\mathbf{x},\mathbf{p})}{\partial p_\alpha}_{|..}$$
$$= \frac{\partial H(t,\mathbf{x},\mathbf{p})}{\partial t}_{|..} - \frac{d\mathscr{X}^\alpha(t)}{dt}\frac{dP_\alpha(t)}{dt} + \frac{d\mathscr{X}^\alpha(t)}{dt}\frac{dP_\alpha(t)}{dt}$$
$$= \frac{\partial H(t,\mathbf{x},\mathbf{p})}{\partial t}_{|..} = -\frac{\partial L(t,\mathbf{x},\mathbf{v})}{\partial t}_{|..}. \tag{5.1.12}$$

In case $\partial L(t,\mathbf{x},\mathbf{v})/\partial t_{|..} \equiv 0$, we have the integral

$$H(\mathbf{x},\mathbf{p}) = E, \tag{5.1.13}$$

where E is the integration constant representing the total energy of the particle.

A *conservative system* is characterized by the work function satisfying

$$\frac{\partial U(t,\mathbf{x},\mathbf{v})}{\partial t} \equiv 0,$$
$$\frac{\partial U(t,\mathbf{x},\mathbf{v})}{\partial v^\alpha} \equiv 0, \tag{5.1.14}$$
$$U(t,\mathbf{x},\mathbf{v}) = -V(\mathbf{x}),$$

where $V(\mathbf{x})$ is called the *potential energy* of the particle. For such a system [using (5.1.6), (5.1.8), (5.1.13), and $p_\alpha = m\delta_{\alpha\beta}v^\beta$] we obtain

$$H(\mathbf{x},\mathbf{p}) = p_\alpha v^\alpha - [(1/2)m\delta_{\alpha\beta}v^\alpha v^\beta - V(\mathbf{x})]$$
$$= (1/2m)(p_1^2 + p_2^2 + p_3^2) + V(\mathbf{x})$$
$$= (1/2m)\delta^{\alpha\beta}p_\alpha p_\beta + V(\mathbf{x})$$
$$= E. \tag{5.1.15}$$

The above relation yields a five-dimensional submanifold in the six-dimensional phase space. Such a submanifold is called an *energy hypersurface* (or *energy shell*). The motion curve $\gamma^{\#}$ of a particle in a conservative system with total energy E must lie wholly in the corresponding energy hypersurface. We can also express the constraining energy hypersurface by

$$\omega(\mathbf{x}, \mathbf{p}) \equiv H(\mathbf{x}, \mathbf{p}) - E = 0. \tag{5.1.16}$$

The variational principle satisfying a constraint can be accommodated by the method of Lagrange multipliers. To illustrate this technique, the *canonical action integral* for the motion in the constraining hypersurface (5.1.16) is taken to be

$$\begin{aligned} J_h[\tilde{\gamma}] &= \int_{t_1}^{t_2} [L_h(\mathbf{x}, \mathbf{p}, \mathbf{v}, \underline{\pi})_{|x^\alpha = \mathscr{X}^\alpha(t),\, p_\alpha = P_\alpha(t),\, v^\alpha = d\mathscr{X}^\alpha(t)/dt,\, \pi_\alpha = dP_\alpha(t)/dt}]\, dt \\ &\equiv \int_{t_1}^{t_2} [p_\alpha v^\alpha - \lambda\omega(\mathbf{x}, \mathbf{p})]_{|..} dt. \end{aligned} \tag{5.1.17}$$

Here $\tilde{\gamma}$ is the motion curve in the twelve-dimensional manifold, which is coordinatized by $(x^1, x^2, x^3, p_1, p_2, p_3, v^1, v^2, v^3, \pi_1, \pi_2, \pi_3)$, such that its projection onto the phase space is the curve $\gamma^{\#}$. The corresponding Euler-Lagrange equations derived from (5.1.17) are (5.1.18)

$$\begin{aligned} &\left[\frac{\partial L_h(\mathbf{x}, \mathbf{p}, \mathbf{v}, \underline{\pi})}{\partial x^\alpha}\right]_{|..} - \frac{d}{dt}\left\{\left[\frac{\partial L_h(\mathbf{x}, \mathbf{p}, \mathbf{v}, \underline{\pi})}{\partial v^\alpha}\right]_{|..}\right\} \\ &\quad = -\lambda \frac{\partial\omega(\mathbf{x}, \mathbf{p})}{\partial x^\alpha}_{|..} - \frac{dP_\alpha(t)}{dt} \\ &\quad = -\lambda \frac{\partial H(\mathbf{x}, \mathbf{p})}{\partial x^\alpha}_{|..} - \frac{dP_\alpha(t)}{dt} \\ &\quad = 0, \\ &\left[\frac{\partial L_h(\mathbf{x}, \mathbf{p}, \mathbf{v}, \underline{\pi})}{\partial p_\alpha}\right]_{|..} - \frac{d}{dt}\left\{\left[\frac{\partial L_h(\mathbf{x}, \mathbf{p}, \mathbf{v}, \underline{\pi})}{\partial \pi_\alpha}\right]_{|..}\right\} \\ &\quad = v^\alpha{}_{|..} - \lambda \frac{\partial\omega(\mathbf{x}, \mathbf{p})}{\partial p_\alpha}_{|..} \\ &\quad = \frac{d\mathscr{X}^\alpha(t)}{dt} - \lambda \frac{\partial H(\mathbf{x}, \mathbf{p})}{\partial p_\alpha}_{|..} \\ &\quad = 0. \end{aligned} \tag{5.1.18}$$

These equations are equivalent to Hamilton's canonical equations (5.1.11) for a conservative system provided that the Lagrange multiplier λ is chosen to be unity.

Example: The problem is to find the prerelativistic motion curve of a three-dimensional oscillator with mass $m > 0$ and spring constant $k > 0$. The corresponding Hamiltonian function is

$$H(\mathbf{x}, \mathbf{p}) = (1/2m)(p_1^2 + p_2^2 + p_3^2) + (k/2)[(x^1)^2 + (x^2)^2 + (x^3)^2] \geq 0.$$

For the case of strictly positive energy, the energy hypersurface is given by

$$\Sigma_5 \equiv \{(\mathbf{x}, \mathbf{p}): (1/2m)(p_1^2 + p_2^2 + p_3^2) + (k/2)[(x^1)^2 + (x^2)^2 + (x^3)^2] = E > 0\}.$$

The Hamilton's canonical equations are given by:

$$\frac{dP_\alpha(t)}{dt} = -\frac{\partial H(\mathbf{x}, \mathbf{p})}{\partial x^\alpha}\bigg|_{..} = -k\mathscr{X}^\alpha(t),$$

$$\frac{d}{dt}\mathscr{X}^\alpha(t) = \frac{\partial H(\mathbf{x}, \mathbf{p})}{\partial p_\alpha}\bigg|_{..} = (1/m)P_\alpha(t).$$

Putting the second system of equations into the first, we get a system of second-order decoupled ordinary differential equations. The general solutions of this system, yielding all possible $\gamma^\#$, are given by

$$\mathscr{X}^\alpha(t) = A^\alpha \cos[\sqrt{(k/m)}t + c^\alpha], \qquad P_\alpha(t) = -\sqrt{mk}A^\alpha \sin[\sqrt{(k/m)}t + c^\alpha],$$

where A^α, c^α are six constants of integration. The total energy of the oscillator is constant and is given by

$$E = (1/2m)[\delta^{\alpha\beta}P_\alpha(t)P_\beta(t)] + (k/2)[\delta_{\alpha\beta}\mathscr{X}^\alpha(t)\mathscr{X}^\beta(t)]$$
$$= (k/2)\delta_{\alpha\beta}A^\alpha A^\beta \geq 0.$$

In case $E > 0$ and the energy-hypersurface Σ_5 is prescribed, three constants of integration A^α must satisfy the above condition. Therefore, two of the three A^α and the three c^α can be adjusted to five prescribed initial conditions to obtain a *unique* motion curve $\gamma^\#$ on the prescribed energy hypersurface in the phase space. □

EXERCISES 5.1

1. For a plane pendulum with mass $m > 0$, length $l > 0$, the Lagrangian in polar coordinates is given by

$$L(\theta, \dot{\theta}) = (1/2)ml^2\dot{\theta}^2 - mgl(1 - \cos\theta),$$

where g is the constant of gravity.
(i) Obtain the corresponding Hamiltonian as given in (5.1.9).
(ii) Solve the corresponding canonical equations of Hamilton as given in (5.1.11).

2. Suppose that the coordinate x^3 is *ignorable*, i.e., $\partial\omega(\mathbf{x}, \mathbf{p})/\partial x^3 \equiv 0$, and thus $p_3 = c_3$, a constant by the canonical equations (5.1.18). Prove that the remaining canonical equations can be derived from the *reduced* action inte-

gral (*Routh's procedure*):

$$J_h^{(0)}[\tilde{\gamma}_0] = \int_{t_1}^{t_2} [p_1 v^1 + p_2 v^2 - \lambda\omega(x^1, x^2, 0, p_1, p_2, c_3)]_{|..} \, dt$$

where $\tilde{\gamma}_0$ is the curve represented by

$$x^\alpha = \mathscr{X}^\alpha(t), \quad p_\alpha = P_\alpha(t), \quad v^\alpha = \frac{d\mathscr{X}^\alpha(t)}{dt} \qquad \text{for } \alpha = 1, 2.$$

5.2. Prerelativistic Particle Mechanics in Space and Time $\mathbb{E}_3 \times \mathbb{R}$

Before passing to the relativistic mechanics, we shall discuss an *intermediate stage*, where the preceding prerelativistic mechanics is developed in terms of a mathematical parameter τ as the independent variable instead of the *usual time* parameter t. The motion curve γ in the product manifold $\mathbb{E}_3 \times \mathbb{R}$, which represents space and time, is parametrized by

$$\begin{gathered} t = \mathscr{T}(\tau), \qquad x^\alpha = \mathscr{X}^\alpha(t) \equiv \xi^\alpha(\tau), \\ \frac{d\mathscr{T}(\tau)}{dt} \neq 0, \qquad \frac{d\mathscr{X}^\alpha(t)}{dt} = \frac{d\xi^\alpha(\tau)}{d\tau}\left[\frac{d\mathscr{T}(\tau)}{d\tau}\right]^{-1}. \end{gathered} \tag{5.2.1}$$

The action integral in (5.1.4) yields

$$\begin{aligned} &\int_{t_1}^{t_2} [L(t, \mathbf{x}, \mathbf{v})_{|x^\alpha = \mathscr{X}^\alpha(t),\, v^\alpha = d\mathscr{X}^\alpha(t)/dt}] \, dt \\ &= \int_{\tau_1}^{\tau_2} \left[\frac{d\mathscr{T}(\tau)}{d\tau} L(t, \mathbf{x}, \mathbf{v})_{|t = \mathscr{T}(\tau),\, x^\alpha = \xi^\alpha(\tau),\, v^\alpha = (d\xi^\alpha(\tau)/d\tau)/(d\mathscr{T}(\tau)/d\tau)}\right] d\tau \\ &\equiv \int_{\tau_1}^{\tau_2} [L'(t, \mathbf{x}, t', \mathbf{x}')_{|t = \mathscr{T}(\tau),\, x^\alpha = \xi^\alpha(\tau),\, t' = d\mathscr{T}(\tau)/d\tau,\, x'^\alpha = d\xi^\alpha(\tau)/d\tau}] \, d\tau. \end{aligned}$$

The modified Lagrangian is

$$L'(t, \mathbf{x}, t', \mathbf{x}') \equiv t'[L(t, \mathbf{x}, \mathbf{v})_{|v^\alpha = (x'^\alpha)/t'}]. \tag{5.2.2}$$

The Euler–Lagrange equations derived from both Lagrangians should be equivalent. We can check that explicitly using (5.2.2) in the following steps.

$$\frac{\partial L'(t, \mathbf{x}, t', \mathbf{x}')}{\partial x^\alpha} = t' \frac{\partial L(t, \mathbf{x}, \mathbf{v})}{\partial x^\alpha}_{|v^\alpha = (x'^\alpha)/t'},$$

$$\frac{\partial L'(t, \mathbf{x}, t', \mathbf{x}')}{\partial t} = t' \frac{\partial L(t, \mathbf{x}, \mathbf{v})}{\partial t}_{|v^\alpha = (x'^\alpha)/t'},$$

$$p'_\alpha \equiv \frac{\partial L'(t, \mathbf{x}, t', \mathbf{x}')}{\partial x'^\alpha} = t' \left[\frac{\partial L(t, \mathbf{x}, \mathbf{v})}{\partial v^\alpha} \right]_{|v^\alpha = (x'^\alpha)/t'} (t')^{-1}$$

$$= \frac{\partial L(t, \mathbf{x}, \mathbf{v})}{\partial v^\alpha}_{|v^\alpha = (x'^\alpha)/t'},$$

$$p'_t \equiv \frac{\partial L'(t, \mathbf{x}, t', \mathbf{x}')}{\partial t'} \tag{5.2.3}$$

$$= L(t, \mathbf{x}, \mathbf{v})_{|v^\alpha = (x'^\alpha)/t'} + t' \left[\frac{\partial L(t, \mathbf{x}, \mathbf{v})}{\partial v^\alpha} \right]_{|v^\alpha = (x'^\alpha)/t'} [-x'^\alpha (t')^{-2}]$$

$$= \left[L(t, \mathbf{x}, \mathbf{v}) - v^\alpha \frac{\partial L(t, \mathbf{x}, \mathbf{v})}{\partial v^\alpha} \right]_{|v^\alpha = (x'^\alpha)/t'} = -H(t, \mathbf{x}, \mathbf{p}').$$

The modified Euler–Lagrange equations are

$$\frac{\partial L'(t, \mathbf{x}, t', \mathbf{x}')}{\partial x^\alpha}_{|..} - \left\{ \frac{d}{d\tau} \left[\frac{\partial L'(t, \mathbf{x}, t', \mathbf{x}')}{\partial x'^\alpha} \right]_{|..} \right\}$$

$$= \frac{d\mathcal{T}(\tau)}{d\tau} \left\{ \frac{\partial L(t, \mathbf{x}, \mathbf{v})}{\partial x^\alpha}_{|..} - \frac{d}{dt} \left[\frac{\partial L(t, \mathbf{x}, \mathbf{v})}{\partial v^\alpha} \right]_{|..} \right\} = 0,$$

and thus the modified Euler–Lagrange equations hold. However, there exists the fourth Euler–Lagrange equation:

$$\frac{\partial L'(t, \mathbf{x}, t', \mathbf{x}')}{\partial t}_{|..} - \frac{d}{d\tau} \left\{ \left[\frac{\partial L'(t, \mathbf{x}, t', \mathbf{x}')}{\partial t'} \right]_{|..} \right\}$$

$$= \frac{d\mathcal{T}(\tau)}{d\tau} \left\{ \frac{\partial L(t, \mathbf{x}, \mathbf{v})}{\partial t}_{|..} - \frac{d}{dt} \left[L(t, \mathbf{x}, \mathbf{v}) - v^\alpha \frac{\partial L(t, \mathbf{x}, \mathbf{v})}{\partial v^\alpha} \right]_{|..} \right\}$$

$$= \frac{d\mathcal{T}(\tau)}{d\tau} \left\{ \frac{\partial L(t, \mathbf{x}, \mathbf{v})}{\partial t}_{|..} + \frac{d}{dt} [H(t, \mathbf{x}, \mathbf{p})]_{|..} \right\} \equiv 0.$$

The last expression vanishes by (5.1.12). Thus the two systems of Euler–Lagrange equations are equivalent.

Now we notice an interesting property of the modified Lagrangian defined in equation (5.2.2). For an arbitrary $v \neq 0$,

$$L'(t, \mathbf{x}, vt', v\mathbf{x}') = vt' L(t, \mathbf{x}, \mathbf{v})_{|v^\alpha = (vx'^\alpha)/vt' = (x'^\alpha)/t'}$$

$$= vL'(t, \mathbf{x}, t', \mathbf{x}'). \tag{5.2.4}$$

Thus the modified Lagrangian function L' is *homogeneous of degree one* in the generalized velocity variables t', $\mathbf{x}'$. By Euler's theorem on homogeneous functions we must have

$$t' \frac{\partial L'(t, \mathbf{x}, t', \mathbf{x}')}{\partial t'} + x'^\alpha \frac{\partial L'(t, \mathbf{x}, t', \mathbf{x}')}{\partial x'^\alpha} = L'(t, \mathbf{x}, t', \mathbf{x}'). \tag{5.2.5}$$

But the above equation implies that the generalized Hamiltonian

$$\begin{aligned} H'(t, \mathbf{x}, p'_t, \mathbf{p}') &\equiv p'_t t' + p'_\alpha x'^\alpha - L'(t, \mathbf{x}, t', \mathbf{x}') \\ &= t' \frac{\partial L'(t, \mathbf{x}, t', \mathbf{x}')}{\partial t'} + x'^\alpha \cdot \frac{\partial L'(t, \mathbf{x}, t', \mathbf{x}')}{\partial x'^\alpha} \\ &\quad - L'(t, \mathbf{x}, t', \mathbf{x}') \\ &\equiv 0. \end{aligned} \tag{5.2.6}$$

Hence the *generalized Hamiltonian is identically zero*. But we need some other function to replace the generalized Hamiltonian in order to obtain the generalized canonical equations. We can treat the relation

$$\omega'(t, \mathbf{x}, p'_t, \mathbf{p}') \equiv p'_t + H(t, \mathbf{x}, \mathbf{p}') = 0 \tag{5.2.7}$$

as a seven-dimensional *constraining* hypersurface in the eight-dimensional *extended phase-space* manifold coordinatized by $(t, \mathbf{x}, p'_t, \mathbf{p}')$. Now we can generalize the canonical action integral (5.1.17) (involving the Lagrange multiplier λ) to

$$\begin{aligned} J'_h[\tilde{\gamma}] &= \int_{\tau_1}^{\tau_2} L'_h(t, \mathbf{x}, p'_t, \mathbf{p}', t', \mathbf{x}', \pi'_t, \underline{\pi}')_{|..}\, d\tau \\ &\equiv \int_{\tau_1}^{\tau_2} [p'_t t' + p'_\alpha x'^\alpha - \lambda\omega'(t, \mathbf{x}, p'_t, \mathbf{p}')_{|..}]\, d\tau, \end{aligned} \tag{5.2.8}$$

where $|..$ denotes evaluation along the curve $\tilde{\gamma}$ in the sixteen-dimensional generalized extended phase space. The corresponding Euler–Lagrange equations are given by

$$\begin{aligned} &\left[\frac{\partial L'_h(..)}{\partial t}\right]_{|..} - \frac{d}{d\tau}\left[\frac{\partial L'_h(..)}{\partial t'}_{|..}\right] \\ &\quad = -\lambda\left[\frac{\partial \omega'(..)}{\partial t}\right]_{|..} - \frac{d}{d\tau}[p'_t]_{|..} \\ &\quad = -\left\{\frac{dP'_t(\tau)}{d\tau} + \lambda\left[\frac{\partial H(..)}{\partial t}_{|..}\right]\right\} = 0, \\ &\left[\frac{\partial L'_h(..)}{\partial x^\alpha}\right]_{|..} - \frac{d}{d\tau}\left\{\left[\frac{\partial L'_h(..)}{\partial x'^\alpha}\right]_{|..}\right. \\ &\quad = -\lambda\left[\frac{\partial \omega'(..)}{\partial x^\alpha}\right]_{|..} - \frac{dp'_\alpha}{d\tau}_{|..} \\ &\quad = -\left\{\frac{dP'_\alpha(\tau)}{d\tau} + \lambda\left[\frac{\partial H(..)}{\partial x^\alpha}\right]_{|..}\right\} = 0, \end{aligned}$$

$$\left[\frac{\partial L'_h(..)}{\partial p'_t}\right]_{|..} - \frac{d}{d\tau}\left\{\left[\frac{\partial L'_h(..)}{\partial \pi'_t}\right]_{|..}\right\}$$

$$= t' - \lambda\left[\frac{\partial \omega'(..)}{\partial p'_t}\right]_{|..} - 0 = [t' - \lambda]_{|..}$$

$$= \frac{d\mathcal{T}(\tau)}{d\tau} - l(\tau) = 0,$$

$$\left[\frac{\partial L'_h(..)}{\partial p'_\alpha}\right]_{|..} - \frac{d}{d\tau}\left\{\left[\frac{\partial L'_h(..)}{\partial \pi'_\alpha}\right]_{|..}\right.$$

$$= \left\{x'^\alpha - \lambda\left[\frac{\partial \omega'(..)}{\partial p'_\alpha}\right]\right\}_{|..} - 0$$

$$= \frac{d\xi^\alpha(\tau)}{d\tau} - \lambda\left[\frac{\partial H(..)}{\partial p'_\alpha}\right]_{|..} = 0. \tag{5.2.9}$$

As a special choice we can have $t = \mathcal{T}(\tau) \equiv \tau$. In that case $\lambda = l(\tau) = 1$ and Hamilton's canonical equations (5.1.11) and (5.1.12) are *exactly* recovered in (5.2.9).

Example: Consider the prerelativistic mechanics of planetary motions. The Lagrangian is given by (outside the origin)

$$L(t, \mathbf{x}, \mathbf{v}) = (m/2)\delta_{\alpha\beta}v^\alpha v^\beta + Mm(\delta_{\mu\nu}x^\mu x^\nu)^{-1/2},$$

where $m > 0$ is the planetary mass and $M > 0$ is the solar mass times the Newtonian gravitational constant. The corresponding Hamiltonian is

$$H(t, \mathbf{x}, \mathbf{p}) = (1/2m)\delta^{\alpha\beta}p_\alpha p_\beta - Mm(\delta_{\mu\nu}x^\mu x^\nu)^{-1/2}.$$

Therefore, the energy constraint is

$$\omega'(t, \mathbf{x}, p'_t, \mathbf{p}') \equiv p'_t + H(t, \mathbf{x}, \mathbf{p}')$$

$$= p'_t + (1/2m)\delta^{\alpha\beta}p'_\alpha p'_\beta - Mm(\delta_{\mu\nu}x^\mu x^\nu)^{-1/2} = 0,$$

$$\frac{\partial \omega'(..)}{\partial t} = \frac{\partial H(..)}{\partial t} \equiv 0,$$

$$\frac{\partial \omega'(..)}{\partial x^\alpha} = \frac{\partial H(..)}{\partial x^\alpha} = Mm\delta_{\alpha\beta}x^\beta(\delta_{\mu\nu}x^\mu x^\nu)^{-3/2},$$

$$\frac{\partial \omega'(..)}{\partial p'_t} = 1,$$

$$\frac{\partial \omega'(..)}{\partial p'_\alpha} = \frac{\partial H(..)}{\partial p'_\alpha} = (1/m)\delta^{\alpha\beta}p'_\beta.$$

Therefore, the Euler–Lagrange equations derived from the corresponding action principle [see equation (5.2.8)] are

$$\frac{dP'_t(\tau)}{d\tau} = 0,$$

$$\frac{dP'_\alpha(\tau)}{d\tau} + \lambda Mm\delta_{\alpha\beta}x^\beta(\delta_{\mu\nu}x^\mu x^\nu)^{-3/2} = 0,$$

$$\frac{d\mathcal{T}(\tau)}{d\tau} - \lambda = \frac{d\mathcal{T}(\tau)}{d\tau} - l(\tau) = 0,$$

$$\frac{d\xi^\alpha(\tau)}{d\tau} - l(\tau)(1/m)\delta^{\alpha\beta}P'_\beta(\tau) = 0.$$

Integrating the differential equation on $\mathcal{T}(\tau)$ we get

$$t = \int dt = \int \frac{d\mathcal{T}(\tau)}{d\tau}\,d\tau = \int l(\tau)\,d\tau + C,$$

where C is a constant. Recalling $\xi^\alpha(\tau) = \mathcal{X}^\alpha(t)$, $P'_\alpha(\tau) = P_\alpha(t)$ and noticing that $P'_t(\tau) = -E$, the negative energy constant, we get

$$P_\alpha(t) = m\delta_{\alpha\beta}\frac{d\mathcal{X}^\beta(t)}{dt},$$

$$\frac{dP_\alpha(t)}{dt} + Mm\delta_{\alpha\beta}\mathcal{X}^\beta(t)[\delta_{\mu\nu}\mathcal{X}^\mu(t)\mathcal{X}^\nu(t)]^{-3/2} = 0,$$

$$-E + (1/2m)\delta^{\alpha\beta}P_\alpha(t)P_\beta(t) - Mm[\delta_{\mu\nu}\mathcal{X}^\mu(t)\mathcal{X}^\nu(t)]^{-1/2} = 0$$

Eliminating $P_\alpha(t)$ we get the following second-order equations of motion and the first integral (indicating the conservation of energy),

$$\frac{d^2\mathcal{X}^\alpha(t)}{dt^2} + M\mathcal{X}^\alpha(t)[\delta_{\mu\nu}\mathcal{X}^\mu(t)\mathcal{X}^\nu(t)]^{-3/2} = 0,$$

$$(m/2)\delta_{\alpha\beta}\frac{d\mathcal{X}^\alpha(t)}{dt}\frac{d\mathcal{X}^\beta(t)}{dt} - Mm[\delta_{\mu\nu}\mathcal{X}^\mu(t)\mathcal{X}^\nu(t)]^{-1/2} = E.$$

The second-order equations of motion could have been obtained directly from the original Lagrangian $L(t, \mathbf{x}, \mathbf{v})$ as the Euler–Lagrange equations. This fact shows the inner consistency of this mathematical method. The equations of motion permit three other first integrals (indicating the conservation of angular momentum), namely,

$$\mathcal{X}^\alpha(t)\frac{d\mathcal{X}^\beta(t)}{dt} - \mathcal{X}^\beta(t)\frac{d\mathcal{X}^\alpha(t)}{dt} = h^{\alpha\beta} = -h^{\beta\alpha} = \text{const.}$$

The relativistic version of the planetary motion will be completely solved in the next section. □

EXERCISES 5.2

1. Consider the motion curve on the energy hypersurface

$$H(\mathbf{x},\mathbf{p}) \equiv (1/2m)\delta^{\alpha\beta}p_\alpha p_\beta + V(\mathbf{x}) = E.$$

Prove that the same motion curves are obtained by the variation of the action integral (*Jacobi's Principle*)

$$J'[\gamma'] = \int_{\tau_1}^{\tau_2} \left\{ [E - V(\mathbf{x})]_{|x^\alpha = \xi^\alpha(\tau)}\, \delta_{\alpha\beta} \frac{d\xi^\alpha(\tau)}{d\tau}\frac{d\xi^\beta(\tau)}{d\tau} \right\}^{1/2} d\tau.$$

5.3. The Relativistic Equation of Motion of a Particle

In this section we shall discuss the special relativistic mechanics of a single particle. For that purpose it is convenient to use the proper time s along a motion curve as in Figure 22.

The parametric equations of a motion curve are given by

$$\begin{aligned} x^k &= (\pi^k \circ \chi \circ \gamma)(s) = \mathscr{X}^k(s), \\ t \equiv x^4 &= \mathscr{X}^4(s), \end{aligned} \tag{5.3.1}$$

where $s \in [0, s_1]$. It is assumed that the functions $\mathscr{X}^k \in \mathscr{C}^2[0, s_1]$. Furthermore, it is assumed that the motion curve in $\mathbf{M}_4$ is *future-pointing* and time-like, i.e.,

$$\frac{dt}{ds} = \frac{d\mathscr{X}^4(s)}{ds} > 0, \qquad d_{kl}\frac{d\mathscr{X}^k(s)}{ds}\frac{d\mathscr{X}^l(s)}{ds} < 0. \tag{5.3.2}$$

From the first inequality it follows that the function $\mathscr{X}^4$ is invertible in $[0, s_1]$; hence, we can write

$$s = \mathscr{S}(x^4) = \mathscr{S}(t) \tag{5.3.3a}$$

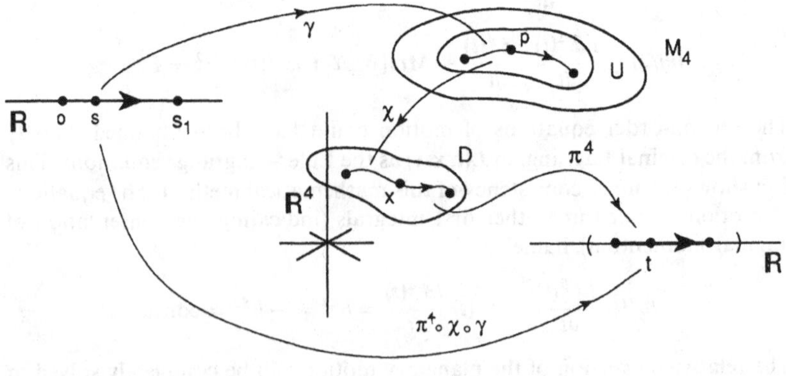

FIGURE 22. Proper time of a motion curve.

for $t \in [t_0, t_1]$. We define the *four-velocity vector* along the curve as

$$u^k(s) \equiv \frac{d\mathscr{X}^k(s)}{ds}, \tag{5.3.3b}$$

which is also the *unit tangent vector* along the curve. It follows from (5.3.3b) and (5.3.2) that

$$d_{kl} u^k(s) u^l(s) = -1. \tag{5.3.4}$$

But the three-velocity components, which are actually measured, are *different* from these components. To express the measurable components of velocity, we reparameterize the motion curve by

$$x^\alpha = \mathscr{X}^\alpha(s) = \mathscr{X}^\alpha[\mathscr{S}(t)] \equiv \mathscr{X}^{\#\alpha}(t), \qquad x^4 = \mathscr{X}^{\#4}(t) \equiv t,$$
$$v^\alpha(t) \equiv \frac{d\mathscr{X}^{\#\alpha}(t)}{dt}. \tag{5.3.5}$$

Note that these components of the measurable velocity are *not* components of any Minkowski vector.

Theorem (5.3.1): *The components of a Minkowski velocity vector and the corresponding components of a measurable velocity vector satisfy the inequality*:

$$|v^\alpha(t)| \le |u^\alpha(s)|. \tag{5.3.6a}$$

Proof: Equation (5.3.4) yields

$$\left[\frac{d\mathscr{X}^4(s)}{ds}\right]^2 - \delta_{\alpha\beta}\frac{d\mathscr{X}^\alpha(s)}{ds}\frac{d\mathscr{X}^\beta(s)}{ds} = 1,$$

so

$$\left[\frac{d\mathscr{X}^4(s)}{ds}\right]^2 = \left\{1 - \delta_{\alpha\beta}\left[\frac{d\mathscr{X}^\alpha(s)}{ds}\right]\left[\frac{d\mathscr{X}^4(s)}{ds}\right]^{-1}\left[\frac{d\mathscr{X}^\beta(s)}{ds}\right]\left[\frac{d\mathscr{X}^4(s)}{ds}\right]^{-1}\right\}^{-1}$$
$$= [1 - \delta_{\alpha\beta} v^\alpha(t) v^\beta(t)]^{-1}.$$

Thus

$$\left|\frac{d\mathscr{X}^4(s)}{ds}\right| = |[1 - \delta_{\alpha\beta} v^\alpha(t) v^\beta(t)]^{-1/2}| \ge 1. \tag{5.3.6b}$$

Hence,

$$|v^\alpha(t)| = \left|\frac{d\mathscr{X}^\alpha(s)}{ds}\right|\left|\frac{d\mathscr{X}^4(s)}{ds}\right|^{-1} = |u^\alpha(s)[1 - \delta_{\alpha\beta} v^\alpha(t) v^\beta(t)]^{1/2}|,$$

which gives $|v^\alpha(t)| \le |u^\alpha(s)|$. ∎

For a photon, one can choose a *special parameter* τ along the null world-line such that

$$x^k = \mathscr{X}^k(\tau), \qquad \tau \in [\tau_1, \tau_2],$$
$$u^k(\tau) \equiv \frac{d\mathscr{X}^k(\tau)}{d\tau}, \tag{5.3.7}$$
$$d_{kl}u^k(\tau)u^l(\tau) = 0.$$

The *proper* (*rest*) *mass* of a massive particle $m = M(s) > 0$ for $s \in [0, s_1]$. For a photon it is assumed that $m \equiv 0$.

The *four-momentum* vector of a free massive particle (in this case the mass is taken to be a constant) is defined as

$$p_k(s) \equiv M(s)d_{kl}u^l(s) = md_{kl}u^l(s).$$

By (5.3.4), one can obtain

$$d^{kl}p_k(s)p_l(s) = -m^2. \tag{5.3.8a}$$

For a photon one assumes that

$$d^{kl}p_k(\tau)p_l(\tau) = 0. \tag{5.3.8b}$$

The four-dimensional space of all possible four-momenta is the dual Minkowski space $\tilde{\mathbf{M}}_4$, which allows the Minkowski coordinate chart $(\tilde{\chi}, \tilde{\mathbf{M}}_4)$ such that $\tilde{\chi}: \tilde{\mathbf{M}}_4 \to \mathbb{R}^4$. In $\mathbb{R}^4$ all possible timelike four-momenta for a given particle of proper constant mass $m \neq 0$ are constrained into the mass hypersurface in four-momentum space given by

$$\tilde{\Sigma}_3 \equiv \{p: d^{kl}p_kp_l + m^2 = 0\}. \tag{5.3.9}$$

This hypersurface is called a *pseudo-three-sphere* and is plotted in Figure 23.

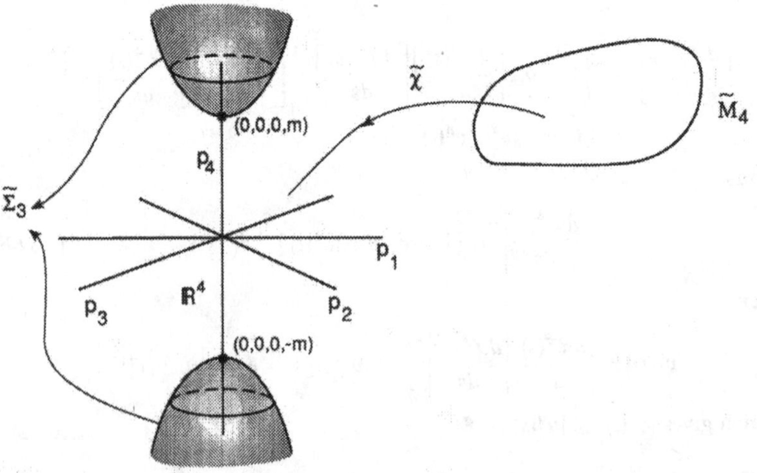

FIGURE 23. Pseudo-three-sphere representing a mass hypersurface.

This hypersurface has two disconnected sheets. For the upper sheet we have $p_4 = \sqrt{p_1^2 + p_2^2 + p_3^2 + m^2} > 0$, and for the lower sheet we have $p_4 = -\sqrt{p_1^2 + p_2^2 + p_3^2 + m^2} < 0$. Usually the upper sheet is ignored from physical consideration.

Along a twice differentiable world-line of a massive particle, *four-acceleration* vector components are defined to be $d^2\mathscr{X}^k(s)/ds^2$. Note that

$$\begin{aligned} d_{kj}\frac{d\mathscr{X}^k(s)}{ds}\frac{d^2\mathscr{X}^j(s)}{ds^2} &= \frac{d}{ds}\left[(1/2)d_{kj}\frac{d\mathscr{X}^k(s)}{ds}\frac{d\mathscr{X}^j(s)}{ds}\right] \\ &= \frac{d}{ds}(-1/2) \equiv 0. \end{aligned}$$

Therefore, the four-velocity is *always M-orthogonal* to the four-acceleration vector.

Now we shall postulate the relativistic equation of motion. For a particle under external force, the *axiom of four-vector equation of motion* (in terms of components) is

$$\frac{d}{ds}\left[M(s)\frac{d\mathscr{X}^k(s)}{ds}\right] = F^k(x,u)_{|x^k=\mathscr{X}^k(s),\, u^k=d\mathscr{X}^k(s)/ds}, \tag{5.3.10}$$

where $x \equiv (x^1, x^2, x^3, x^4)$, $u \equiv (u^1, u^2, u^3, u^4)$, and $F^k(x,u)$ stand for the components of the external four-force field. This law is the natural generalization of Newton's equations of motion (5.1.1).

Theorem (5.3.2): *The proper mass $M(s)$ is a constant along the massive world-line if and only if*

$$d_{kl}\frac{d\mathscr{X}^k(s)}{ds}F^l(x,u,)_{|\ldots} \equiv 0. \tag{5.3.11}$$

Proof: From (5.3.10) we get

$$\frac{dM(s)}{ds}\frac{d\mathscr{X}^k(s)}{ds} + M(s)\frac{d^2\mathscr{X}^k(s)}{ds^2} = F^k(x,u)_{|\ldots}.$$

Multiplying the above equation with $d_{kl}(d\mathscr{X}^l(s)/ds)$ and summing we have

$$\begin{aligned} \frac{dM(s)}{ds}d_{kl}\frac{d\mathscr{X}^k(s)}{ds}\frac{d\mathscr{X}^l(s)}{ds} + M(s)d_{kl}\frac{d^2\mathscr{X}^k(s)}{ds^2}\frac{d\mathscr{X}^l(s)}{ds} \\ = d_{kl}F^k(x,u)_{|\ldots}\frac{d\mathscr{X}^l(s)}{ds}, \end{aligned}$$

which reduces to [via (5.3.4)]

$$\begin{aligned} \frac{dM(s)}{ds}(-1) + (1/2)M(s)\frac{d}{ds}\left[d_{kl}\frac{d\mathscr{X}^k(s)}{ds}\frac{d\mathscr{X}^l(s)}{ds}\right] \\ = d_{kl}\frac{d\mathscr{X}^k(s)}{ds}F^l(x,u)_{|\ldots}. \end{aligned}$$

Thus

$$-\frac{dM(s)}{ds} = d_{kl}\frac{d\mathscr{X}^k(s)}{ds}F^l(x,u)_{|..}.$$

Therefore, $dM(s)/ds \equiv 0$ if and only if $d_{kl}(d\mathscr{X}^k(s)/ds)F^l(x,u)_{|..} \equiv 0$. ■

Corollary (5.3.1): *If $M(s)$ is a positive constant along the world-line of a particle and the external force components $F^k(x,u)_{|..}$ are not identically zero, then $F^k(..)_{|..}$ represent a spacelike vector along the world-line.*

Proof: From the preceding theorem we conclude that

$$d_{kl}\frac{d\mathscr{X}^k(s)}{ds}F^l(x,u)_{|..} \equiv 0.$$

Therefore, $d\mathscr{X}^k(s)/ds$ and $F^l(x,u)_{|..}$ are M-orthogonal along the world-line. But $d\mathscr{X}^k(s)/ds$ for a massive particle is timelike. Thus by Theorems (1.1.4) and (1.1.6) it follows that $F^l(x,u)_{|..}$ is spacelike. ■

Now we shall physically interpret the relativistic equation of motion (5.3.10) by comparing and contrasting it with the Newtonian equation of motion (5.1.1). Suppose that we have a differentiable function f along the world-line. By (5.3.5), (5.3.6*b*), and the chain rule, we get

$$f(s) = f[\mathscr{S}(t)] \equiv f^{\#}(t),$$
$$\frac{df(s)}{ds} = \frac{d\mathscr{X}^4(s)}{ds}\frac{df^{\#}(t)}{dt} = [1 - \delta_{\alpha\beta}v^\alpha(t)v^\beta(t)]^{-1/2}\frac{df^{\#}(t)}{dt}. \tag{5.3.12}$$

Putting the above equation into the relativistic motion law (5.3.10) we can have

$$\frac{d}{dt}[M^{\#}(t)v^\alpha(t)[1 - \delta_{\beta\gamma}v^\beta(t)v^\gamma(t)]^{-1/2}]$$
$$= [1 - \delta_{\beta\gamma}v^\beta(t)v^\gamma(t)]^{1/2}F^\alpha(x,u,)_{|..}$$
$$\equiv f^{\#\alpha}(t,\mathbf{x},\mathbf{v},)_{|x^\mu = \mathscr{X}^{\#\mu}(t),\, v^\mu = d\mathscr{X}^{\#\mu}(t)/dt}, \tag{5.3.13a}$$

and also

$$\frac{d}{dt}[M^{\#}(t)[1 - \delta_{\beta\gamma}v^\beta(t)v^\gamma(t)]^{-1/2}]$$
$$= [1 - \delta_{\beta\gamma}v^\beta(t)v^\gamma(t)]^{1/2}F^4(x,u,)_{|..}. \tag{5.3.13b}$$

To interpret physically (5.3.13*a*) we introduce the following concepts:

$$\text{Moving mass} \equiv M_v(t) \equiv M^{\#}(t)[1 - \delta_{\beta\gamma}v^\beta(t)v^\gamma(t)]^{-1/2},$$
$$\text{Relative three-force} \equiv f^{\#\alpha}(..)_{|..} \tag{5.3.14}$$
$$\equiv [1 - \delta_{\beta\gamma}v^\beta(t)v^\gamma(t)]^{1/2}F^\alpha(..)_{|..}.$$

Therefore, equation (5.3.13*a*) can be written as

$$\frac{d}{dt}[M_v(t)v^\alpha(t)] = f^{\#\alpha}(..)|_{..}.$$

which can be recognized as the appropriate relativistic generalization of the Newtonian motion law (5.1.1). To interpret the other equation (5.3.13*b*), let us restrict the case in which $M(s) = M^{\#}(t) = m$, where m is constant. By Theorem (5.3.2), we have

$$\delta_{\alpha\beta}u^\alpha(s)F^\beta(..)|_{..} - u^4(s)F^4(..)|_{..} = 0,$$

so

$$F^4(..)|_{..} = \delta_{\alpha\beta}v^\alpha(t)F^\beta(..)|_{..}.$$

Hence,

$$[1 - \delta_{\beta\gamma}v^\beta(t)v^\gamma(t)]^{1/2}F^4(..)|_{..} = \delta_{\alpha\beta}v^\alpha(t)f^{\#\beta}(..)|_{..}.$$

Putting the above equation into (5.3.13*b*) we get

$$m\frac{d}{dt}[1 - \delta_{\beta\gamma}v^\beta(t)v^\gamma(t)]^{-1/2} = \delta_{\alpha\beta}v^\alpha(t)f^{\#\beta}(..)|_{..}. \tag{5.3.15}$$

Comparing the above equation with (5.1.2), we can interpret (5.3.15) as the statement "*the rate of increase in the relative energy equals the rate of work done by the external relative force.*" As a result of the preceding statement, it is reasonable to define the following quantities:

$$\begin{aligned}
\text{Relative energy} &\equiv E(v) \equiv M_v \\
&\equiv m[1 - \delta_{\beta\gamma}v^\beta v^\gamma]^{-1/2} \\
&\geq m, \\
\text{Relative kinetic energy} &\equiv T(v) \equiv E(v) - E(0) \\
&= m\{[1 - \delta_{\beta\gamma}v^\beta v^\gamma]^{-1/2} - 1\} \\
&= (1/2)m\delta_{\alpha\beta}v^\alpha v^\beta + \mathcal{O}(|v|^4), \\
\text{Rest mass} &\equiv E(0) \equiv M_0 = m.
\end{aligned} \tag{5.3.16}$$

Modern experimental physics confirms the notions inherent in (5.3.16) and (5.3.14). Note that, if we reinstate c as the speed of light, then equation (5.3.16) can be expressed as $E(v) = M_v c^2$ so that $E(0) = M_0 c^2$, or simply $E = Mc^2$. *This last formula is perhaps the most popularized formula in the history of science.*

Example 1: We choose a free massive particle. For this case $F^\beta(..)|_{..} \equiv 0$. Therefore, by Theorem (5.3.2), $M(s) = m$ is a positive constant. So the relativistic equations of motion (5.3.10) yield that the *acceleration four-vector* is zero, or

$$d^2\mathcal{X}^k(s)/ds^2 = 0. \tag{5.3.17}$$

This equation was *assumed* in (2.2.2) (with $t = s$). Integrating (5.3.17) we have $x^k = \mathscr{X}^k(s) = t^k s + b^k$ where the eight constants of integration t^k, b^k satisfy $d_{kl}t^k t^l = -1$, $t^4 > 0$, and otherwise are arbitrary. The constants t^k correspond to an arbitrary point on the upper sheet of a hypersurface like that in Fig. 23. □

Example 2: Consider the case of a constant relative force $K > 0$ acting on a massive point particle in the direction of the x^1-axis. We make further simplifying assumptions, namely, $M(s) = m$, and $v^2(t) = v^3(t) = f^{\#2}(..)_{|..} = f^{\#3}(..)_{|..} \equiv 0$. The equations of motion (5.3.13*a*) yield

$$\frac{d}{dt}[v^1(t)[1 - (v^1(t))^2]^{-1/2}] = (K/m) > 0.$$

Integrating, with the initial condition $v^1(0) = (d\mathscr{X}^{\#1}(t)/dt)_{|t=0} = 0$, we get

$$v^1(t)[1 - (v^1(t))^2]^{-1/2} = (Kt/m).$$

Thus

$$v^1(t) = Kt[m^2 + (Kt)^2]^{-1/2}.$$

It can be noted $0 \leq v^1(t) < 1$ for $t \in [0, \infty)$ and $\lim_{t\to\infty} v^1(t) = 1$, the speed of light. Therefore, even a constant forward thrust on a massive particle does not allow it to actually attain the speed of light! Integrating once more with the initial condition $\mathscr{X}^{\#1}(0) = 0$ we have for $0 \leq x$ and $0 \leq t$

$$[(K/m)x^1 + 1]^2 - [(K/m)t]^2 = 1.$$

The above conditions yield a part of a hyperbola in the 2-flat given by $x^2 = x^3 = 0$. The world-line of the motion curve approaches a null-line asymptotically. □

The equation of motion of a massive charged particle in an external electromagnetic field is governed by the *Lorentz equations of motion.* The Lorentz four-force field is given by

$$F^i(x, u) \equiv eF^i{}_j(x)u^j, \tag{5.3.18}$$

where $F_{ij}(x) = -F_{ji}(x)$ and e is the charge parameter. Note that

$$u^i F_i(x, u)_{|..} = eu^i u^j F_{ij}(x)_{|..} \equiv 0.$$

Thus by Theorem (5.3.2) the proper mass $M(s) = m$ is a constant along the motion curve. The Lorentz equations of motion by (5.3.10), (5.3.18) are

$$m\frac{d^2\mathscr{X}^k(s)}{ds^2} = eu^j F^k{}_j(x)_{|..}\,. \tag{5.3.19}$$

The three components of the above equations, by (5.3.13*a*), yield (5.3.20)

$$m\frac{d}{dt}[v^\alpha(t)[1 - \delta_{\beta\gamma}v^\beta(t)v^\gamma(t)]^{-1/2}]$$

$$= e[E^\alpha(x) + \varepsilon^\alpha{}_{\beta\gamma}v^\beta(t)H^\gamma(x)]_{|x^\alpha = \mathscr{X}^{\#\alpha}(t),\, x^4 = t}, \tag{5.3.20}$$

where $E^\alpha(x) \equiv F^\alpha{}_4(x)$, $H_\alpha(x) \equiv (1/2)\varepsilon_{\alpha\beta\gamma}F^{\beta\gamma}(x)$, and $\varepsilon_{\alpha\beta\gamma}$ is the totally antisymmetric permutation symbol with $\varepsilon_{123} = 1$ [compare (1.3.12)]. The components $E^\alpha(x)$ and $H^\alpha(x)$ are those of the electric and magnetic vector fields respectively. Equation (5.3.13*b*) in this case yields

$$m\frac{d}{dt}[[1 - \delta_{\beta\gamma}v^\beta(t)v^\gamma(t)]^{-1/2}]$$
$$= e\delta_{\alpha\beta}v^\alpha(t)E^\beta(x)_{|x^\alpha = \mathscr{X}^{*\alpha}(t),\, x^4 = t}. \tag{5.3.21}$$

The right-hand side is the rate of work done by the electric field on the charged particle. *The rate of work done by the magnetic field is exactly zero.*

Example: We shall work out a particular problem for motion curves of a charged elementary particle in a *cyclotron*. There exists a *constant magnetic field* along the x^3-axis. Only uniplanar motions in the $(x^1\text{-}x^2)$-plane are considered. Thus we choose

$$H^1(x) = H^2(x) \equiv 0, \qquad H^3(x) = h, \quad \text{a constant},$$
$$E^\alpha(x) \equiv 0, \qquad v^3(t) \equiv 0.$$

Putting the above conditions in (5.3.20) and defining the speed

$$v(t) \equiv \{[v^1(t)]^2 + [v^2(t)]^2\}^{1/2},$$

we obtain the following two equations:

$$m\frac{d}{dt}[v^1(t)[1 - (v(t))^2]^{-1/2}] = ehv^2(t) = ehv(t)\sin\Theta(t),$$

$$m\frac{d}{dt}[v^2(t)[1 - (v(t))^2]^{-1/2}] = -ehv^1(t) = -ehv(t)\cos\Theta(t).$$

Here $\tan\Theta(t) \equiv v^2(t)/v^1(t)$. Multiplying the first equation by $\cos\Phi(t)$ and the second by $\sin\Theta(t)$ and adding we get

$$\cos\Theta(t)\frac{d}{dt}[v(t)\cos\Theta(t)[1 - (v(t))^2]^{-1/2}]$$
$$+ \sin\Theta(t)\frac{d}{dt}[v(t)\sin\Theta(t)[1 - (v(t))^2]^{-1/2}] = 0.$$

Hence,

$$\frac{d}{dt}[v(t)[1 - (v(t))^2]^{-1/2}] = 0,$$

which implies that $v(t) = v_0$, a constant. Putting the above result in either of the equations of motion we obtain

$$\frac{d\Theta(t)}{dt} = -(eh/m)[1 - (v_0)^2]^{1/2} \equiv \omega.$$

Thus the particle describes a circular orbit in the $(x^1\text{-}x^2)$-plane with constant angular velocity ω. Measuring experimentally the radius of the circular orbit $r_0 \equiv (v_0)/\omega$ and the time period $\tau_0 \equiv 2\pi/\omega$ of a revolution, one can determine an important property of the elementary particle:

$$(e/m) = -(\omega/h)[1 - (v_0)^2]^{-1/2}$$
$$= -(2\pi/h)[(\tau_0)^2 - (2\pi r_0)^2]^{-1/2}. \quad \square$$

EXERCISES 5.3

1. (i) Express the equation of motion (5.3.13*a*) for $M^{\#}(t) = m$, a constant, in the form

$$M^{\alpha}{}_{\beta}(v)_{|..}\frac{d^2\mathscr{X}^{\#\beta}(t)}{dt^2} = f^{\#\alpha}(..)_{|..},$$

where $v \equiv (v^1, v^2, v^3)$.

(ii) Obtain the eigenvalues of $[M^{\alpha\beta}(v)]$, and discuss the physical meaning of these eigenvalues.

2. Consider the Lorentz equations of motion (5.3.20) in *absence* of any electric field ($E^{\alpha}(x) \equiv 0$). Prove that these equations can be reduced to

$$[1 - v^2]^{-1/2}\frac{dv^{\alpha}(t)}{dt} = (e/m)\varepsilon^{\alpha}{}_{\beta\gamma}v^{\beta}(t)H^{\gamma}(x)_{|x^{\alpha} = \mathscr{X}^{\#\alpha}(t),\, x^4 = t},$$

where $v \equiv +[\delta_{\alpha\beta}v^{\alpha}(t)v^{\beta}(t)]^{1/2}$.

5.4. The Relativistic Lagrangian and Hamiltonian Mechanics of a Particle

We shall now study the relativistic mechanics of a particle in both the Lagrangian and the Hamiltonian formulations. For that purpose we *cannot* use the Lagrangian $L'(t, \mathbf{x}, t', \mathbf{x}')$ from equation (5.2.2), since it is not Lorentz invariant (a scalar under Lorentz transformations). We need a new Lagrangian $\Lambda(x, u)$ in terms of the space–time variables $x \equiv (x^1, x^2, x^3, x^4)$ and $u \equiv (u^1, u^2, u^3, u^4)$. $\Lambda(x, u)$ is *assumed* to have the following properties:

(i) $\Lambda \in \mathscr{C}^2(D)$, where $D \subset \mathbb{R}^8$ is a domain corresponding to that of *an extended space.*
(ii) Λ is a *Lorentz invariant*, i.e., a scalar under Lorentz transformations.
(iii) Λ is a *positive, homogeneous function of degree one* in the variables u^k.

We may define a metric $g_{ij}(x, u)$ on the extended space, which is associated with the Lagrangian $\Lambda(x, u)$, by the requirement

$$g_{ij}(x, u) \equiv -(1/2)\frac{\partial^2\Lambda^2(x, u)}{\partial u^i \partial u^j} = g_{ji}(x, u).$$

It is assumed that $\det[g_{ij}] \neq 0$ [compare the inequality in (5.1.7)]. Note that from property (iii) of $\Lambda(x, u)$ one can show that $g_{ij}(x, u)$ is a *Finsler metric* (see the book by Lovelock and Rund referred at the end).

By property (iii) and Euler's theorem on homogeneous functions we obtain [cf. equation (5.2.6)]

$$u^k \frac{\partial \Lambda(x, u)}{\partial u^k} = \Lambda(x, u). \tag{5.4.1}$$

The relativistic invariant action integral for a smooth curve γ is

$$\begin{aligned} J_L[\gamma] &\equiv \int_{\tau_1}^{\tau_2} [\Lambda(x, u)_{|x^k = \mathscr{X}^k(\tau),\, u^k = d\mathscr{X}^k(\tau)/d\tau}]\, d\tau \\ &= \int_{\tau_1}^{\tau_2} [\Lambda(x, u)_{|..}]\, d\tau. \end{aligned} \tag{5.4.2}$$

This action integral is invariant under the *reparametrization*

$$\hat{\tau} = \hat{\mathscr{T}}(\tau), \qquad \frac{d\hat{\mathscr{T}}(\tau)}{d\tau} > 0, \qquad \hat{\mathscr{X}}^k(\hat{\tau}) \equiv \mathscr{X}^k(\tau). \tag{5.4.3}$$

The Euler–Lagrange equations from (5.4.2) are

$$\left[\frac{\partial \Lambda(x, u)}{\partial x^k}\right]_{|..} - \frac{d}{d\tau}\left\{\left[\frac{\partial \Lambda(x, u)}{\partial u^k}\right]_{|..}\right\} = 0. \tag{5.4.4}$$

For a timelike curve $d_{kl}[d\mathscr{X}^k(\tau)/d\tau][d\mathscr{X}^l(\tau)/d\tau] < 0$, one can choose $\tau = s$, the proper time parameter, in (5.4.4). *The four equations* (5.4.4) *are not independent* as there exists one identity among them. Using (5.4.1) we find

$$\begin{aligned} &\frac{d\mathscr{X}^k(\tau)}{d\tau}\left\{\left[\frac{\partial \Lambda(x, u)}{\partial x^k}\right]_{|..} - \frac{d}{d\tau}\left[\frac{\partial \Lambda(x, u)}{\partial u^k}\right]_{|..}\right\} \\ &= \frac{d\mathscr{X}^k(\tau)}{d\tau}\left[\frac{\partial \Lambda(x, u)}{\partial x^k}\right]_{|..} + \frac{d^2\mathscr{X}^k(\tau)}{d\tau^2}\left[\frac{\partial \Lambda(x, u)}{\partial u^k}\right]_{|..} - \frac{d}{d\tau}\left[u^k \frac{\partial \Lambda(x, u)}{\partial u^k}\right]_{|..} \\ &= \frac{d\Lambda(x, u)}{d\tau}_{|..} - \frac{d\Lambda(x, u)}{d\tau}_{|..} \\ &\equiv 0. \end{aligned}$$

Example: Consider the simple case of a massive ($m > 0$) free particle. The relativistic Lagrangian is chosen to be

$$\begin{aligned} \Lambda(x, u) &\equiv -m|d_{kl}u^k u^l|^{1/2}. \\ &= -m(-d_{kl}u^k u^l)^{1/2}. \end{aligned} \tag{5.4.5}$$

Thus

$$\frac{\partial \Lambda(x,u)}{\partial x^k} \equiv 0,$$

$$\frac{\partial \Lambda(x,u)}{\partial u^k} = mu^k |d_{jl}u^j u^l|^{-1/2},$$

$$g_{kl}(x,u) \equiv -(1/2)\frac{\partial^2 \Lambda^2(x,u)}{\partial u^k \partial u^l} = m^2 d_{kl},$$

$$\det[g_{kl}] \equiv -m^8 \neq 0.$$

The Euler–Lagrange equations (5.4.4) yield

$$0 - \frac{d}{d\tau}\{[mu^k|d_{jl}u^j u^l|^{-1/2}]_{|..}\} = 0.$$

After putting $\tau = s$ and $[d_{kl}u^k u^l]_{|u^k = d\mathscr{X}^k(s)/ds} = -1$, we obtain

$$m\frac{d^2\mathscr{X}^k(s)}{ds^2} = 0,$$

which are the same equations as in (5.3.17). □

A very general class of the relativistic invariant Lagrangians is provided in the following theorem.

Theorem (5.4.1): *Let $V(x)$, $A_i(x)$, $S_{ij}(x)$, $\phi_{ijk}(x)$, $\psi_{ijkl}(x)$ be various totally symmetric Minkowski tensor fields of class $\mathscr{C}^2$ on their domains. Then, for arbitrary constants c_0, c_1, c_2, c_3, c_4, the Lagrangian*

$$\begin{aligned}\Lambda(x,u) \equiv [1 + c_0 V(x)] \\ \times [c_1|A_i(x)u^i| + c_2|S_{ij}(x)u^i u^j|^{1/2} \\ + c_3|\phi_{ijk}(x)u^i u^j u^k|^{1/3} + c_4|\psi_{ijkl}(x)u^i u^j u^k u^l|^{1/4}],\end{aligned} \quad (5.4.6)$$

is a relativistic invariant and a positive homogeneous function of degree one in the variables u^k.

Proof: The proof is straightforward and is left to the reader. ■

Now we shall develop the relativistic Hamiltonian particle mechanics. First we define the generalized four-momentum variables by

$$p_i \equiv \frac{\partial \Lambda(x,u)}{\partial u^i}. \quad (5.4.7)$$

The coresponding relativistic "Hamiltonian" is

$$p_i u^i - \Lambda(x,u) = u^i\frac{\partial \Lambda(x,u)}{\partial u^i} - \Lambda(x,u) \equiv 0,$$

by (5.4.1) [analogous to the identity (5.2.6)]. So we need another function in the place of this "Hamiltonian." To obtain that function, we first notice that by differentiating (5.4.1) we get

$$u^i \frac{\partial^2 \Lambda(x,u)}{\partial u^j \partial u^i} + \frac{\partial u^i}{\partial u^j} \cdot \frac{\partial \Lambda(x,u)}{\partial u^i} = \frac{\partial \Lambda(x,u)}{\partial u^j}.$$

This is equivalent to

$$\left[\frac{\partial^2 \Lambda(x,u)}{\partial u^j \partial u^i}\right] u^i = 0$$

for all $u \in D \subset \mathbb{R}^4$. Thus we see that for nonzero solutions of above linear homogeneous equations on u^i's we must have

$$\det\left[\frac{\partial^2 \Lambda(x,u)}{\partial u^j \partial u^i}\right] = \det\left[\frac{\partial p_j}{\partial u^i}\right] = 0.$$

Thus all the p_i are *not independent* and *there must exist at least one functional relationship among them*, i.e.,

$$\Omega(x,p) = 0, \tag{5.4.8}$$

where we write $p \equiv (p_1, p_2, p_3, p_4)$. The above relationship denotes a constraining seven-dimensional *mass hypersurface* in the eight-dimensional *extended phase space* M_8. The function Ω is called the "Super-Hamiltonian" function.

The relativistic canonical action integral that generalizes (5.1.17) and (5.2.8) can be expressed involving a Lagrange multiplier as

$$J_H[\hat{\gamma}] = \int_{\tau_1}^{\tau_2} [p_i u^i - \lambda \Omega(x,p)]_{|..}\, d\tau. \tag{5.4.9}$$

The corresponding Euler–Lagrange equations yield

$$\begin{aligned} \frac{dP_i(\tau)}{d\tau} &= -\left[\lambda \frac{\partial \Omega(x,p)}{\partial x^i}\right]_{|..} = -l(\tau)\left[\frac{\partial \Omega(x,p)}{\partial x^i}\right]_{|..}, \\ \frac{d\mathscr{X}^i(\tau)}{d\tau} &= \left[\lambda \frac{\partial \Omega(x,p)}{\partial p_i}\right]_{|..} = l(\tau)\left[\frac{\partial \Omega(x,p)}{\partial p_i}\right]_{|..}. \end{aligned} \tag{5.4.10}$$

If we reparameterize the curve by the *special parameter* (arbitrary up to addition of a constant) defined by the indefinite integral

$$\begin{aligned} \theta = \Theta(\tau) &\equiv \int^{\tau} l(\xi)\, d\xi, \\ \bar{\mathscr{X}}^i(\theta) &\equiv \mathscr{X}^i(\tau), \\ \bar{P}_i(\theta) &\equiv P_i(\tau), \end{aligned} \tag{5.4.11}$$

then the equations (5.4.10) transform into the relativistic canonical form:

$$\frac{d\bar{P}_i(\theta)}{d\theta} = -\left[\frac{\partial\Omega(x,p)}{\partial x^i}\right]_{|\ldots}$$
$$\frac{d\bar{\mathcal{X}}^i(\theta)}{d\theta} = \left[\frac{\partial\Omega(x,p)}{\partial p_i}\right]_{|\ldots}. \tag{5.4.12}$$

Now we shall illustrate this method by some examples.

Example 1: We shall consider the simple case of a free massive particle with the Lagrangian given in (5.4.5) [which is a special case of (5.4.6)]. In that case the four-momentum variables are given by

$$\begin{aligned} p_i &= m d_{ij} u^j |d_{kl} u^k u^l|^{-1/2}, \\ p_\alpha &= m\delta_{\alpha\beta} v^\beta |1 - \delta_{\tau\sigma} v^\tau v^\sigma|^{-1/2}, \\ p_4 &= -m|1 - \delta_{\tau\sigma} v^\tau v^\sigma|^{-1/2}, \\ v^\alpha &\equiv u^\alpha / u^4. \end{aligned} \tag{5.4.13}$$

The p_i and the v^α are homogeneous functions of degree zero of the u^k. Eliminating three ratios v^α we obtain

$$\begin{aligned} d^{kl} p_k p_l &= \delta^{\alpha\beta} p_\alpha p_\beta - (p_4)^2 \\ &= m^2 |1 - \delta_{\tau\sigma} v^\tau v^\sigma|^{-1} [\delta_{\alpha\beta} v^\alpha v^\beta - 1] \\ &= -m^2, \end{aligned} \tag{5.4.14}$$

$$\Omega(x,p) \equiv (1/2m)[d^{kl} p_k p_l + m^2] = 0.$$

The corresponding canonical equations (5.4.12) yield:

$$\frac{d\bar{P}_i(\theta)}{d\theta} \equiv 0,$$

$$\frac{d\bar{\mathcal{X}}^i(\theta)}{d\theta} = d^{ij}\bar{P}_j(\theta)/m,$$

$$\frac{d^2\bar{\mathcal{X}}^i(\theta)}{d\theta^2} = (1/m) d^{ij} \frac{d\bar{P}_j(\theta)}{d\theta} = 0.$$

Comparing the above equations with (5.3.17) we conclude that

$$\theta = s + c, \tag{5.4.15}$$

where s is the proper time parameter along the curve γ and c is an arbitrary constant. □

Example 2: We choose the case of a massive charged particle in the external electromagnetic field. In this case, the Lagrangian is assumed to be

$$\Lambda(x,u) = eA_k(x)u^k - m|d_{kl}u^k u^l|^{1/2}, \tag{5.4.16}$$

which is a kind of special case of (5.4.6).

The corresponding four-momentum components are

$$\begin{aligned} p_i &= eA_i(x) + md_{ij}u^j|d_{kl}u^k u^l|^{-1/2}, \\ p_\alpha &= m\delta_{\alpha\beta}v^\beta|1 - \delta_{\tau\sigma}v^\tau v^\sigma|^{-1/2} + eA_\alpha(x), \\ p_4 &= -m|1 - \delta_{\tau\sigma}v^\tau v^\sigma|^{-1/2} + eA_4(x), \\ v^\alpha &\equiv u^\alpha/u^4. \end{aligned} \tag{5.4.17}$$

Eliminating three ratios v^α we get

$$\begin{aligned} &d^{kl}[p_k - eA_k(x)][p_l - eA_l(x)] \\ &\qquad = m^2[\delta_{\alpha\beta}v^\alpha v^\beta - 1]/[1 - \delta_{\tau\sigma}v^\tau v^\sigma] = -m^2, \\ &\Omega(x, p) \equiv (1/2m)\{d^{kl}[p_k - eA_k(x)][p_l - eA_l(x)] + m^2\} = 0. \end{aligned} \tag{5.4.18}$$

The corresponding canonical equations (5.4.12) yield

$$\begin{aligned} \frac{d\bar{P}_a(\theta)}{d\theta} &= -\left[\frac{\partial\Omega(x, p)}{\partial x^a}\right]_{|..} = (e/m)A_{k,a}[p^k - eA^k(x)]_{|..}, \\ \frac{d\mathscr{X}^a(\theta)}{d\theta} &= \left[\frac{\partial\Omega(x, p)}{\partial p_a}\right]_{|..} = (1/m)[p^a - eA^a(x)]_{|..}\,. \end{aligned} \tag{5.4.19}$$

Eliminating $\bar{P}_a(\theta)$ in the above equations we obtain

$$\begin{aligned} \frac{d\bar{P}^a(\theta)}{d\theta} &= m\frac{d^2\bar{\mathscr{X}}^a(\theta)}{d\theta^2} + e\frac{d\bar{\mathscr{X}}^k(\theta)}{d\theta}A^a{}_{,k|..} \\ &= ed^{ab}\frac{d\bar{\mathscr{X}}^k(\theta)}{d\theta}A_{k,b|..}. \end{aligned}$$

Thus we find

$$m\frac{d^2\bar{\mathscr{X}}^a(\theta)}{d\theta^2} = ed^{ab}\frac{d\bar{\mathscr{X}}^k(\theta)}{d\theta}[A_{k,b} - A_{b,k}]_{|..}\,.$$

The above equation is equivalent to (5.3.19) provided that

$$F_{bk}(x) = A_{k,b} - A_{b,k}, \qquad \theta = s + c,$$

where s is the proper time parameter and c is a constant. □

Example 3: In this example we study the motion of a massive particle ($m > 0$) in a gravitational field according to a special relativistic model equivalent to *Einstein's linearized gravitational theory.* In this theory, the gravity is generated by a symmetric Minkowski tensor field $\gamma_{ab}(x)$ such that

$$\begin{aligned} \gamma_{ab}(x) &= \gamma_{ba}(x), \\ |\gamma_{ab}(x)| &< 1, \\ \det[d_{ab} + \gamma_{ab}(x)] &< 0. \end{aligned} \tag{5.4.20}$$

From the last condition we can conclude the existence of the inverse matrix $[d_{ab} + \gamma_{ab}(x)]^{-1}$. Thus there is a contravariant symmetric Minkowski tensor field $\hat{\gamma}^{ab}(x) = \hat{\gamma}^{ba}(x)$ such that

$$|\hat{\gamma}^{ab}(x)| < 1,$$
$$[d^{ab} + \hat{\gamma}^{ab}(x)][d_{bc} + \gamma_{bc}(x)] = \delta^a{}_c. \tag{5.4.21}$$

The relativistic Lagrangian is assumed to be

$$\Lambda(x, u) = -m|[d_{ab} + \gamma_{ab}(x)]u^a u^b|^{1/2}; \tag{5.4.22}$$

hence, we have the Finsler metric

$$g_{ab}(x, u) \equiv -(1/2)\frac{\partial^2 \Lambda^2(x, u)}{\partial u^a \partial u^b} = m^2[d_{ab} + \gamma_{ab}(x)],$$

with $\det[g_{ab}(x, u)] < 0$. The above equations are again a special case of equation (5.4.6).

For brevity we now introduce a new function $\hat{S}$ by the solution of the ordinary differential equation

$$\frac{d\hat{S}(\tau)}{d\tau} = [|[d_{ab} + \gamma_{ab}(x)]u^a u^b|^{1/2}]_{|x^a = \mathscr{X}^a(\tau),\, u^a = d\mathscr{X}^a(\tau)/d\tau}$$
$$> 0. \tag{5.4.23}$$

With this notation we find the Euler–Lagrange equations to be

$$(1/2)[\gamma_{ab,k}(x)u^a u^b]_{|..}\left[\frac{d\hat{S}(\tau)}{d\tau}\right]^{-1}$$
$$- \frac{d}{d\tau}\left(\{[d_{kl} + \gamma_{kl}(x)]u^l\}_{|..}\left[\frac{d\hat{S}(\tau)}{d\tau}\right]^{-1}\right) = 0.$$

Thus

$$(1/2)([\gamma_{ab,k} - \gamma_{ka,b} - \gamma_{kb,a}]u^a u^b)_{|..} - [d_{kl} + \gamma_{kl}(x)]_{|..}\frac{d^2\mathscr{X}^l(\tau)}{d\tau^2}$$
$$+ \left\{([d_{kl} + \gamma_{kl}(x)]u^l)_{|..}\frac{d^2\hat{S}(\tau)}{d\tau^2}\left[\frac{d\hat{S}(\tau)}{d\tau}\right]^{-1}\right\} = 0.$$

The above equation is equivalent to

$$\frac{d^2\mathscr{X}^c(\tau)}{d\tau^2} - (1/2)\{[d^{ck} + \hat{\gamma}^{ck}(x)][\gamma_{ab,k} - \gamma_{ka,b} - \gamma_{kb,a}]u^a u^b\}_{|..}$$
$$= \frac{d\mathscr{X}^c(\tau)}{d\tau}\frac{d^2\hat{S}(\tau)}{d\tau^2}\left[\frac{d\hat{S}(\tau)}{d\tau}\right]^{-1}. \tag{5.4.24}$$

Now we choose the parameter τ such that

$$\frac{d^2\hat{S}(\tau)}{d\tau^2} = 0,$$

$$\frac{d\hat{S}(\tau)}{d\tau} = c_1 > 0, \tag{5.4.25}$$

$$\hat{s} = \hat{S}(\tau) = c_1\tau + c_2,$$

where c_1, c_2 are constants of integration. This class of parameters is a *special class*. The simplest choice of the special parameters is $\tau = \hat{s}$. Using this parameter, the equations of motion (5.4.24) become

$$\frac{d^2\mathscr{X}^c(\hat{s})}{d\hat{s}} - (1/2)\{[d^{ck} + \hat{\gamma}^{ck}(x)][\gamma_{ab,k} - \gamma_{ka,b} - \gamma_{kb,a}]u^a u^b\}_{|..} = 0. \tag{5.4.26a}$$

A first integral of (5.4.26a), which follows from (5.4.23) and $\dfrac{d\hat{S}(\tau)}{d\tau} = 1$, is

$$[d_{ab} + \gamma_{ab}(x)]_{|x^a = \mathscr{X}^a(\hat{s})} \frac{d\mathscr{X}^a(\hat{s})}{d\hat{s}}\frac{d\mathscr{X}^b(\hat{s})}{d\hat{s}} = -1. \tag{5.4.26b}$$

We note that the Euler–Lagrange equations (5.4.26a) can also be derived from the *squared Lagrangian* (using $\tau = \hat{s}$)

$$\begin{aligned} \Lambda_2(x, u) &\equiv -(1/2m)[\Lambda(x, u)]^2 \\ &= (m/2)[d_{ab} + \gamma_{ab}(x)]u^a u^b. \quad \square \end{aligned} \tag{5.4.27}$$

Example 4: Now we shall work out a *specific problem* in the linearized gravitation, namely, the problem of planetary motion in the gravitational field of the sun. For that purpose, we use the squared Lagrangian (5.4.27) corresponding to the *linearized Schwarzschild field*:

$$\begin{aligned} \Lambda_2(x, u) &\equiv (m/2)\{[\delta_{\alpha\beta} + \gamma_{\alpha\beta}(x)]u^\alpha u^\beta - [1 - \gamma_{44}(x)](u^4)^2\}, \\ \gamma_{\alpha\beta}(x) &\equiv 2M\delta_{\alpha\beta}[\delta_{\mu\nu}x^\mu x^\nu]^{-1/2}, \\ \gamma_{44}(x) &\equiv 2M[\delta_{\mu\nu}x^\mu x^\nu]^{-1/2}. \end{aligned} \tag{5.4.28}$$

Here $m > 0$ denotes the planetary mass and $M > 0$ denotes the solar mass multiplied by the Newtonian gravitational constant.

To make the problem easier to solve, we transform from the Minkowski chart to spherical polar and time coordinates by

$$\begin{aligned} x^1 &= X^1(\hat{x}) \equiv \hat{x}^1 \sin\hat{x}^2 \cos\hat{x}^3 \equiv r\cdot\sin\theta\cdot\cos\phi, \\ x^2 &= X^2(\hat{x}) \equiv \hat{x}^1 \sin\hat{x}^2 \sin\hat{x}^3 \equiv r\cdot\sin\theta\cdot\sin\phi, \\ x^3 &= X^3(\hat{x}) \equiv \hat{x}^1 \cos\hat{x}^2 \equiv r\cdot\cos\theta, \\ x^4 &= X^4(\hat{x}) \equiv \hat{x}^4 \equiv t. \end{aligned} \tag{5.4.29}$$

Using (5.4.29), the squared Lagrangian (5.4.28) transforms (see problem 2 of Exercises 5.4) to

$$\begin{aligned}\Lambda_2(x,u) &= \hat{\Lambda}_2(r,\theta,\phi,t,u^r,u^\theta,u^\phi,u^t)\\ &= (m/2)\{(1+2M/r)[(u^r)^2 + r^2(u^\theta)^2 + r^2\sin^2\theta(u^\phi)^2]\\ &\quad -(1-2M/r)(u^t)^2\}. \end{aligned} \tag{5.4.30}$$

The Euler–Lagrange equation corresponding to the θ-coordinate is

$$\begin{aligned}&\left[\frac{\partial\hat{\Lambda}_2(..)}{\partial\theta}\right]_{|..} - \frac{d}{d\hat{s}}\left\{\left[\frac{\partial\hat{\Lambda}_2(..)}{\partial u^\theta}\right]_{|..}\right\}\\ &\quad = m\{[(r^2+2Mr)\sin\theta\cos\theta(u^\phi)^2]_{|..}\\ &\qquad -(r^2+2Mr)_{|..}\frac{d^2\Theta(\hat{s})}{d\hat{s}^2} - \frac{d\Theta(\hat{s})}{d\hat{s}}\frac{d}{d\hat{s}}(r^2+2Mr)_{|..}\Bigg\}\\ &\quad = 0. \end{aligned} \tag{5.4.31}$$

The above differential equation yields a *particular solution*

$$\theta = \Theta(\hat{s}) = \pi/2, \tag{5.4.32}$$

which corresponds to a special *uniplanar motion* in space. We shall choose this solution subsequently. We note that coordinates ϕ and t do not appear explicitly in (5.4.30). So the coordinates ϕ and t are *ignorable*. The momenta corresponding to ϕ and t are conserved. We can deduce explicitly from the Euler–Lagrange equations

$$\begin{aligned}-\frac{d}{d\hat{s}}\left\{\left[\frac{\partial\hat{\Lambda}_2(..)}{\partial u^\phi}\right]_{|..}\right\} &= -\frac{d}{d\hat{s}}\{m[(1+2M/r)r^2\sin^2\theta u^\phi]_{|..}\}\\ &= 0\end{aligned}$$

that

$$m[r^2(1+2M/r)]_{|..}\frac{d\Phi(\hat{s})}{d\hat{s}} = mh, \tag{5.4.33a}$$

where h is a constant. Similarly we have

$$\begin{aligned}-\frac{d}{d\hat{s}}\left\{\left[\frac{\partial\hat{\Lambda}_2(..)}{\partial u^t}\right]_{|..}\right\} &= \frac{d}{d\hat{s}}\left\{m[(1-2M/r)]_{|..}\cdot\frac{d\mathscr{X}^4(\hat{s})}{d\hat{s}}\right\}\\ &= 0;\end{aligned}$$

thus,

$$m(1-2M/r)_{|..}\frac{d\mathscr{X}^4(\hat{s})}{d\hat{s}} = E, \tag{5.4.33b}$$

a constant. Now instead of a second-order equation of motion for the r-coordinate, we can use the first integral (5.4.26b) together with (5.4.33a) to get

$$(1 + 2M/r)_{|\ldots}\left[\frac{d\mathscr{R}(\hat{s})}{d\hat{s}}\right]^2 + [h^2/(r^2 + 2Mr)_{|\ldots}] - [E^2/m^2(1 - 2M/r)_{|\ldots}] = -1. \tag{5.4.34}$$

Now we shall assume the constant "areal velocity" $h > 0$, which implies by (5.4.33*a*) that $d\Phi(\hat{s})/d\hat{s} > 0$. Therefore, there exists an inverse function $\hat{\mathscr{S}}$ such that

$$\hat{s} = \hat{\mathscr{S}}(\phi),$$
$$\frac{d\mathscr{R}(\hat{s})}{d\hat{s}} = \frac{d\Phi(\hat{s})}{d\hat{s}}\frac{d\mathscr{R}(\hat{\mathscr{S}}(\phi))}{d\phi} = h\frac{d\hat{\mathscr{R}}(\phi)}{d\phi}\bigg/(r^2 + 2Mr)_{|\ldots}. \tag{5.4.35}$$

Putting (5.4.35) into (5.4.34), we obtain

$$[h^2/(1 + 2M/r)_{|\ldots}]\left\{\left[\frac{d}{d\phi}(1/\hat{\mathscr{R}}(\phi))\right]^2 + [1/\hat{\mathscr{R}}(\phi)]^2\right\} - [E^2/m^2(1 - 2M/r)_{|\ldots}] = -1. \tag{5.4.36}$$

Making the inversion

$$u = 1/r > 0,$$
$$u = U(\phi) \equiv 1/\hat{\mathscr{R}}(\phi), \tag{5.4.37}$$

equation (5.4.36) transform into

$$\{[1 - 2MU(\phi)]/[1 + 2MU(\phi)]\}\left\{\left[\frac{dU(\phi)}{d\phi}\right]^2 + [U(\phi)]^2\right\} + [1 - 2MU(\phi)]/h^2 = E^2/m^2h^2. \tag{5.4.38}$$

We can solve this nonlinear first-order equation exactly in terms of elliptic functions. But it is easier to solve by the perturbation method, the second-order equation obtained by differentiating (5.4.38). It is not difficult to extract the physically relevant effects from the second-order equation

$$\begin{aligned} u'' + u &= (M/h^2)[1 + 2Mu]/[1 - 2Mu] + 2M[(u')^2 + u^2]/(1 - 4M^2u^2) \\ &= (M/h^2) + (4M^2u/h^2) + 2M(1 + 4M^2u^2)[(u')^2 + u^2] \\ &\quad + \mathscr{O}[M^3u^2/h^2] + \mathscr{O}\{M^5u^4[(u')^2 + u^2]\}. \end{aligned} \tag{5.4.39}$$

For brevity we have written $u' \equiv dU(\phi)/d\phi$, $u'' \equiv d^2U(\phi)/d\phi^2$. For all practical purposes, we can *ignore* the higher-order terms since (M/h) is a small number for all the planets in the solar system. We shall solve (5.4.39) by the method of perturbations, namely,

$$\begin{aligned} u &= u_0 + u_1 + \cdots, \\ u_0'' + u_0 &= M/h^2, \\ u_1'' + u_1 &= 4M^2u_0/h^2 + 2M[1 + 4M^2u_0^2][(u_0')^2 + u_0^2], \\ &\cdots \end{aligned} \tag{5.4.40}$$

The general solution for u_0 (the Newtonian approximation) is

$$u_0 = \mathscr{U}_0(\phi) = (M/h^2)[1 + e \cdot \cos(\phi - \tilde{\omega})]. \tag{5.4.41}$$

Here e, $\tilde{\omega}$ are constants of integration and represent the eccentricity and angle of perihelion of the orbit respectively. The orbit is circular, elliptic, parabolic, or hyperbolic if $e = 0$, $0 < e < 1$, $e = 1$, $1 < e$ respectively. We shall consider the elliptic case $0 < e < 1$, which corresponds to the planetary orbit. Putting (5.4.41) into (5.4.40) we get the differential equation for the first-order correction as

$$u_1'' + u_1 = 2(3 + e^2)M^3h^{-4} + 8M^3h^{-4}e \cdot \cos(\phi - \tilde{\omega}) + \mathcal{O}(M^7h^{-8}). \tag{5.4.42}$$

Neglecting the $\mathcal{O}(M^7h^{-8})$ term, the particular integral of (5.4.42) is

$$u_1 = \mathscr{U}_1(\phi) = 2(3 + e^2)M^3h^{-4} + 4M^3h^{-4}e\phi \sin(\phi - \tilde{\omega}). \tag{5.4.43}$$

Putting (5.4.41), (5.4.43) into (5.4.40) we have (neglecting higher-order terms)

$$\begin{aligned} u &= u_0 + u_1 \\ &= (M/h^2)[1 + 2(3 + e^2)M^2h^{-2} + e \cdot \cos(\phi - \tilde{\omega}) + 4M^2h^{-2}e\phi \sin(\phi - \tilde{\omega})] \\ &= (M/h^2)[1 + 2(3 + e^2)M^2h^{-2}][1 + \hat{e} \cos(\phi - \tilde{\omega} - \tilde{\delta\omega})], \end{aligned} \tag{5.4.44}$$

where

$$\hat{e} \equiv e[1 + (4M^2h^{-2}\phi)^2]^{1/2}[1 + 2(3 + e^2)M^2h^{-2}]^{-1},$$

$$\tilde{\delta\omega} \equiv \operatorname{Arctan}(4M^2h^{-2}\phi) = 4M^2h^{-2}\phi + \mathcal{O}[(4M^2h^{-2}\phi)^3].$$

The elliptic orbit is distorted in the sense that the eccentricity $\hat{e}$ and the perihelion angle $\tilde{\omega}$ change as the planet moves around the orbit. The change of the eccentricity is too small to measure. The perihelion angle $\tilde{\omega}$ changes at the rate $\tilde{\delta\omega}$ per radian of revolution. Therefore, the perihelion shift per full revolution is

$$\Delta\tilde{\omega} = (2\pi)(4M^2h^{-2}) = 32\pi^3a^2\tau_0^{-2}(1 - e^2)^{-1}, \tag{5.4.45}$$

where

$a \equiv$ semimajor axis of the orbit,
$e \equiv$ eccentricity of the orbit,
$\tau_0 \equiv$ period of revolution of the orbit (planet's year).

For the planet Mercury, the above amount is a little over 57″ per century, which is slightly larger than the observed effect. Einstein's theory of general relativity, without linearization, yields the exact perihelion shift 43″ per century (within the limit of present day observations). □

EXERCISES 5.4

1. Consider the relativistic Lagrangian

$$\Lambda(x, u) = -m|[d_{ab} + \gamma_{ab}(x)]u^a u^b|^{1/2}.$$

Set up explicitly the corresponding relativistic action integral $J_H[\tilde{\gamma}]$ as given in equation (5.4.9).

2. Consider a general coordinate transformation from one chart to another given by

$$\hat{x}^k = \hat{X}^k(x^1, x^2, x^3, x^4),$$

$$\hat{u}^k = u^l \frac{\partial \hat{X}^k(x)}{\partial x^l},$$

$$\frac{\partial(\hat{x}^1, \ldots, \hat{x}^4)}{\partial(x^1, \ldots, x^4)} \neq 0.$$

Suppose that the Lagrangian remains invariant, i.e., $\hat{\Lambda}(\hat{x}, \hat{u}) = \Lambda(x, u)$. Prove that the Euler–Lagrange equations from both Lagrangians are equivalent.

3. Consider the relativistic canonical equations

$$\frac{dP_k(s)}{ds} = -\left[\frac{\partial \Omega(x, p)}{\partial x^k}\right]_{|..},$$

$$\frac{d\mathscr{X}^k(s)}{ds} = \left[\frac{\partial \Omega(x, p)}{\partial p_k}\right]_{|..}.$$

Consider a general coordinate transformation in the extended phase space M_8,

$$\hat{x}^k = \xi^k(x, p), \qquad \hat{p}^k = \eta^k(x, p), \qquad J \equiv \left[\begin{array}{c|c} \partial\hat{x}^k/\partial x^l & \partial\hat{x}^k/\partial p^l \\ \hline \partial\hat{p}^k/\partial x^l & \partial\hat{p}^k/\partial p^l \end{array}\right]$$

for all $(x, p) \in D \subset \mathbb{R}^8$. Prove that the relativistic canonical equations transform covariantly provided that $J^{\mathrm{T}}\Gamma J = \Gamma$, where Γ is the 8×8 matrix whose block decomposition is

$$\Gamma = \left[\begin{array}{c|c} 0 & D \\ \hline -D & 0 \end{array}\right],$$

$D \equiv [d_{ij}]$.

References

1. P. G. Bergman, *Introduction to the Theory of Relativity*, Prentice-Hall, New Jersey, 1942. [pp. 134, 158]
2. P. A. M. Dirac, Canad. J. Math. **2** (1950), 129–148.
3. I. M. Gelfand and S. V. Fomin, *Calculus of variations*, Prentice-Hall, New Jersey, 1963. [p. 128]
4. C. Lanczos, *The variational principles of mechanics*, University of Toronto Press, Toronto, 1977. [pp. 126, 128, 131, 134, 136, 137]
5. J. L. Synge, *Relativity: The special theory*, North-Holland, Amsterdam, 1964. [pp. 141, 145, 147, 150]
6. ———, *Classical dynamics*, Reprint from *Handbuch der Physik*, Springer-Verlag, Berlin, 1960. [p. 157]

6 The Special Relativistic Classical Field Theory

6.1. Variational Formalism for Relativistic Classical Fields

We shall define the Lagrangian function for a relativistic classical field. Let $\bar{D} \subset \mathbb{R}^4$ be a bounded region corresponding to a region $\bar{U}$ in space–time $\mathbf{M}_4$. Let $F: \bar{D} \to \mathbb{R}^p$ be a twice differentiable function, i.e., $F \in \mathscr{C}^2(\bar{D}; \mathbb{R}^p)$ (see Figure 24).

Let $\phi^R \equiv \pi^R \circ F$, where R is a multi-index taking values in $\{1, \ldots, p\}$. We shall use capital indices R, S, T, U, V, W as multi-indices in the sequel. A multi-index is simply the concatenation of some finite number of indices taking values in some fixed range. Since $\phi^R: D \to \mathbb{R}$, we can associate $y^R = \phi^R(x)$ with components of a Minkowski tensor field.

Let the *Lagrangian* function $\mathscr{L}: \bar{D} \times \mathbb{R}^p \times \mathbb{R}^{4p} \to \mathbb{R}$, and furthermore assume that $\mathscr{L}$ is twice differentiable. The real value of this function is denoted by $\mathscr{L}(x, y^R, y^R{}_{,i})$. Let the *action functional* $\mathscr{A}$ be a function $\mathscr{A}: \mathscr{C}^2(\bar{D}; \mathbb{R}^p) \to \mathbb{R}$ such that

$$\mathscr{A}(F) \equiv \int_{\bar{D}} \mathscr{L}(x, y^R, y^R{}_{,i})_{|y^R = \phi^R(x),\, y^R{}_{,i} = \phi^R{}_{,i}}\, d^4x. \tag{6.1.1}$$

It is further assumed that $\mathscr{A}$ is a *totally differentiable* function. A "slightly varied" function $F + \varepsilon h$, which has the same boundary value, is defined by

$$\begin{aligned} y^R &= \phi^R(x) + \varepsilon h^R(x), \\ h^R(x)_{|\partial D} &\equiv 0. \end{aligned} \tag{6.1.2}$$

Here h^R is assumed to be twice differentiable and $\varepsilon > 0$ is a small positive number. The *variation* of the action integral (6.1.1) is given by (summation convention on index R is used),

$$\begin{aligned} \delta\mathscr{A}(F) \equiv \mathscr{A}(F + \varepsilon h) - \mathscr{A}(F) &= \int_{\bar{D}} [\mathscr{L}(x, y^R, y^R{}_{,i})_{|\hat{\Sigma}} - \mathscr{L}(x, y^R, y^R{}_{,i})_{|\Sigma}]\, d^4x \\ &= \varepsilon \int_{\bar{D}} \left[h^R(x) \frac{\partial \mathscr{L}(..)}{\partial y^R}_{|\Sigma} + h^R{}_{,i}(x) \frac{\partial \mathscr{L}(..)}{\partial y^R{}_{,i}}_{|\Sigma} \right] d^4x + \mathcal{O}(\varepsilon^2), \end{aligned} \tag{6.1.3}$$

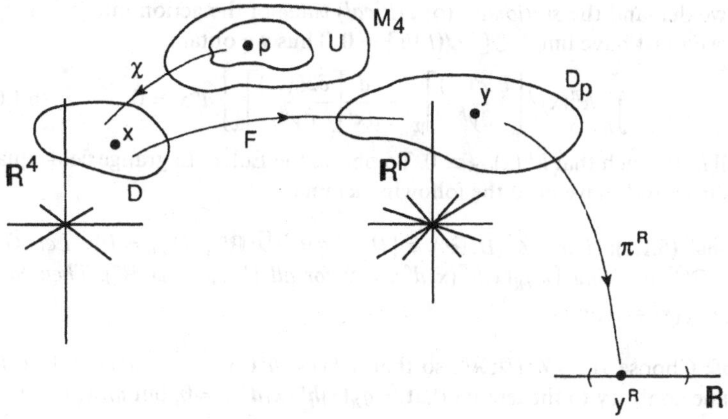

FIGURE 24. The mapping corresponding to a tensor field $\phi^R(x)$.

where $\hat{\Sigma}$ is the submanifold $y^R = \phi^R(x) + \varepsilon h^R(x)$, $y^R{}_i = \phi^R{}_{,i} + \varepsilon h^R{}_{,i}$, and where Σ is the submanifold $y^R = \phi^R(x)$, $y^R{}_i = \phi^R{}_{,i}$. We introduce now the notation for a *total partial derivative*:

$$\frac{d}{dx^i}[f(x, y^R, y^R{}_j)_{|\Sigma}]$$
$$\equiv \frac{\partial f(..)}{\partial x^i}_{|\Sigma} + \phi^R{}_{,i}\left[\frac{\partial f(..)}{\partial y^R}\right]_{|\Sigma} + \phi^R{}_{,ji}\left[\frac{\partial f(..)}{\partial y^R{}_j}\right]_{|\Sigma}, \qquad (6.1.4)$$

where Σ is the submanifold $y^R = \phi^R(x)$, $y^R{}_i = \phi^R{}_{,i}$. With this notation, (6.1.3) becomes

$$\delta\mathscr{A}(F) = \varepsilon \int_{\bar{D}} h^R(x)\left[\frac{\partial \mathscr{L}(..)}{\partial y^R}\right]_{|\Sigma} d^4x$$
$$+ \varepsilon \int_{\bar{D}} \left\{\frac{d}{dx^i}\left[h^R(x)\frac{\partial \mathscr{L}(..)}{\partial y^R{}_i}{}_{|\Sigma}\right] - h^R(x)\frac{d}{dx^i}\left[\frac{\partial \mathscr{L}(..)}{\partial y^R{}_i}\right]_{|\Sigma}\right\} d^4x + \mathscr{O}(\varepsilon^2).$$
$$= \varepsilon \int_{\bar{D}} h^R(x)\left\{\left[\frac{\partial \mathscr{L}(..)}{\partial y^R}\right]_{|\Sigma} - \frac{d}{dx^i}\left[\frac{\partial \mathscr{L}(..)}{\partial y^R{}_i}\right]_{|\Sigma}\right\} d^4x$$
$$+ \varepsilon \int_{\partial D}\left[\frac{\partial \mathscr{L}(.,)}{\partial y^R{}_i}{}_{|\Sigma}\right] h^R(x) n^i d_3\sigma + \mathscr{O}(\varepsilon^2).$$

Here we have used Gauss's Theorem (2.4.5). Thus we find that

$$\delta\mathscr{A}(F) = \varepsilon \int_{\bar{D}} h^R(x)\left\{\left[\frac{\partial \mathscr{L}(..)}{\partial y^R}\right]_{|\Sigma} - \frac{d}{dx^i}\left[\frac{\partial \mathscr{L}(..)}{\partial y^R{}_i}\right]_{|\Sigma}\right\} d^4x + \mathscr{O}(\varepsilon^2) \qquad (6.1.5)$$

since $h^R(x)_{|\partial D} \equiv 0$.

If we demand the *stationary* (or *critical*) *value* of the action integral $\mathscr{A}(F)$, then we must have $\lim_{\varepsilon \to 0+} [\delta\mathscr{A}(F)/\varepsilon] = 0$. Thus we obtain

$$\int_{\bar{D}} h^R(x) \left\{ \left[\frac{\partial \mathscr{L}(..)}{\partial y^R} \right]_{|\Sigma} - \frac{d}{dx^i} \left[\frac{\partial \mathscr{L}(..)}{\partial y^R{}_i} \right]_{|\Sigma} \right\} d^4x = 0 \tag{6.1.6}$$

for all $h^R(x)$ such that $h^R(x)_{|\partial D} \equiv 0$. To obtain the Euler–Lagrange field equations from (6.1.6) we need the following lemma.

Lemma (6.1.1): *Let* ${}_0\mathscr{C}^2(\bar{D}; \mathbb{R}^p) \equiv \{H \colon H \in \mathscr{C}^2(\bar{D}; \mathbb{R}^p),\ H_{|\partial D} = 0\}$. *Let* $G \in \mathscr{C}^2(\bar{D}; \mathbb{R}^p)$ *such that* $\int_{\bar{D}} g_R(x) h^R(x)\, d^4x = 0$ *for all* $H \in {}_0\mathscr{C}^2(\bar{D}; \mathbb{R}^p)$. *Then each of the* $g_R(x) \equiv 0$ *on* $\bar{D}$.

Proof: Choose $H \in {}_0\mathscr{C}^2(\bar{D}; \mathbb{R}^p)$ so that $h^2(x) = h^3(x) = \cdots = h^p(x) \equiv 0$ in $\bar{D}$. Assume contrary to the lemma that $\int_{\bar{D}} g_R(x) h^R(x)\, d^4x = 0$, but $g_{(1)}(\xi) \neq 0$ for some $\xi \in \bar{D}$. Suppose that $g_{(1)}(\xi) > 0$ (the case of a negative value can be treated similarly). From the assumption of continuity, there exists a neighbourhood $N_\delta(\xi) \subset D$ such that $g_{(1)}(\xi) > 0$ in $N_\delta(\xi)$. The neighbourhood $N_\delta(\xi)$ is exterior to the hypercube

$$C_{(4)} \equiv \{x \colon -\delta/2 \leq x^k - \xi^k \leq \delta/2,\ k = 1, 2, 3, 4\}.$$

We define a twice differentiable function $h^{(1)}(x)$ on $\bar{D}$ by

$$h^{(1)}(x) \equiv \begin{cases} \prod_{k=1}^{4} \left(x^k - \xi^k - \dfrac{\delta}{2} \right)^4 \left(x^k - \xi^k + \dfrac{\delta}{2} \right)^4 & \text{for } x \text{ in } C_{(4)}, \\ 0 & \text{for } x \text{ in } \bar{D} - C_{(4)}. \end{cases}$$

Therefore, the resulting test function H belongs to ${}_0\mathscr{C}^2(\bar{D}; \mathbb{R}^p)$. Thus

$$\begin{aligned} \int_{\bar{D}} g_R(x) h^R(x)\, d^4x &= \int_{\bar{D}} g_{(1)}(x) h^{(1)}(x)\, d^4x \\ &= \int_{C_{(4)}} g_{(1)}(x) h^{(1)}(x)\, d^4x \\ &> 0. \end{aligned}$$

The above inequality contradicts the starting hypothesis that the above integral is zero. ■

Theorem (6.1.1) (Euler–Lagrange): *Let the action integral* $\mathscr{A}(F)$ *be given by* (6.1.1). *If the stationary condition* $\lim_{\varepsilon \to 0+} [\delta\mathscr{A}(F)/\varepsilon] = 0$ *holds, then the Euler–Lagrange equations*:

$$\left[\frac{\partial \mathscr{L}(..)}{\partial y^R} \right]_{|..} - \frac{d}{dx^i} \left\{ \left[\frac{\partial \mathscr{L}(..)}{\partial y^R{}_i} \right]_{|..} \right\} = 0 \tag{6.1.7}$$

must be satisfied in the domain $D \subset \mathbb{R}^4$.

Proof: The proof is straightforward from equation (6.1.6) and the preceding lemma. ■

Example: We consider the case of a real scalar wave field $\phi(x)$. The relativistic Lagrangian $\mathscr{L}: \bar{D} \times \mathbb{R}^5 \to \mathbb{R}$ is defined by

$$\mathscr{L}(x, y, y_i) \equiv -(1/2)d^{kl}y_k y_l,$$
$$\frac{\partial \mathscr{L}(x, y, y_i)}{\partial y} \equiv 0,$$
$$\frac{\partial \mathscr{L}(x, y, y_i)}{\partial y_k} = -d^{kl}y_l,$$
$$\det\left[\frac{\partial^2 \mathscr{L}(x, y, y_i)}{\partial y_k \partial y_l}\right] = -1 \neq 0.$$

The Euler-Lagrange equations (6.1.7) in D yield

$$\left[\frac{\partial \mathscr{L}(x, y)}{\partial y}\right]_{|y=\phi(x), y_j=\phi_{,j}} - \frac{d}{dx^i}\left\{\left[\frac{\partial \mathscr{L}(x, y)}{\partial y_i}\right]_{|y=\phi(x), y_j=\phi_{,j}}\right\}$$
$$= -\frac{d}{dx^i}\{-d^{ij}\phi_{,j}\} = d^{ij}\phi_{,ij} = \Box\phi(x) = 0.$$

The complex fields $\psi^R(x)$ describe the wave mechanics of charged particles in nature (like Dirac's bispinor field). The variational method has to be slightly extended for such a field. Suppose that

$$\psi^R(x) = \phi^R(x) + i\eta^R(x),$$
$$\phi^R(x) \equiv \text{Re}[\psi^R(x)],$$
$$\eta^R(x) \equiv \text{Im}[\psi^R(x)].$$

We introduce the Lagrangian function $\mathscr{L}: \bar{D} \times \mathbb{R}^p \times \mathbb{R}^p \times \mathbb{R}^{4p} \times \mathbb{R}^{4p} \to \mathbb{R}$ so that the Lagrangian field $\mathscr{L}(x, y^R, w^R, y^R{}_k, w^R{}_k)$ is real-valued.

We introduce complex-conjugate coordinates so that

$$\zeta^R = y^R + iw^R,$$
$$\bar{\zeta}^R = y^R - iw^R,$$
$$y^R = (1/2)[\zeta^R + \bar{\zeta}^R],$$
$$w^R = -(i/2)[\zeta^R - \bar{\zeta}^R],$$
$$\frac{\partial}{\partial \zeta^R} \equiv (1/2)\left[\frac{\partial}{\partial y^R} - i\frac{\partial}{\partial w^R}\right],$$
$$\frac{\partial}{\partial \bar{\zeta}^R} \equiv (1/2)\left[\frac{\partial}{\partial y^R} + i\frac{\partial}{\partial w^R}\right], \tag{6.1.8}$$
$$\frac{\partial}{\partial y^R} \equiv \frac{\partial}{\partial \zeta^R} + \frac{\partial}{\partial \bar{\zeta}^R},$$
$$\frac{\partial}{\partial w^R} \equiv i\left[\frac{\partial}{\partial \zeta^R} - \frac{\partial}{\partial \bar{\zeta}^R}\right],$$
$$\hat{\mathscr{L}}(x, \zeta^R, \bar{\zeta}^R, \zeta^R{}_k, \bar{\zeta}^R{}_k) \equiv \mathscr{L}(x, y^A, w^A, y^A{}_k, w^A{}_{,k}).$$

Consider the Euler–Lagrange equations derived from the Lagrangian function $\mathscr{L}$, namely,

$$\left[\frac{\partial \mathscr{L}(..)}{\partial y^R}\right]_{|..} - \frac{d}{dx^i}\left\{\left[\frac{\partial \mathscr{L}(..)}{\partial y^R{}_i}\right]_{|..}\right\} = 0,$$

$$\left[\frac{\partial \mathscr{L}(..)}{\partial w^R}\right]_{|..} - \frac{d}{dx^i}\left\{\left[\frac{\partial \mathscr{L}(..)}{\partial w^R{}_i}\right]_{|..}\right\} = 0.$$

A complex combination of the above equations can be written as

$$\left\{\left[\frac{\partial \mathscr{L}(..)}{\partial y^R}\right]_{|..} - i\left[\frac{\partial \mathscr{L}(..)}{\partial w^R}\right]_{|..}\right\} - \frac{d}{dx^k}\left\{\left[\frac{\partial \mathscr{L}(..)}{\partial y^R{}_k}\right]_{|..} - i\left[\frac{\partial \mathscr{L}(..)}{\partial w^R{}_k}\right]_{|..}\right\} = 0.$$

Therefore multiplying by 1/2 and using the definitions above we get

$$\left[\frac{\partial \hat{\mathscr{L}}(..)}{\partial \zeta^R}\right]_{|..} - \frac{d}{dx^k}\left\{\left[\frac{\partial \hat{\mathscr{L}}(..)}{\partial \zeta^R{}_k}\right]_{|..}\right\} = 0. \tag{6.1.9}$$

Example: Consider a relativistic Lagrangian function for a complex scalar field $\psi(x)$ such that

$$\hat{\mathscr{L}}(x, \zeta, \bar{\zeta}, \zeta_k, \bar{\zeta}_k) \equiv -d^{kl}\zeta_k\bar{\zeta}_l + g(\zeta\bar{\zeta})^n,$$

where g is a real constant and n is a positive integer. Thus we have

$$\frac{\partial \hat{\mathscr{L}}(..)}{\partial \zeta} = gn(\zeta\bar{\zeta})^{n-1}\bar{\zeta},$$

$$\frac{\partial \hat{\mathscr{L}}(..)}{\partial \zeta_k} = -d^{kl}\bar{\zeta}_l.$$

Consequently the Euler-Lagrange equations (6.1.9) yield

$$\left[\frac{\partial \hat{\mathscr{L}}(..)}{\partial \zeta}\right]_{|..} - \frac{d}{dx_k}\left\{\left[\frac{\partial \hat{\mathscr{L}}(..)}{\partial \zeta_k}\right]_{|..}\right\} = gn|\psi(x)|^{2(n-1)}\bar{\psi}(x) + \Box\bar{\psi}(x)$$
$$= 0.$$

Taking the complex conjugation of the last equation, we have the nonlinear wave equation

$$\Box\psi(x) + gn|\psi(x)|^{2(n-1)}\psi(x) = 0. \quad \Box$$

Subsequently the hat on a Lagrangian $\hat{\mathscr{L}}(..)$ for complex fields will be *omitted.*

Now we shall discuss the *invariance* of the action integral (6.1.1) under a coordinate transformation and the consequent *conservation equations of*

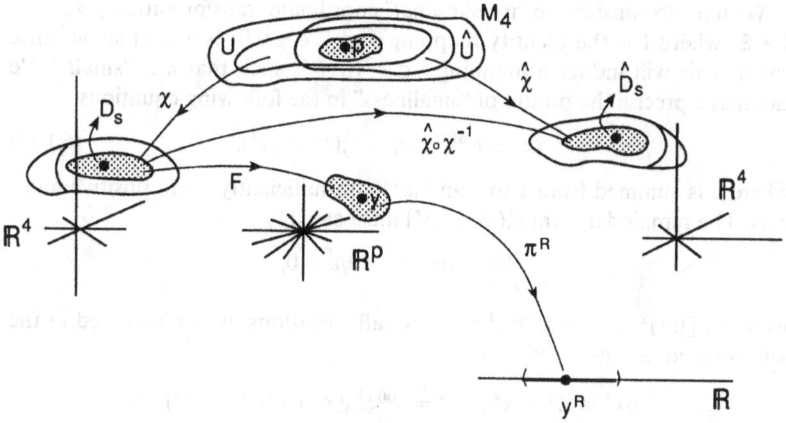

FIGURE 25. Two coordinate charts and a tensor field mapping.

Noether. For that purpose consider two charts in $\mathbf{M}_4$ and the corresponding two expressions of the action integral (6.1.1) (see Figure 25).

From these mappings we conclude that

$$\hat{x}^k = [\pi^k \circ \hat{\chi} \circ \chi^{-1}](x) \equiv \hat{X}^k(x) \equiv x^k + \xi^k(x),$$

$$\frac{\partial(\hat{x}^1, \hat{x}^2, \hat{x}^3, \hat{x}^4)}{\partial(x^1, x^2, x^3, x^4)} = \det[\delta^k{}_l + \xi^k{}_{,l}] \neq 0,$$

$$y^R = [\pi^R \circ F](x) \equiv \phi^R(x),$$

$$\begin{aligned} \hat{\phi}^{k_1\cdots}(\hat{x}) &\equiv \hat{\phi}^R(\hat{x}) \\ &= [\pi^R \circ F \circ \hat{\chi} \circ \chi^{-1}](\hat{x}) \\ &= \left[\frac{\partial \hat{X}^{k_1}\cdots}{\partial x^{j_1}}\right] \phi^{j_1}{}^{\cdots}_{\cdots}(x) \\ &= [(\delta^{k_1}{}_{j_1} + \xi^{k_1}{}_{,j_1})\cdots]\phi^{j_1}{}^{\cdots}_{\cdots}(x) \\ &= \phi^{k_1}{}^{\cdots}_{\cdots}(x) + h^{k_1}{}^{\cdots}_{\cdots}(x) \\ &= \phi^R(x) + h^R(x). \end{aligned} \tag{6.1.10}$$

$$\mathscr{A}(F + h) - \mathscr{A}(F) = \int_{\hat{\bar{D}}} \mathscr{L}(\hat{x}, y^R, y^R{}_i)_{|\hat{\Sigma}}\, d^4\hat{x} - \int_{\bar{D}} \mathscr{L}(x, y^R, y^R{}_i)_{|\Sigma}\, d^4x,$$

where $\hat{\Sigma}$ is the submanifold $y^R = \phi^R(x) + h^R(x)$, $y^R{}_i = \phi^R{}_{,i} + h^R{}_{,i}$, and Σ is the submanifold $y^R = \phi^R(x)$, $y^R{}_i = \phi^R{}_{,i}$.

We have to consider an "infinitesimal" coordinate transformation $\hat{\chi} \circ \chi^{-1} = \mathrm{I} + \Xi$, where I is the identity mapping and $\Xi \in \mathscr{C}^2(D; \mathbb{R}^4)$ is in some sense small. This will induce a mapping $h \in \mathscr{C}^2(D; R^p)$ such that h is "small." We can make precise the notion of "smallness" in the following equations:

$$\xi^k(x) = \varepsilon^{(\mu)}\xi^k_{(\mu)}(x) + r^k_0(\varepsilon^1, \ldots, \varepsilon^r), \tag{6.1.11}$$

where μ is summed from 1 to r and $|\varepsilon^{(\mu)}|$ are sufficiently small positive numbers. The remainder term $r^k_0(\varepsilon^1, \ldots, \varepsilon^r)$ must satisfy

$$\lim_{\varepsilon \to 0+} r^k_0(\varepsilon^1, \ldots, \varepsilon^r)/\varepsilon = 0,$$

with $\varepsilon \equiv [(\varepsilon^1)^2 + \cdots + (\varepsilon^r)^2]^{1/2}$. The small variations are summarized in the following equations:

$$\begin{aligned}
\delta x^k &\equiv \hat{x}^k - x^k = \xi^k = \varepsilon^{(\mu)}\xi^k_{(\mu)}(x) + r^k_0(\varepsilon^1, \ldots, \varepsilon^r), \\
\frac{\partial(\hat{x}^1, \hat{x}^2, \hat{x}^3, \hat{x}^4)}{\partial(x^1, x^2, x^3, x^4)} &= 1 + (\delta x^j)_{,j} + r^k_0(\varepsilon^1, \ldots, \varepsilon^r), \\
\delta\phi^R(x) &\equiv \hat{\phi}^R(\hat{x}) - \phi^R(x) = h^R(x) \\
&= \varepsilon^{(\mu)}h^R_{(\mu)}(x) + r^R(\varepsilon^1, \ldots, \varepsilon^r), \\
[\delta\phi^R(x)]_{,i} &= \varepsilon^{(\mu)}h^R_{(\mu),i} + r^R{}_{,i}(\varepsilon^1, \ldots, \varepsilon^r), \\
\delta_L\phi^R(x) &\equiv \hat{\phi}^R(x) - \phi^R(x) \\
&= [\hat{\phi}^R(\hat{x}) - \phi^R(x)] - [\hat{\phi}^R(\hat{x}) - \hat{\phi}^R(x)], \\
&= \delta\phi^R(x) - \phi^R{}_{,i}\delta x^i - \hat{r}^R(\varepsilon^1, \ldots, \varepsilon^r), \\
[\delta_L\phi^R(x)]_{,i} &\equiv \frac{\partial\hat{\phi}^R(x)}{\partial x^i} - \frac{\partial\phi^R(x)}{\partial x^i} = \delta_L[\phi^R{}_{,i}], \\
\delta[\phi^R{}_{,i}] &= \delta_L[\phi^R{}_{,i}] + \phi^R{}_{,ij}\delta x^j + \hat{r}^R{}_{,i}(\varepsilon^1, \ldots, \varepsilon^r).
\end{aligned} \tag{6.1.12}$$

Here all the remainder terms must satisfy an equation like

$$\lim_{\varepsilon \to 0+} r(\varepsilon^1, \ldots, \varepsilon^r)/\varepsilon = 0.$$

Now we are ready to discuss the Noether's theorem.

Theorem (6.1.2) (Noether's theorem): *Suppose that the action integral $\mathscr{A}(F)$, as defined in* (6.1.1), *is invariant under the transformation of coordinates in every domain $D \subset D_s \subset \mathbb{R}^4$. Furthermore, let the field functions $\phi^R \equiv \pi^R \circ F$ satisfy the Euler-Lagrange equations* (6.1.7). *Then the following partial differential equations will hold in D*:

$$\frac{d}{dx^i}\left\{\xi^i_{(\mu)}(x)[\mathscr{L}(\ldots)]_{|\ldots} + \left[\frac{\partial\mathscr{L}(\ldots)}{\partial y^R{}_i}\right]_{|\ldots}[h^R_{(\mu)}(x) - \xi^j_{(\mu)}(x)\phi^R{}_{,j}(x)]\right\} = 0. \tag{6.1.13}$$

Proof: From the invariance of the action integral and equations (6.1.10), (6.1.12), it follows that

$$
\begin{aligned}
0 &= \mathscr{A}(F+h) - \mathscr{A}(F) \\
&= \int_{\bar{D}} \left\{ \mathscr{L}(x+\delta x, y^R + \delta y^R, y^R{}_{,i} + \delta y^R{}_{,i})_{|..} \frac{\partial(\hat{x}^1, \hat{x}^2, \hat{x}^3, \hat{x}^4)}{\partial(x^1, x^2, x^3, x^4)} \right. \\
&\qquad \left. - \mathscr{L}(x, y^R, y^R{}_{,i})_{|..} \right\} d^4x \\
&= \int_{\bar{D}} \{ [\mathscr{L}(x+\delta x, y^R + \delta y^R, y^R{}_{,i} + \delta y^R{}_{,i})_{|..}][1 + (\delta x^j)_{,j}] \\
&\qquad - \mathscr{L}(x, y^R, y^R{}_{,i})_{|..} \} d^4x + r_1(\varepsilon^1, \ldots, \varepsilon^r) \\
&= \int_{\bar{D}} \{ \mathscr{L}(x+\delta x, y^R + \delta y^R, y^R{}_{,i} + \delta y^R{}_{,i})_{|..} \\
&\qquad - \mathscr{L}(x, y^R, y^R{}_{,i})_{|..} + (\delta x^j)_{,j} \mathscr{L}(x, y^R, y^R{}_{,i})_{|..} \} d^4x + \hat{r}_1(\varepsilon^1, \ldots, \varepsilon^r) \\
&= \int_{\bar{D}} \left\{ \left[\frac{\partial \mathscr{L}(..)}{\partial x^i} \right]_{|..} \delta x^i + \left[\frac{\partial \mathscr{L}(..)}{\partial y^R} \right]_{|..} \delta \phi^R(x) \right. \\
&\qquad \left. + \left[\frac{\partial \mathscr{L}(..)}{\partial y^R{}_{,i}} \right]_{|..} \delta[\phi^R{}_{,i}] + (\delta x^j)_{,j} \mathscr{L}(x, y^R, y^R{}_{,i})_{|..} \right\} d^4x + \hat{r}_1'(\varepsilon^1, \ldots, \varepsilon^r) \\
&= \int_{\bar{D}} \left\{ [\delta x^i \mathscr{L}(..)]_{,i} + \left[\frac{\partial \mathscr{L}(..)}{\partial y^R} \right]_{|..} [\delta_L \phi^R + \delta x^i \phi^R{}_{,i}] \right. \\
&\qquad \left. + \left[\frac{\partial \mathscr{L}(..)}{\partial y^R{}_{,j}} \right]_{|..} [\delta_L(\phi^R{}_{,j}) + \phi^R{}_{,ji} \delta x^i] \right\} d^4x + \hat{\hat{r}}_1(\varepsilon^1, \ldots, \varepsilon^r) \\
&= \int_{\bar{D}} \left\{ \frac{d}{dx^i}[\delta x^i \mathscr{L}(..)] + \left[\frac{\partial \mathscr{L}(..)}{\partial y^R} \right]_{|..} \delta_L \phi^R + \left[\frac{\partial \mathscr{L}(..)}{\partial y^R{}_{,i}} \right]_{|..} \delta_L(\phi^R{}_{,i}) \right\} d^4x \\
&\quad + \hat{\hat{r}}_1(\varepsilon^1, \ldots, \varepsilon^r) \\
&= \int_{\bar{D}} \left\{ \frac{d}{dx^i} \left([\delta x^i \mathscr{L}(..)] + \left[\frac{\partial \mathscr{L}(..)}{\partial y^R{}_{,i}} \right]_{|..} \delta_L \phi^R \right) \right. \\
&\qquad \left. + \left(\left[\frac{\partial \mathscr{L}(..)}{\partial y^R} \right]_{|..} - \frac{d}{dx^i} \left[\frac{\partial \mathscr{L}(..)}{\partial y^R{}_{,i}} \right]_{|..} \right) \delta_L \phi^R \right\} d^4x + \hat{\hat{r}}_1(\varepsilon^1, \ldots, \varepsilon^r).
\end{aligned}
\tag{6.1.14}
$$

Using the Euler-Lagrange equations (6.1.7) we have

$$
\begin{aligned}
0 = \varepsilon^{(\mu)} \int_{\bar{D}} \left\{ \frac{d}{dx^i} \left([\xi^i_{(\mu)}(x) \mathscr{L}(..)]_{|..} \right. \right. \\
\left. \left. + \left[\frac{\partial \mathscr{L}(..)}{\partial y^R{}_{,i}} \right]_{|..} [h^R_{(\mu)}(x) - \xi^j_{(\mu)}(x) \phi^R{}_{,j}] \right) \right\} d^4x + \hat{r}_2(\varepsilon^1, \ldots, \varepsilon^r).
\end{aligned}
$$

From the assumption that the action integral $\mathscr{A}$ is totally differentiable, it now follows that

$$0 = \lim_{\varepsilon \to 0+} \left[\varepsilon^{(\mu)} \varepsilon^{-1} \int_{\bar{D}} \left\{ \frac{d}{dx^i} \left([\xi^i_{(\mu)}(x) \mathscr{L}(..)]_{|..} + \left[\frac{\partial \mathscr{L}(..)}{\partial y^R{}_i} \right]_{|..} [h^R_{(\mu)}(x) - \xi^j_{(\mu)}(x) \phi^R{}_{,j}] \right) \right\} d^4x + \hat{r}_2(\varepsilon^1, \ldots, \varepsilon^r)/\varepsilon \right],$$

or equivalently

$$\int_{\bar{D}} \left\{ \frac{d}{dx^i} \left([\xi^i_{(\mu)}(x) \mathscr{L}(..)]_{|..} + \left[\frac{\partial \mathscr{L}(..)}{\partial y^R{}_i} \right]_{|..} [h^R_{(\mu)}(x) - \xi^j_{(\mu)}(x) \phi^R{}_{,j}] \right) \right\} d^4x = 0, \quad (6.1.15)$$

for $\mu = 1, \ldots, r$. Since $\bar{D}$ is an arbitrary contractible domain, we must conclude that equations (6.1.13) hold. ■

Now we shall discuss three examples on Noether's theorem.

Example 1: Let us consider the "infinitesimal" translation of Minkowski coordinates, that is,

$$\begin{gathered}
\hat{x}^k = x^k + \varepsilon^k, \quad \delta x^k = \varepsilon^j \xi^k{}_j(x) = \varepsilon^k, \qquad \xi^k{}_j(x) = \delta^k{}_j, \\
\frac{\partial(\hat{x}^1, \hat{x}^2, \hat{x}^3, \hat{x}^4)}{\partial(x^1, x^2, x^3, x^4)} = \det[\delta^k{}_j + 0] = 1, \\
\hat{\phi}^R(\hat{x}) = \phi^R(x), \qquad \delta\phi^R \equiv 0, \qquad h^R{}_k(x) \equiv 0, \\
\delta_L \phi^R = 0 - \phi^R{}_{,i} \delta x^i - r^R(\varepsilon^1, \ldots, \varepsilon^r) = -\varepsilon^i \phi^R{}_{,i} - \hat{r}^R(\varepsilon^1, \ldots, \varepsilon^r).
\end{gathered} \quad (6.1.16)$$

Noether's theorem summarized in (6.1.13) yields:

$$\frac{d}{dx^i} \left\{ \delta^i{}_j [\mathscr{L}(..)]_{|..} + \left[\frac{\partial \mathscr{L}(..)}{\partial y^R{}_i} \right]_{|..} [0 - \delta^k{}_j \phi^R{}_{,k}] \right\} = 0.$$

Thus $T^i{}_{j,i} = 0$, where we define

$$T^i{}_j(x) \equiv \delta^i{}_j [\mathscr{L}(..)]_{|..} - \left[\frac{\partial \mathscr{L}(..)}{\partial y^R{}_i} \right]_{|..} \phi^R{}_{,j}. \quad (6.1.17)$$

The second-order tensor field $T^i{}_j(x)$ for the field $\phi^R(x)$ is called the *canonical energy-momentum-stress tensor density*. It is not necessarily a symmetric tensor. Equation (6.1.17) physically represents the differential conservation of energy and momentum of the field $\phi^R(x)$. □

Example 2: Let us consider an "infinitesimal" proper, orthochronous Lorentz transformation:

$$l^i{}_j = \delta^i{}_j + \varepsilon^i{}_j.$$

From the condition (1.2.10) we obtain

$$d_{ij}(\delta^i{}_k + \varepsilon^i{}_k)(\delta^j{}_m + \varepsilon^j{}_m) = d_{km},$$

so that

$$d_{ij}\varepsilon^i{}_k\delta^j{}_m + d_{ij}\delta^i{}_k\varepsilon^j{}_m = -d_{ij}\varepsilon^i{}_k\varepsilon^j{}_m.$$

Therefore,

$$\varepsilon_{mk} + \varepsilon_{km} = -d^{ij}\varepsilon_{ik}\varepsilon_{jm} \equiv r_{km}(\varepsilon_{ab}).$$

Therefore, we can write

$$\begin{aligned}
\delta x^k &= \varepsilon^k{}_j x^j = \varepsilon^{ab}\delta^k{}_a d_{jb} x^j \\
&= (1/2)\varepsilon^{ab}[(\delta^k{}_a d_{jb} - \delta^k{}_b d_{ja})x^j] + r_0^k(\varepsilon_{cd}),
\end{aligned}$$

$$\xi^k_{ab}(x) = (1/2)(\delta^k{}_a d_{jb} - \delta^k{}_b d_{ja})x^j = -\xi^k{}_{ba}(x),$$

$$\frac{\partial(\hat{x}^1, \hat{x}^2, \hat{x}^3, \hat{x}^4)}{\partial(x^1, x^2, x^3, x^4)} = \det[\delta^k{}_j + \varepsilon^k{}_j] = 1,$$

$$\begin{aligned}
\hat{\phi}^{k\cdots}_{\cdots}(\hat{x}) &= (\delta^k{}_i + \varepsilon^k{}_i)\cdots\phi^{i\cdots}_{\cdots}(x), \\
\delta\phi^{k\cdots}_{\cdots} &\equiv \varepsilon^k{}_i\cdots\phi^{i\cdots}_{\cdots}(x) + \cdots \\
&= \varepsilon^{ab}\delta^k{}_a d_{bi}\cdots\phi^{i\cdots}_{\cdots}(x) + \cdots \\
&= (1/2)\varepsilon^{ab}[(\delta^k{}_a d_{bi} - d^k{}_b d_{ai})\cdots]\phi^{i\cdots}_{\cdots} + r_1^k(\varepsilon_{cd}) \\
&\equiv (1/2)\varepsilon^{ab}[S^{k\cdots}_{abi\cdots}\phi^{i\cdots}_{\cdots}] + r_1^k(\varepsilon_{cd}), \\
h^k{}_{ab}(x) &\equiv (1/2)[S^{k\cdots}_{abi\cdots}\phi^{i\cdots}_{\cdots}] = -h^k{}_{ba}(x), \\
\delta_L\phi^{k\cdots}_{\cdots} &= \delta\phi^k - \delta x^i \phi^{k\cdots}_{\cdots,i} - \hat{r}^k(\varepsilon_{cd}) \\
&= (1/2)\varepsilon^{ab}[S^{k\cdots}_{abi\cdots}\phi^{i\cdots}_{\cdots} - (d_{jb}\phi^{k\cdots}_{\cdots,a} - d_{ja}\phi^{k\cdots}_{\cdots,b})x^j] + r^k(\varepsilon_{cd}).
\end{aligned} \tag{6.1.18}$$

Putting the above equations into (6.1.13) we obtain the following six equations:

$$\begin{aligned}
0 &= \frac{d}{dx^i}\left\{\xi^i{}_{ab}(x)[\mathscr{L}(..)]_{|..} + \left[\frac{\partial\mathscr{L}(..)}{\partial y^{k\cdots}_{\cdots i}}\right]_{|..}[h^{k\cdots}{}_{ab}(x) - \xi^j{}_{ab}(x)\phi^{k\cdots}_{\cdots,j}]\right\}, \\
0 &= \frac{d}{dx^i}\left\{(\delta^i{}_a d_{jb}x^j - d^i{}_b d_{ja}x^j)[\mathscr{L}(..)]_{|..}\right. \\
&\quad \left. + \left[\frac{\partial\mathscr{L}(..)}{\partial y^{k\cdots}_{\cdots i}}\right]_{|..}[S^{k\cdots}_{abj\cdots}\phi^{j\cdots}_{\cdots} - (\delta^j{}_a d_{lb} - \delta^j{}_b d_{la})x^l\phi^{k\cdots}_{\cdots,j}]\right\} \\
&= \mathscr{J}^i{}_{ab,i}.
\end{aligned} \tag{6.1.19}$$

Here we define $\mathcal{J}^i{}_{ab}(x)$ by

$$\mathcal{J}^i{}_{ab}(x) \equiv \{T^i{}_a(x)d_{bj}x^j - T^i{}_b(x)d_{aj}x^j\} + \left\{\left[\frac{\partial \mathcal{L}(..)}{\partial y^{k\cdots}{}_{\cdots i}}\right]_{|..} [S^{k}_{abj}{}^{\cdots}_{\cdots}\phi^{j\cdots}_{\cdots}(x)\right\}$$

$$= -\mathcal{J}^i{}_{ba}(x).$$

The third-order tensor field $\mathcal{J}^i{}_{ab}(x)$ represents the *relativistic angular-momentum tensor density* for the field $\phi^{i\cdots}_{\cdots}(x)$. It contains the *orbital angular-momentum tensor density* plus the *spin angular-momentum tensor density* of the field [each enclosed in curly parentheses in the definition of $\mathcal{J}^i{}_{ab}(x)$]. □

Example 3: This example involves a complex scalar field $\psi^R(x)$. We consider the "infinitesimal" global gauge transformation characterized by the following equations:

$$\begin{gathered}
\delta x^i \equiv 0, \qquad \xi^i(x) \equiv 0, \\
\frac{\partial(\hat{x}^1, \hat{x}^2, \hat{x}^3, \hat{x}^4)}{\partial(x^1, x^2, x^3, x^4)} = 1, \\
\hat{\psi}^R(x) = e^{i\varepsilon}\psi^R(x) = \psi^R(x) + i\varepsilon\psi^R(x) + r_1^R(\varepsilon), \\
\delta\psi^R(x) = \eta^R(x) + r_1^R(x), \\
\eta^R(x) \equiv i\varepsilon\psi^R(x), \qquad \overline{\eta^R(x)} \equiv -i\varepsilon\overline{\psi^R(x)}.
\end{gathered} \tag{6.1.20}$$

The complex generalization of Noether's equations (6.1.13) for this case is

$$0 = \frac{d}{dx^k}\left\{\xi^k(x)[\mathcal{L}(..)]_{|..} + \left[\frac{\partial \mathcal{L}(..)}{\partial \zeta^R{}_k}\right]_{|..} [\eta^R(x) - \xi^j(x)\psi^R{}_{,j}(x)] + (\text{c.c})\right\},$$

where "c.c" means the complex conjugate terms corresponding to the previous terms in this expression. Putting (6.1.20) into the above equations we obtain

$$\begin{gathered}
j^k{}_{,k} = 0, \\
j^k(x) \equiv -ie\left\{\left[\frac{\partial \mathcal{L}(..)}{\partial \zeta^R{}_k}\right]_{|..} \psi^R(x) - \left[\frac{\partial \mathcal{L}(..)}{\partial \bar{\zeta}^R{}_k}\right]_{|..} \bar{\psi}^R(x)\right\},
\end{gathered} \tag{6.1.21}$$

where e is the charge parameter and $j^k(x)$ represent components of the *electric charge-current vector density* for the complex scalar field $\psi^R(x)$. □

Now we shall discuss the concepts of integral or total conservation of various properties of a field. For that purpose we express Noether's equations (6.1.13) in the following form:

$$J^i_{(\mu),i} = 0, \tag{6.1.22}$$

where we define $J^i_{(\mu)}(x)$ by

$$J^i_{(\mu)}(x) \equiv \xi^i_{(\mu)}(x)[\mathcal{L}(..)]_{|..} + \left[\frac{\partial \mathcal{L}(..)}{\partial y^R{}_i}\right]_{|..} [h^R_{(\mu)}(x) - \xi^j_{(\mu)}(x)\phi^R{}_{,j}].$$

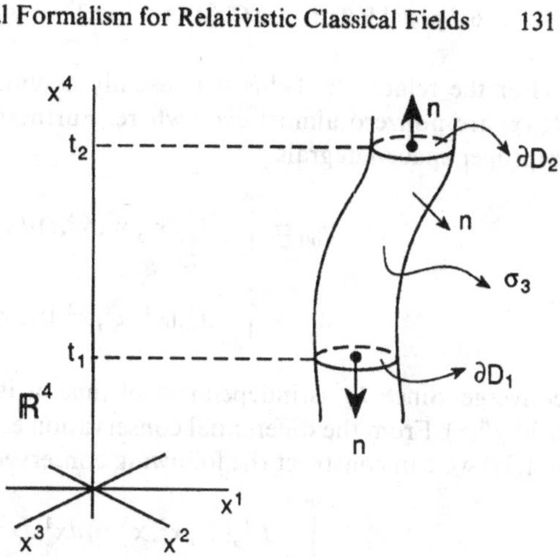

FIGURE 26. A world-tube containing the field $\phi^R(x)$.

Theorem (6.1.3) (Integral conservation laws): *Let $J^i_{(\mu)}(x) \not\equiv 0$ inside a world-tube (see Figure 26) and $J^i_{(\mu)}(x) \equiv 0$ outside the world-tube. Let $x^4 = t_1$ and $x^4 = t_2$ be three-dimensional cross-sections of the tube denoted by $\partial_1 D$ and $\partial_2 D$ respectively and σ_3 the "cylindrical wall" around the domain D of the tube. Furthermore, let $J^i_{(\mu)} n_{i|\sigma_3} = 0$, where n^i denotes the unit outer normal on the continuous, piecewise smooth, orientable boundary ∂D. Then the following integral conservation law holds:*

$$J_{(\mu)} \equiv \int_{\partial_1 D} J^4_{(\mu)}(x^1, x^2, x^3, t_1)\, dx^1\, dx^2\, dx^3$$
$$= \int_{\partial_2 D} J^4_{(\mu)}(x^1, x^2, x^3, t_2)\, dx^1\, dx^2\, dx^3. \tag{6.1.23a}$$

Proof: From Fig. 26 it is clear that $\partial D = \partial_1 D \cup \partial_2 D \cup \sigma_3$. Applying the generalized Gauss's theorem (2.4.5) and equation (6.1.22) we have

$$0 = \int_{\bar{D}} J^i_{(\mu),i}(x)\, dx^1\, dx^2\, dx^3\, dx^4$$
$$= \int_{\partial_1 D} J^i_{(\mu)}(x) n_i\, dx^1\, dx^2\, dx^3 + \int_{\partial_2 D} J^i_{(\mu)}(x) n_i\, dx^1\, dx^2\, dx^3$$
$$+ \int_{\sigma_3} J^i_{(\mu)}(x) n_i\, d_3\sigma$$
$$= \int_{\partial_1 D} J^4_{(\mu)}(x) n_4\, dx^1\, dx^2\, dx^3 + \int_{\partial_2 D} J^4_{(\mu)}(x) n_4\, dx^1\, dx^2\, dx^3 + 0$$
$$= \int_{\partial_1 D} J^4_{(\mu)}(x^1, x^2, x^3, t_2)(1)\, dx^1\, dx^2\, dx^3.$$
$$+ \int_{\partial_2 D} J^4_{(\mu)}(x^1, x^2, x^3, t_2)(-1)\, dx^1\, dx^2\, dx^3. \quad \blacksquare$$

For the relativistic fields it is usually assumed that the fields $\phi^R(x)$ and $J^4_{(\mu)}(x)$ are nonzero almost everywhere. Furthermore, it is assumed that the improper space-integrals

$$J_{(\mu)} \equiv \int_{\mathbb{R}^3} J^4_{(\mu)}(x^1, x^2, x^3, t)\, dx^1\, dx^2\, dx^3$$
$$= \int_{\mathbb{R}^3} J^4_{(\mu)}(x^1, x^2, x^3, 0)\, dx^1\, dx^2\, dx^3 \qquad (6.1.23b)$$

converge. Since $J_{(\mu)}$ is independent of time, it is a *conserved quantity* of the field $\phi^R(x)$. From the differential conservation equations (6.1.17), (6.1.19), and (6.1.21) we can construct the following conserved quantities:

$$P_a \equiv \int_{\mathbb{R}^3} T^4{}_a(x^1, x^2, x^3, 0)\, dx^1\, dx^2\, dx^3,$$
$$J_{ab} \equiv \int_{\mathbb{R}^3} \mathcal{J}^4{}_{ab}(x^1, x^2, x^3, 0)\, dx^1\, dx^2\, dx^3 = -J_{ba}, \qquad (6.1.24)$$
$$\mathcal{Q} \equiv \int_{\mathbb{R}^3} j^4(x^1, x^2, x^3, 0)\, dx^1\, dx^2\, dx^3.$$

The quantities P_a represent components of the *total energy-momentum four-vector* of the field. The quantities J_{ab} represent the components of the *total relativistic angular-momentum tensor* of the field. The quantity $\mathcal{Q}$ represents the *total electric charge* of the field.

EXERCISES 6.1

1. Consider the relativistic sine-Gordon field equation:

$$\Box\psi = k \sin\psi,$$

where $k \neq 0$ is a real constant. Construct an action integral such that the corresponding Euler-Lagrange equation is the sine-Gordon equation.

2. Let

$$\mathcal{L}(x, y^R, y^R{}_i)|_{..} \equiv \sum_{j=1}^{4} \frac{d}{dx^j}\{S^j(x, y^R)|_{..}\}.$$

Show that the Euler-Lagrange equations vanish identically. Explain this result.

3. Consider the action integral

$$\mathcal{A}(F) \equiv \int_{\bar{D}} \mathcal{L}(x, y^R, y^R{}_i, y^R{}_{ij})_{|y^R = \phi^R(x), y^R{}_i = \phi^R{}_{,i}, y^R{}_{ij} = \phi^R{}_{,ij}}\, d^4x.$$

(i) Obtain the requisite equation for $\mathcal{A}$ to have a critical value for $y^R = \phi^R(x)$.

(ii) Prove that, if the above action integral is invariant under a general coordinate transformation

$$\hat{x}^i = \hat{X}^i(x),$$
$$\frac{\partial(\hat{x}^1, \hat{x}^2, \hat{x}^3, \hat{x}^4)}{\partial(x^1, x^2, x^3, x^4)} \neq 0,$$

for all $x \in D \subset \mathbb{R}^4$, then $\mathscr{L}$ must satisfy

$$\frac{\partial \mathscr{L}(..)}{\partial x^a} \equiv 0,$$

$$\frac{\partial \mathscr{L}(..)}{\partial y^R{}_i} y^R{}_j + 2\frac{\partial \mathscr{L}(..)}{\partial y^R{}_{ik}} y^R{}_{jk} = \delta^i{}_j \mathscr{L}(..),$$

$$\frac{\partial \mathscr{L}(..)}{\partial y^R{}_{ij}} y^R{}_k \equiv 0.$$

4. Suppose that the parameterized world-sheet of a string in M_4 be given by the twice differentiable functions ξ^i:

$$x^i = \xi^i(u), \qquad u = (u^1, u^2) \in \bar{D} \subseteq \mathbb{R}^2,$$

$$\operatorname{rank}\left[\frac{\partial \xi^i(u)}{\partial u^A}\right] = 2, \qquad A, B \in \{1, 2\}.$$

Assume the metric tensor on the world-sheet is differentiable and satisfies:

$$g_{AB}(u) = g_{BA}(u),$$
$$g(u) \equiv \det[g_{AB}(u)] < 0.$$

Furthermore, the action integral for the string is assumed to be

$$\int_{\bar{D}} d_{ij} g^{AB}(u) \frac{\partial \xi^i(u)}{\partial u^A} \frac{\partial \xi^j(u)}{\partial u^B} [-g(u)]^{1/2} \, du^1 \, du^2.$$

(Here the summation convention operates for lower-case and capital indices). Show that the Euler-Lagrange equations are

$$\frac{\partial}{\partial u^A}\left[[-g(u)]^{1/2} g^{AB}(u) \frac{\partial \xi^k(u)}{\partial u^B}\right] = 0,$$

$$d_{ij}\left[\frac{\partial \xi^i(u)}{\partial u^A}\frac{\partial \xi^j(u)}{\partial u^B} - (1/2) g_{AB}(u) g^{CD}(u) \frac{\partial \xi^i(u)}{\partial u^C}\frac{\partial \xi^j(u)}{\partial u^D}\right] = 0.$$

6.2. The Klein–Gordon Scalar Field

Let us consider a complex scalar field $\psi(x)$ that satisfies the *Klein–Gordon equation*:

$$\Box\psi(x) - \mu^2\psi(x) = d^{kl}\psi_{,kl} - \mu^2\psi(x) = 0, \tag{6.2.1}$$

where μ is a real mass parameter and $\psi \in \mathscr{C}^2(\mathbf{M}_4; \mathbb{C})$.

In this section and subsequently we *choose units* such that $c = \hbar = 1$, where $\hbar$ is the Planck constant divided by 2π.

The real-valued Lagrangian function $\mathscr{L}: \mathbb{C} \times \mathbb{C} \times \mathbb{C}^4 \times \mathbb{C}^4 \to \mathbb{R}$ is given by:

$$\begin{aligned}
\mathscr{L}(\zeta, \bar{\zeta}, \zeta_k, \bar{\zeta}_l) &\equiv -d^{kl}\bar{\zeta}_k\zeta_l - \mu^2\bar{\zeta}\zeta, \\
\frac{\partial \mathscr{L}(..)}{\partial \bar{\zeta}} &= -\mu^2\zeta, \\
\frac{\partial \mathscr{L}(..)}{\partial \bar{\zeta}_k} &= -d^{kl}\zeta_l.
\end{aligned} \tag{6.2.2}$$

The Euler–Lagrange equation is

$$\left[\frac{\partial \mathscr{L}(..)}{\partial \bar{\zeta}}\right]_{|..} - \frac{d}{dx^k}\left\{\left[\frac{\partial \mathscr{L}(..)}{\partial \bar{\zeta}_k}\right]_{|..}\right\} = -\mu^2\psi(x) + d^{kl}\psi_{,kl} = 0, \tag{6.2.3}$$

which is the Klein–Gordon equation (6.2.1).

Example: We shall obtain a special class of solution of (6.2.1). Assume that the solution is nontrivial and separable, that is,

$$\psi(x^1, x^2, x^3, x^4) = \xi(x^1)\eta(x^2)\zeta(x^3)\mathscr{T}(x^4) \not\equiv 0.$$

Equation (6.2.1) yields

$$\begin{aligned}
&\xi''(x^1)/\xi(x^1) + \eta''(x^2)/\eta(x^2) \\
&\quad + \zeta''(x^3)/\zeta(x^3) - \mathscr{T}''(x^4)/\mathscr{T}(x^4) - \mu^2 = 0.
\end{aligned}$$

From the separation of variables it follows that

$$\begin{aligned}
\xi''(x^1)/\xi(x^1) &= \alpha, \\
\eta''(x^2)/\eta(x^2) &= \beta, \\
\zeta''(x^3)/\zeta(x^3) &= \gamma, \\
\mathscr{T}''(x^4)/\mathscr{T}(x^4) &= \delta.
\end{aligned}$$

Here α, β, γ, δ are complex constants that satisfy the relations:

$$\begin{aligned}
\mathrm{Re}(\alpha + \beta + \gamma - \delta) &= \mu^2, \\
\mathrm{Im}(\alpha + \beta + \gamma - \delta) &= 0.
\end{aligned}$$

These constants are otherwise arbitrary. The most general separable solution of (6.2.1) is

$$\begin{aligned}
\psi(x) = \exp\{&\sqrt{\alpha}x^1 + \sqrt{\beta}x^2 + \sqrt{\gamma}x^3 \\
&+ [\mathrm{Re}(\alpha + \beta + \gamma) - \mu^2 + i\,\mathrm{Im}(\alpha + \beta + \gamma)]^{1/2}x^4\},
\end{aligned} \tag{6.2.4a}$$

where α, β, γ are arbitrary complex constants. In case we choose α, β, γ to be nonpositive real numbers, we obtain the special case of (6.2.4*a*), namely,

$$\psi(x) = \exp\{i[k_\alpha x^\alpha + \omega(\mathbf{k})x^4]\} \tag{6.2.4b}$$

where $\omega(\mathbf{k}) \equiv \omega(k_1, k_2, k_3) \equiv [(k_1)^2 + (k_1)^2 + (k_1)^2 + \mu^2]^{1/2}$. The real and the imaginary parts of the solution in (6.2.4*b*) represent *simple harmonic plane waves with frequency* $\omega(\mathbf{k})$ *and wave number* $\mathbf{k} \equiv (k_1, k_2, k_3)$. □

The Lagrangian (6.2.2) is invariant under Poincaré transformations as well as the global gauge transformation (6.1.20). Thus the Noether theorems (6.1.17), (6.1.19), and (6.1.21) hold. It will be instructive to work out explicitly the tensors $T^k{}_l(x)$, $\mathscr{J}^k{}_{ab}(x)$, $j^k(x)$ from the Lagrangian (6.2.2). These are computed to be

$$\begin{aligned}
T^k{}_j(x) &= \delta^k{}_j \mathscr{L}(\ldots)|_{\ldots} - \left[\frac{\partial \mathscr{L}(\ldots)}{\partial \zeta_k}\right]_{|\ldots} \psi_{,j} - \left[\frac{\partial \mathscr{L}(\ldots)}{\partial \bar{\zeta}_k}\right]_{|\ldots} \bar{\psi}_{,j} \\
&= d^{kl}[\bar{\psi}_{,l}\psi_{,j} + \bar{\psi}_{,j}\psi_{,l}] - \delta^k{}_j[d^{ab}\bar{\psi}_{,a}\psi_{,b} + \mu^2\bar{\psi}(x)\psi(x)], \\
T_{kj}(x) &= \bar{\psi}_{,k}\psi_{,j} + \bar{\psi}_{,j}\psi_{,k} + d_{kj}\mathscr{L}(\ldots)|_{\ldots} = T_{jk}(x), \qquad (6.2.5) \\
\mathscr{J}^k{}_{ab}(x) &= [T^k{}_a(x)d_{bj} - T^k{}_b(x)d_{aj}]x^j = -\mathscr{J}^k{}_{ba}(x), \\
j^k(x) &= -ie\left\{\left[\frac{\partial \mathscr{L}(\ldots)}{\partial \zeta_k}\right]_{|\ldots} \psi(x) - \left[\frac{\partial \mathscr{L}(\ldots)}{\partial \bar{\zeta}_k}\right]_{|\ldots} \bar{\psi}(x)\right\} \\
&= -ied^{ka}[\bar{\psi}(x)\psi_{,a} - \bar{\psi}_{,a}\psi(x)].
\end{aligned}$$

Suppose that ψ is a complex-valued function such that the improper integral $\int_{\mathbb{R}^4} |\psi(x)|\, dx^1\, dx^2\, dx^3\, dx^4$ converges. Then ψ possesses the Fourier transform

$$\begin{aligned}
\psi(x) &= (1/4\pi^2) \int_{\mathbb{R}^4} \Gamma(k_1, k_2, k_3, k_4) e^{ik_a x^a}\, dk_1\, dk_2\, dk_3\, dk_4 \\
&\equiv (1/4\pi^2) \int_{\mathbb{R}^4} \Gamma(k) e^{ik_a x^a}\, d^4k, \qquad (6.2.6)
\end{aligned}$$

where Γ is an appropriate complex-valued function. We shall now seek *the class of general solutions* $\psi(x)$ of the Klein–Gordon equation (6.2.1) such that ψ admits a Fourier transform.

Theorem (6.2.1): *Let the complex-valued field* $\psi(x)$ *be given by the uniformly convergent improper integral (plane wave decomposition):*

$$\begin{aligned}
\psi(x) = (1/2\pi)^{3/2} \int_{\mathbb{R}^3} [2\omega(\mathbf{k})]^{-1/2} \{&\alpha(\mathbf{k}) e^{i[k_\mu x^\mu - \omega(k)x^4]} \\
&+ \bar{\beta}(\mathbf{k}) e^{-i[k_\mu x^\mu - \omega(k)x^4]}\}\, d^3\mathbf{k}, \qquad (6.2.7)
\end{aligned}$$

$$\begin{aligned}
\omega(\mathbf{k}) &\equiv \omega(k_1, k_2, k_3) = [k_\alpha k_\alpha + \mu^2]^{1/2}, \\
d^3\mathbf{k} &\equiv dk_1\, dk_2\, dk_3.
\end{aligned}$$

Furthermore let the following improper integrals converge uniformly

$$\int_{\mathbb{R}^3} k_\mu [2\omega(\mathbf{k})]^{-1/2} \{\alpha(\mathbf{k}) e^{i[k_\mu x^\mu - \omega(k)x^4]} + \bar{\beta}(\mathbf{k}) e^{-i[k_\mu x^\mu - \omega(k)x^4]}\} \, d^3\mathbf{k},$$

$$\int_{\mathbb{R}^3} [\omega(\mathbf{k})/2]^{1/2} \{\alpha(\mathbf{k}) e^{i[k_\mu x^\mu - \omega(k)x^4]} + \bar{\beta}(\mathbf{k}) e^{-i[k_\mu x^\mu - \omega(k)x^4]}\} \, d^3\mathbf{k},$$

$$\int_{\mathbb{R}^3} k_\mu k_\mu [2\omega(\mathbf{k})]^{-1/2} \{\alpha(\mathbf{k}) e^{i[k_\mu x^\mu - \omega(k)x^4]} + \bar{\beta}(\mathbf{k}) e^{-i[k_\mu x^\mu - \omega(k)x^4]}\} \, d^3\mathbf{k},$$

$$\int_{\mathbb{R}^3} \{[\omega(\mathbf{k})]^3/2\}^{1/2} \{\alpha(\mathbf{k}) e^{i[k_\mu x^\mu - \omega(k)x^4]} + \bar{\beta}(\mathbf{k}) e^{-i[k_\mu x^\mu - \omega(k)x^4]}\} \, d^3\mathbf{k}.$$

Let the continuous complex-valued functions α, β be otherwise arbitrary. Then the equation (6.2.7) *yields a class of general solutions of the Klein–Gordon equation* (6.2.1).

Proof: By the assumption that the above improper integrals converge uniformly we may differentiate under the integral signs. Thus

$$\begin{aligned}
&\Box\psi(x) - \mu^2\psi(x) \\
&\quad = (1/2\pi)^{3/2} \int_{\mathbb{R}^3} [2\omega(\mathbf{k})]^{-1/2} \{-k_\nu k_\nu + [\omega(\mathbf{k})]^2 - \mu^2\} \\
&\qquad \times \{\alpha(\mathbf{k}) e^{i[k_\mu x^\mu - \omega(k)x^4]} + \bar{\beta}(\mathbf{k}) e^{-i[k_\mu x^\mu - \omega(k)x^4]}\} \, d^3\mathbf{k}, \\
&\quad \equiv 0. \quad \blacksquare
\end{aligned}$$

Example 1: With the general solution in (6.2.7) we shall compute some of the integral constants given in (6.1.24). Let $(\mathbf{x}, 0) = (x^1, x^2, x^3, 0)$ and $d^3\mathbf{x} \equiv dx^1\,dx^2\,dx^3$. By (6.2.5), (6.2.7), and (6.1.24) we have for the total electric charge of the field

$$\mathcal{Q} = -ie \int_{\mathbb{R}^3} d^{44}\{\bar{\psi}(\mathbf{x},0)\psi_{,4}(\mathbf{x},0) - \bar{\psi}_{,4}(\mathbf{x},0)\psi(\mathbf{x},0)\} \, d^3\mathbf{x}. \tag{6.2.8}$$

By (6.2.7), we get

$$\begin{aligned}
\bar{\psi}(\mathbf{x},0)\psi_{,4}(\mathbf{x},0) &= (-i/16\pi^3) \\
&\quad \times \int_{\mathbb{R}^3}\int_{\mathbb{R}^3} [\omega(\mathbf{k})/\omega(\mathbf{k}')]^{1/2} \{\alpha(\mathbf{k}) e^{ik_\mu x^\mu} - \bar{\beta}(\mathbf{k}) e^{ik_\mu x^\mu}\} \\
&\quad \times \{\bar{\alpha}(\mathbf{k}') e^{-ik'_\mu x^\mu} + \beta(\mathbf{k}') e^{ik'_\mu x^\mu}\} \, d^3\mathbf{k}' \, d^3\mathbf{k}.
\end{aligned}$$

Putting the above expression in (6.2.8), we obtain

$$
\begin{aligned}
\mathscr{Q} = (e/16\pi^3) &\int_{\mathbb{R}^3}\int_{\mathbb{R}^3}\int_{\mathbb{R}^3} [\omega(\mathbf{k})/\omega(\mathbf{k}')]^{1/2} \\
&\times \{[\alpha(\mathbf{k})e^{ik_\mu x^\mu} - \bar{\beta}(\mathbf{k})e^{-ik_\mu x^\mu}]\cdot[\bar{\alpha}(\mathbf{k}')e^{-ik'_\mu x^\mu} + \beta(\mathbf{k}')e^{ik'_\mu x^\mu}] \\
&+ (\text{c.c.})\}\, d^3\mathbf{x}\, d^3\mathbf{k}'\, d^3\mathbf{k} \\
= (e/16\pi^3) &\int_{\mathbb{R}^3}\int_{\mathbb{R}^3}\int_{\mathbb{R}^3} [\omega(\mathbf{k})/\omega(\mathbf{k}')]^{1/2} \\
&\times \{[\alpha(\mathbf{k})\bar{\alpha}(\mathbf{k}')e^{i(k_\mu - k'_\mu)x^\mu} - \bar{\beta}(\mathbf{k})\beta(\mathbf{k}')e^{-i(k_\mu - k'_\mu)x^\mu} \\
&+ \alpha(\mathbf{k})\beta(\mathbf{k}')e^{i(k_\mu + k'_\mu)x^\mu} - \bar{\alpha}(\mathbf{k}')\bar{\beta}(\mathbf{k})e^{-i(k_\mu + k'_\mu)x^\mu}] + (\text{c.c.})\}\, d^3\mathbf{x}\, d^3\mathbf{k}'\, d^3\mathbf{k}.
\end{aligned} \tag{6.2.9}
$$

We shall digress slightly to introduce the Dirac "delta function," which will be used to simplify the nine-fold integral in (6.2.9). Let $f(k, k')$ be a real-valued continuous function from $\mathbb{R}^2$ such that the Cauchy principal value of the improper integral C.P.V. $\int_{-\infty}^{+\infty} f(k, k)\, dk$ converges. In that case

$$
\begin{aligned}
\text{C.P.V.}\ (1/2\pi) &\int_{-\infty}^{+\infty}\int_{-\infty}^{+\infty}\int_{-\infty}^{+\infty} f(k, k')e^{i(k-k')x}\, dx\, dk'\, dk \\
&\equiv \lim_{L_3\to\infty}\ \lim_{L_2\to\infty}\ \lim_{L_1\to\infty}\ (1/2\pi) \\
&\quad \times \int_{-L_3}^{+L_3}\int_{-L_2}^{+L_2}\left[f(k, k')\int_{-L_1}^{+L_1} e^{i(k-k')x}\, dx\right] dk'\, dk \\
&\equiv \lim_{L_3\to\infty}\ \lim_{L_2\to\infty}\ \lim_{L_1\to\infty}\ (1/\pi) \\
&\quad \times \int_{-L_3}^{+L_3}\int_{-L_2}^{+L_2} [f(k, k')\sin[L_1(k - k')]/(k - k')]\, dk'\, dk.
\end{aligned} \tag{6.2.10a}
$$

Thus we have the delta function defined by the equation

$$
\begin{aligned}
\text{C.P.V.} &\int_{-\infty}^{+\infty}\int_{-\infty}^{+\infty} f(k, k')\delta(k - k')\, dk'\, dk \\
&\equiv \lim_{L_3\to\infty}\ \lim_{L_2\to\infty}\ \lim_{L_1\to\infty}\ (1/\pi) \\
&\quad \times \int_{-L_3}^{+L_3}\int_{-L_2}^{+L_2} [f(k, k')\sin[L_1(k - k')]/(k - k')]\, dk'\, dk \\
&= \text{C.P.V.} \int_{-\infty}^{+\infty} f(k, k)\, dk.
\end{aligned}
$$

We give a proof of the last step under stronger assumptions. Let $g(x)$ be a smooth function that vanishes outside a finite interval. It is well known that C.P.V. $\int_{-\infty}^{+\infty} (\sin Lx/x)\, dx = \pi$ for every positive L. Since g is continuous, one has $|g(x) - g(0)| < \varepsilon$ whenever $|x| < \delta_\varepsilon$. Thus we have

$$\text{C.P.V.} \int_{-\infty}^{+\infty} |g(x) - g(0)|(\sin Lx/x)\,dx$$
$$= \int_{-\delta_\varepsilon}^{\delta_\varepsilon} |g(x) - g(0)|(\sin Lx/x)\,dx + \int_{-\infty}^{-\delta_\varepsilon} |g(x) - g(0)|(\sin Lx/x)\,dx$$
$$+ \int_{\delta_\varepsilon}^{\infty} |g(x) - g(0)|(\sin Lx/x)\,dx.$$

Integration by parts shows that the limit of the last two integrals is zero for every $\delta_\varepsilon > 0$ when L increases without bound. Furthermore, we see that the first integral is bounded by $2\varepsilon L\delta_\varepsilon$. So choosing $\varepsilon = L^{-2}$ we can prove that the limit of the first integral is zero as L increases without bound. Thus we have shown that C.P.V. $\lim_{L\to\infty} (1/\pi)\int_{-\infty}^{+\infty} g(x)(\sin Lx/x)\,dx = g(0)$. (There is a weakness in this proof on account of switching two limiting processes.)

We generalize (6.2.10a) to the three-dimensional case as follows:

$$\text{C.P.V.} \int_{\mathbb{R}^3}\int_{\mathbb{R}^3} f(\mathbf{k},\mathbf{k}')\delta^3(\mathbf{k}-\mathbf{k}')\,d^3\mathbf{k}'\,d^3\mathbf{k}$$
$$\equiv \text{C.P.V.}\,(1/2\pi)^3 \int_{\mathbb{R}^3}\int_{\mathbb{R}^3}\int_{\mathbb{R}^3} f(\mathbf{k},\mathbf{k}')e^{i(k_\mu - k'_\mu)x^\mu}\,d^3\mathbf{x}\,d^3\mathbf{k}'\,d^3\mathbf{k}$$
$$= \text{C.P.V.} \int_{\mathbb{R}^3} f(\mathbf{k},\mathbf{k})\,d^3\mathbf{k}. \tag{6.2.10b}$$

Note that in the sequel we shall suppress explicit mention of the Cauchy principal value of an integral; i.e., "C.P.V." *will be omitted*. Using (6.2.10b) in (6.2.9) we get

$$\mathcal{Q} = (e/2)\int_{\mathbb{R}^3}\int_{\mathbb{R}^3} [\omega(\mathbf{k})/\omega(\mathbf{k}')]^{1/2}\{[\alpha(\mathbf{k})\bar{\alpha}(\mathbf{k}') - \bar{\beta}(\mathbf{k})\beta(\mathbf{k}')]\delta^3(\mathbf{k}-\mathbf{k}')$$
$$+ [\alpha(\mathbf{k})\beta(\mathbf{k}') - \bar{\alpha}(\mathbf{k}')\bar{\beta}(\mathbf{k})]\delta^3(\mathbf{k}+\mathbf{k}') + \text{c.c.}\}\,d^3\mathbf{k}'\,d^3\mathbf{k}$$
$$= (e/2)\int_{\mathbb{R}^3} \{[|\alpha(\mathbf{k})|^2 - |\beta(\mathbf{k})|^2]$$
$$+ [\alpha(\mathbf{k})\beta(-\mathbf{k}) - \bar{\alpha}(-\mathbf{k})\bar{\beta}(\mathbf{k})] + (\text{c.c.})\}\,d^3\mathbf{k}$$
$$= e\int_{\mathbb{R}^3} [|\alpha(\mathbf{k})|^2 - |\beta(\mathbf{k})|^2]\,d^3\mathbf{k}. \quad \square \tag{6.2.11}$$

Example 2: Next we shall compute the total energy of the field from (6.2.24), (6.2.5), (6.2.7), and (6.2.10b). Thus we have

$$H \equiv -P_4 = -\int_{\mathbb{R}^3} T^4{}_4(\mathbf{x},0)\,d^3\mathbf{x}$$
$$= -\int_{\mathbb{R}^4} \{d^{44}(2\bar{\psi}_{,4}\psi_{,4}) - (\bar{\psi}_{,\nu}\psi_{,\nu} - \bar{\psi}_{,4}\psi_{,4} + \mu^2|\psi|^2)\}_{|(\mathbf{x},0)}\,d^3\mathbf{x}$$

$$= \int_{\mathbb{R}^3} (\bar{\psi}_{,\nu}\psi_{,\nu} + \bar{\psi}_{,4}\psi_{,4} + \mu^2|\psi|^2)_{|(\mathbf{x},0)}\, d^3\mathbf{x}$$

$$= (1/16\pi^3)\int_{\mathbb{R}^3}\int_{\mathbb{R}^3}\int_{\mathbb{R}^3} [\omega(\mathbf{k})\omega(\mathbf{k}')]^{-1/2}$$
$$\times \{[k_\nu k'_\nu + \omega(\mathbf{k})\omega(\mathbf{k}')][\alpha(\mathbf{k})e^{ik_\gamma x^\gamma} - \bar{\beta}(\mathbf{k})e^{-ik_\gamma x^\gamma}]$$
$$\times [\bar{\alpha}(\mathbf{k}')e^{-ik'_\sigma x^\sigma} - \beta(\mathbf{k}')e^{ik'_\sigma x^\sigma}] + \mu^2[\alpha(\mathbf{k})e^{ik_\gamma x^\gamma} + \bar{\beta}(\mathbf{k})e^{-ik_\gamma x^\gamma}]$$
$$\times [\bar{\alpha}(\mathbf{k}')e^{-ik'_\sigma x^\sigma} + \beta(\mathbf{k}')e^{ik'_\sigma x^\sigma}]\}\, d^3\mathbf{x}\, d^3\mathbf{k}'\, d^3\mathbf{k}$$

$$= (1/2)\int_{\mathbb{R}^3} [\omega(\mathbf{k})]^{-1}$$
$$\times \{[k_\nu k_\nu + (\omega(\mathbf{k}))^2 + \mu^2][|\alpha(\mathbf{k})|^2 + |\beta(\mathbf{k})|^2]$$
$$+ [k_\nu k_\nu - (\omega(\mathbf{k}))^2 + \mu^2][\alpha(\mathbf{k})\beta(-\mathbf{k}) + \bar{\beta}(\mathbf{k})\bar{\alpha}(-\mathbf{k})]\}\, d^3\mathbf{k}$$

$$= \int_{\mathbb{R}^3} [|\alpha(\mathbf{k})|^2 + |\beta(\mathbf{k})|^2]\omega(\mathbf{k})\, d^3\mathbf{k}. \quad \square \tag{6.2.12}$$

From equations (6.2.11) and (6.2.12) we can interpret *physically* $|\alpha(\mathbf{k})|^2$ and $|\beta(\mathbf{k})|^2$ as densities (or distributions) of the positively charged and the negatively charged scalar fields in the space of wave-numbers (three-dimensional momentum space).

In this section if we put the mass parameter $\mu = 0$ in all the equations then the corresponding equations for a complex wave field obeying $\Box\psi(x) = 0$ are obtained.

EXERCISES 6.2

1. Show that by equations (6.1.24), (6.2.5), (6.2.7), and (6.2.10*b*) the components of the total three-momentum of a complex scalar field are given by

$$P_\nu = \int_{\mathbb{R}^3} [|\alpha(\mathbf{k})|^2 + |\beta(\mathbf{k})|^2]k_\nu\, d^3\mathbf{k}.$$

2. In the spherical polar and time coordinate chart, the Klein–Gordon equation (6.2.1) becomes

$$r^{-2}(r^2\chi_{,r})_{,r} + (1/r^2\sin\theta)(\sin\theta\cdot\chi_{,\theta})_{,\theta}$$
$$+ (1/r^2\sin^2\theta)\chi_{,\phi\phi} - \chi_{,tt} - \mu^2\chi(..) = 0,$$
$$\chi(r,\theta,\phi,t) \equiv \psi(x^1,x^2,x^3,x^4).$$

Obtain the most general separable solution

$$\chi(r,\theta,\phi,t) = \rho(r)\Theta(\theta)\Phi(\phi)\mathscr{T}(t)$$

of this equation.

3. Consider the relativistic wave equation for the "spinless" electron in the external electrostatic field of a positively charged nucleus. This equation is the modified Klein–Gordon equation:

$$r^{-2}(r^2\chi_{,r})_{,r} + (1/r^2\sin\theta)(\sin\theta\cdot\chi_{,\theta})_{,\theta} + (1/r^2\sin^2\theta)\chi_{,\phi\phi}$$
$$+ [-i(\partial/\partial t) + Ze^2/r]^2\chi(..) - \mu^2\chi(..) = 0,$$

where $-e < 0$ is the charge of the electron and Z is the number of protons in the nucleus. The stationary solutions are characterized by the special separable structure:

$$\chi(r,\theta,\phi,t) = \rho(r)\Theta(\theta)\Phi(\phi)e^{-iEt},$$

where E is the parameter representing the energy eigenvalue. Show that for the case of square-integrable solutions, i.e.,

$$\int |\chi(..)|^2 r^2 \sin\theta\, dr\, d\theta\, d\phi < \infty,$$

the energy eigenvalues are given by

$$E = \mu(n+s)[(n+s)^2 + Z^2e^4]^{-1/2},$$
$$s \equiv (-1/2) + (1/2)[(2l+1)^2 - 4Z^2e^4]^{1/2},$$
$$n \in \{1,2,\ldots\}, \qquad l \in \{0,1,\ldots\}.$$

(Note that these values have small discrepancies with the experimental values. The Dirac equation in sec. 5 can remove these discrepancies.)

4. Suppose that $\Omega\colon \mathbb{R}^3 \to \mathbb{C}$, $\Xi\colon \mathbb{R}^3 \to \mathbb{C}$ are such that

$$\omega = \Omega(x^1,x^2,x^3) \equiv x^1\cos\alpha + x^2\sin\alpha + ix^3,$$
$$\xi = \Xi(x^2,x^3,x^4) \equiv x^2 + ix^3\sin\alpha + x^4\cos\alpha,$$

where α is an arbitrary real parameter. Let $h\colon D \subset \mathbb{C}\times\mathbb{C} \to \mathbb{C}$ be represented by $\eta = h(\omega,\xi)$. Let h be an *arbitrary* holomorphic function of two complex variables ω, ξ. Then show that the complex-valued function $\psi(x^1,x^2,x^3,x^4) \equiv h[\Omega(x^1,x^2,x^3),\Xi(x^2,x^3,x^4)]$ satisfies the wave equation $\Box\psi(x) = 0$.

6.3. The Electromagnetic Tensor Field

The special theory of relativity emerged from investigations into the speed of light propagation, which is also the speed of electromagnetic wave propagation. Therefore, the electromagnetic field equations are of particular interest in the special theory of relativity. Maxwell's equations, as three-dimensional vector field equations, are:

$$\begin{aligned} \nabla \times \mathbf{H}(..) &= \frac{\partial \mathbf{E}(..)}{\partial t}, \\ \nabla \cdot \mathbf{E}(..) &= 0, \\ \nabla \cdot \mathbf{H}(..) &= 0, \\ \nabla \times \mathbf{E}(..) &= -\frac{\partial \mathbf{H}(..)}{\partial t}, \end{aligned} \tag{6.3.1}$$

where t is the time variable, ∇ is the gradient operator, and $\mathbf{E}(..)$ and $\mathbf{H}(..)$ are the electric and magnetic fields respectively. The equations (6.3.1) hold *outside* the charged sources. In the language of three-dimensional Cartesian tensor analysis, (6.3.1) can be expressed as:

$$\begin{aligned} \varepsilon_{\mu\nu\lambda} H_{\nu,\lambda} &= -\frac{\partial E_\mu(..)}{\partial t}, \\ E_{\mu,\mu} &= 0, \\ H_{\mu,\mu} &= 0, \\ \varepsilon_{\mu\nu\lambda} E_{\nu,\lambda} &= \frac{\partial H_\mu(..)}{\partial t}, \end{aligned} \tag{6.3.2}$$

where $\varepsilon_{\mu\nu\lambda}$ is the totally antisymmetric permutation symbol. The above equations can be cast into Minkowski tensor field equations. To achieve that we have to identify electric and magnetic three-vectors with an antisymmetric Minkowski tensor (see (5.3.20)):

$$[F_{ab}(x)] \equiv \begin{bmatrix} 0 & H_3(x) & -H_2(x) & E_1(x) \\ -H_3(x) & 0 & H_1(x) & E_2(x) \\ H_2(x) & -H_1(x) & 0 & E_3(x) \\ -E_1(x) & -E_2(x) & -E_3(x) & 0 \end{bmatrix} = -[F_{ab}(x)]. \tag{6.3.3}$$

The above definition unifies both electric and magnetic vector fields into a single Minkowski electromagnetic tensor field. Transformation properties of the tensor $F_{ab}(x)$ under a Lorentz transformation have physical consequences. In the example following equation (1.3.7) in Chapter 1, we find that a purely electric field in one inertial frame appears to be a mixed electromagnetic field to another inertial observer.

Maxwell's equations (6.3.2) can be written as the equations

$$\begin{aligned} F^{\mu\beta}{}_{,\beta} &= -F^{\mu 4}{}_{,4}, \\ F^{4\mu}{}_{,\mu} &= 0, \\ F_{\mu\nu,\lambda} + F_{\nu\lambda,\mu} + F_{\lambda\mu,\nu} &= 0, \\ F_{\nu 4,\lambda} + F_{4\lambda,\nu} &= -F_{\lambda\nu,4}. \end{aligned} \tag{6.3.4}$$

The above equations can be cast into the following Minkowski tensor field equations:

$$M^a \equiv F^{ab}{}_{,b} = 0,$$
$$M_{abc} \equiv F_{ab,c} + F_{bc,a} + F_{ca,b} = 0. \tag{6.3.5}$$

Since M_{abc} is totally antisymmetric, there are four linearly independent equations $M_{abc} = 0$. The eight linearly independent equations in (6.3.5) have two differential identities (assuming that F_{ab} are twice differentiable):

$$M^a{}_{,a} \equiv 0,$$
$$\eta^{abcd} M_{bcd,a} \equiv 0, \tag{6.3.6}$$

where η_{abcd} is the totally antisymmetric pseudotensor.

Example 1: We shall derive the wave equation for the electromagnetic tensor field $F_{ab}(x)$ assuming that the F_{ab} are twice differentiable functions. From (6.3.5) we obtain

$$\begin{aligned} 0 = M^a{}_{bc,a} &= F^a{}_{b,ca} + d^{ak} F_{bc,ka} - F^a{}_{c,ba} \\ &= [F^a{}_{b,a}]_{,c} + \Box F_{bc} - [F^a{}_{c,a}]_{,b} \\ &= 0 + \Box F_{bc} - 0 \\ &= \Box F_{bc}. \end{aligned}$$

The above equation shows that both the electric and magnetic fields travel with the speed of light. □

Example 2: Introducing a complex-valued antisymmetric field [see equation (1.3.15)],

$$\psi_{ab}(x) \equiv F_{ab}(x) + (i/2)\eta_{abcd} F^{cd}(x) = F_{ab}(x) + iF^{*}_{ab}(x),$$

we can express Maxwell's equations (6.3.6) in a neat form:

$$\psi^{ab}{}_{,b} = 0. \quad \Box \tag{6.3.7}$$

The equations $M_{abc} = 0$ can be solved by putting

$$F_{ab}(x) = \frac{\partial A_b(x)}{\partial x^a} - \frac{\partial A_a(x)}{\partial x^b} \equiv A_{b,a} - A_{a,b}, \tag{6.3.8}$$

where the A_a are *arbitrary* twice differentiable functions. The functions A_a are called the components of the *four-potential field.* We note that the four-potential is not uniquely determined by (6.3.8). We can make a *local gauge transformation* (involving an arbitrary twice differentiable function λ):

$$A'_k(x) = A_k(x) - \lambda_{,k},$$
$$F'_{kj}(x) = F_{kj}(x). \tag{6.3.9}$$

Inserting (6.3.8) into Maxwell's equations (6.3.5), we obtain four equations:

$$d^{ac}(A^b{}_{,b})_{,c} - \Box A^a(x) = 0. \tag{6.3.10}$$

Now we make a local gauge transformation (6.3.9) such that λ is thrice differentiable and satisfies

$$\Box\lambda(x) = A^b{}_{,b}. \tag{6.3.11}$$

Thus from (6.3.9) we have

$$\begin{aligned} A'^b{}_{,b} &= A^b{}_{,b} - d^{ba}\lambda_{,ab} \\ &= A^b{}_{,b} - A^b{}_{,b} \\ &= 0. \end{aligned} \tag{6.3.12}$$

This special condition on $A'^b(x)$ is called the *Lorentz gauge condition* and is preserved under the *restricted* gauge transformations:

$$A''_a(x) = A'_a(x) - \lambda_{,a},$$

$$\Box\lambda(x) = 0.$$

Maxwell's equations (6.3.5) with $F_{ab}(x) = F'_{ab}(x) = A'_{b,a} - A'_{a,b}$ become

$$\Box A'^a(x) = 0. \tag{6.3.13}$$

Subsequently, we shall assume the Lorentz gauge condition and drop the prime on the four-potential and write simply

$$\begin{aligned} \Box A^k(x) &= 0, \\ A^k{}_{,k} &= 0. \end{aligned} \tag{6.3.14}$$

Example: We shall obtain simple harmonic plane-wave solutions of (6.3.14). We assume the solutions of the form:

$$A^j(x) = a^j(k)\cos[k_a x^a + \theta(k)], \tag{6.3.15}$$

where

$$[a^1(k)]^2 + [a^2(k)]^2 + [a^3(k)]^2 + [a^4(k)]^2 \not\equiv 0$$

and

$$[k_1]^2 + [k_2]^2 + [k_3]^2 + [k_4]^2 > 0.$$

The wave equation (6.3.14) implies that

$$\begin{aligned} [k_4]^2 &= [k_1]^2 + [k_2]^2 + [k_3]^2 = k_\mu k_\mu > 0, \\ k_4 &= \pm\nu(\mathbf{k}), \qquad \nu(\mathbf{k}) \equiv [k_\mu k_\mu]^{1/2}. \end{aligned} \tag{6.3.16}$$

Therefore, the four-vector k_a must be a nonzero null vector. The Lorentz gauge condition $A^b{}_{,b} = 0$ implies that

$$k_b a^b(k) = k_\mu a^\mu(k) + k_4 a^4(k) = 0. \tag{6.3.17}$$

Using Theorem (1.1.6), equation (6.3.17) implies that $a^b(k)$ cannot be timelike. Thus we must have either $a^b(k)$ is null and

$$a^b(k) = \lambda(k)k^b,$$

for some $\lambda(k) \not\equiv 0$, or else $a^b(k)$ is spacelike.

In the case $a^b(k)$ is null we have $F_{ab}(x) \equiv 0$ and we shall not consider this case further. In the case $a^b(k)$ is spacelike we get

$$0 \leq [a^4(k)]^2 < a^\mu(k)a^\mu(k).$$

Using the Lorentz gauge condition we obtain

$$[k_\mu a^\mu(k)]^2 < [k_\nu k_\nu a^\mu(k)a^\mu(k)].$$

By the Schwartz inequality (1.1.13) we conclude that

$$a^\mu(k) \neq \lambda(k)k^\mu$$

for any $\lambda(k)$. The three functions are otherwise arbitrary. The function a^4 is given by $a^4(k) = -k_\mu a^\mu(k)/k_4$. In this case

$$\begin{aligned}
-F_{\mu 4}(x) &= [k_\mu k_\nu a_\nu(k)/(k_4)^2 - a_\mu(k)]k_4 \sin[k_a x^a + \theta(k)] \not\equiv 0,\\
-F_{\mu\nu}(x) &= [k_\mu a_\nu(k) - k_\nu a_\mu(k)] \sin[k_a x^a + \theta(k)] \not\equiv 0,\\
k_\mu F_{\mu 4}(x) &= k_\lambda \varepsilon_{\lambda\mu\nu} F_{\mu\nu}(x) = F_{\lambda 4}(x)\varepsilon_{\lambda\mu\nu}F_{\mu\nu}(x) \equiv 0.
\end{aligned}$$

The above equations show that the propagation vector $\mathbf{k} \equiv (k_1, k_2, k_3)$, the electric field vector $\mathbf{E}(x)$, and the magnetic field vector $\mathbf{H}(x)$ are all mutually perpendicular. *Thus the electromagnetic wave is propagated in a transverse fashion.* □

Now we shall derive Maxwell's equations from a variational principle. We can choose the Lagrangian $\mathscr{L}: \mathbb{R}^4 \times \mathbb{R}^4 \times \mathbb{R}^{16} \to \mathbb{R}$ as

$$\mathscr{L}(x, y_a, y_{ak}) \equiv -(1/4)d^{ab}d^{kl}(y_{lb} - y_{bl})(y_{ka} - y_{ak}). \tag{6.3.18}$$

Therefore,

$$\mathscr{L}(\ldots)_{|y_a = A_a(x),\, y_{ak} = A_{a,k}} = -(1/4)F^{ab}(x)F_{ab}(x),$$

$$\frac{\partial \mathscr{L}(\ldots)}{\partial y_a} \equiv 0,$$

$$\begin{aligned}
\frac{\partial \mathscr{L}(\ldots)}{\partial y_{ij}} &= -(1/4)d^{ab}d^{kl}[(\delta^i{}_l\delta^j{}_b - \delta^i{}_b\delta^j{}_l)(y_{ka} - y_{ak})\\
&\qquad + (y_{lb} - y_{bl})(\delta^i{}_k\delta^j{}_a - \delta^i{}_a\delta^j{}_k)]\\
&= y^{ji} - y^{ij}.
\end{aligned}$$

Thus the Euler–Lagrange equations are

$$\left[\frac{\partial \mathscr{L}(\ldots)}{\partial y_i}\right]_{|\ldots} - \frac{d}{dx^j}\left\{\left[\frac{\partial \mathscr{L}(\ldots)}{\partial y_{ij}}\right]_{|\ldots}\right\} = -F^{ij}{}_{,j} = 0, \tag{6.3.19}$$

which are Maxwell's equations (6.3.10). The canonical stress-energy-momentum tensor, computed from (6.1.17) and (6.3.18) is

$$T^i{}_j(x) = \delta^i{}_j \mathscr{L}(\ldots)_{|\ldots} - \left[\frac{\partial \mathscr{L}(\ldots)}{\partial y_{ai}}\right]_{|\ldots} A_{a,j}$$

$$= -(1/4)\delta^i{}_j F^{ab}(x)F_{ab}(x) - F^{ai}(x)A_{a,j}. \tag{6.3.20}$$

The above tensor is neither symmetric nor gauge-invariant. We define the symmetrized gauge-invariant stress-energy-momentum tensor by

$$\theta_{ij}(x) \equiv T_{ij}(x) + [F^a{}_i(x)A_j(x)]_{,a}$$

$$= F^a{}_i(x)F_{aj}(x) - (1/4)d_{ij}F^{ab}(x)F_{ab}(x). \tag{6.3.21}$$

It can be verified (see problem 4 of Exercises 1.3) that

$$\theta^i{}_i \equiv 0,$$

$$\theta^{ij}{}_{,j} = T^{ij}{}_{,j} = 0,$$

$$\int_{\mathbb{R}^3} \theta^4{}_k(\mathbf{x},0)\,d^3\mathbf{x} = \int_{\mathbb{R}^3} T^4{}_k(\mathbf{x},0)\,d^3\mathbf{x} + \int_{\mathbb{R}^3} [F^{\alpha 4}(\mathbf{x},0)A_k(\mathbf{x},0)]_{,\alpha}\,d^3\mathbf{x}.$$

The last improper integral may be replaced by a surface integral by Gauss's Theorem, which converges to zero by the usual hypothesis. The total energy-momentum of the electromagnetic field is given by

$$P_k = \int_{\mathbb{R}^3} \theta^4{}_k(\mathbf{x},0)\,d^3\mathbf{x}. \tag{6.3.22}$$

There is a simpler Lagrangian (which is not gauge-invariant) from which Maxwell's equations (6.3.14) can be derived. It is given by

$$\mathscr{L}_0(x, y_a, y_{ak}) = -(1/2)d^{ij}d^{kl}y_{ki}y_{lj}, \tag{6.3.23}$$

where we have

$$\frac{\partial \mathscr{L}_0(\ldots)}{\partial y_a} \equiv 0,$$

$$\frac{\partial \mathscr{L}_0(\ldots)}{\partial y_{ab}} = -(1/2)d^{ij}d^{kl}[\delta^a{}_k\delta^b{}_i y_{lj} + \delta^a{}_l\delta^b{}_j y_{ki}] = -y^{ab}.$$

Thus the Euler–Lagrange equations are

$$\left[\frac{\partial \mathscr{L}_0(\ldots)}{\partial y_i}\right]_{|\ldots} - \frac{d}{dx^j}\left\{\left[\frac{\partial \mathscr{L}_0(\ldots)}{\partial y_{ij}}\right]_{|\ldots}\right\} = 0 + [d^{jc}A^i{}_{,c}(x)]_{,j}$$

$$= \Box A^i(x)$$

$$= 0.$$

The Lorentz gauge condition has to be put as an *additional equation*. The canonical stress-energy-momentum tensor constructed from $\mathscr{L}_0(..)$ is

$$T_{(0)}{}^i{}_j(x) = \delta^i{}_j\mathscr{L}_0(..)_{|..} - \left[\frac{\partial\mathscr{L}_0(..)}{\partial y_{ai}}\right]_{|..} A_{a,j}$$
$$= d^{ki}A^a{}_{,k}A_{a,j} - (1/2)\delta^i{}_j d^{bc}A^a{}_{,c}A_{a,b}, \tag{6.3.24}$$

so that $T_{(0)ij}(x) = T_{(0)ji}(x)$. The total energy of the electromagnetic field according to (6.1.24) and (6.3.24) is

$$H = -P_4$$
$$= -\int_{\mathbb{R}^3} T_{(0)}{}^4{}_4(\mathbf{x}, 0)\, d^3\mathbf{x}$$
$$= \frac{1}{2}\int_{\mathbb{R}^3} [A^a{}_{,4}A_{a,4} + \delta^{\alpha\beta}A^a{}_{,\alpha}A_{a,\beta}]\, d^3\mathbf{x}. \tag{6.3.25}$$

Example: A class of general solutions of Maxwell's equations can be expressed as a Fourier integral [compare the plane-wave solution in (6.3.15)]:

$$A_b(x) = (2\pi)^{-3/2}\int_{\mathbb{R}^3} [2\nu(\mathbf{k})]^{-1/2}$$
$$\times [\alpha_b(\mathbf{k})e^{i[k_\mu x^\mu - \nu(\mathbf{k})x^4]} + \bar{\alpha}_b(\mathbf{k})e^{-i[k_\mu x^\mu - \nu(\mathbf{k})x^4]}]\, d^3\mathbf{k} \tag{6.3.26}$$

where $\nu(\mathbf{k}) \equiv [k_\mu k_\mu]^{1/2}$. Putting the above solution into (6.3.25) we obtain

$$H = -(32\pi^3)^{-1}\int_{\mathbb{R}^3}\int_{\mathbb{R}^3}\int_{\mathbb{R}^3} [\nu(\mathbf{k})\nu(\mathbf{k}')]^{-1/2}$$
$$\times \{[\nu(\mathbf{k})\nu(\mathbf{k}') + k_\nu k'_\nu][\alpha^b(\mathbf{k})e^{ik_\mu x^\mu} - \bar{\alpha}^b(\mathbf{k})e^{-ik_\mu x^\mu}]$$
$$\times [\alpha_b(\mathbf{k}')e^{ik'_\gamma x^\gamma} - \bar{\alpha}_b(\mathbf{k}')e^{-ik'_\gamma x^\gamma}]\}\, d^3\mathbf{x}\, d^3\mathbf{k}'\, d^3\mathbf{k}$$
$$= -(1/4)\int_{\mathbb{R}^3} [\nu(\mathbf{k})]^{-1}[[\nu(\mathbf{k})^2 - k_\nu k_\nu][\alpha^b(\mathbf{k})\alpha_b(-\mathbf{k}) + \bar{\alpha}^b(\mathbf{k})\bar{\alpha}_b(-\mathbf{k})]$$
$$-[\nu(\mathbf{k})^2 + k_\nu k_\nu][2\alpha^b(\mathbf{k})\bar{\alpha}_b(\mathbf{k})]]\, d^3\mathbf{k}$$
$$= \int_{\mathbb{R}^3} [\alpha^b(\mathbf{k})\bar{\alpha}_b(\mathbf{k})]\,\nu(\mathbf{k})\, d^3\mathbf{k}. \tag{6.3.27}$$

We can interpret $|\alpha^b(\mathbf{k})|^2$ physically as the electromagnetic field density (distribution) in the three-momentum space. Then from (6.3.27) we can conclude that the energy density of the electromagnetic field is the product of frequency with the sum of the field densities (in different modes of polarization). □

EXERCISES 6.3

1. Let the complex electromagnetic field be given by [see (6.3.7)]

$$\psi_{ab}(x) \equiv F_{ab}(x) + iF^*_{ab}(x).$$

Prove that the symmetrized energy-momentum-stress tensor in (6.3.21) can be expressed as

$$2\theta_{ab}(x) = \psi_a{}^k(x)\overline{\psi}_{bk}(x).$$

2. Consider the eigenvalue problem:

$$\det[\theta_{ab}(x) - \lambda(x)d_{ab}] = 0,$$

where $\theta_{ab}(x)$ is the symmetrized energy-momentum-stress tensor of the preceding problem. Prove that there exist two invariant eigenvalues

$$\lambda(x) = \pm(1/4)\{[F_{ab}(x)F^{ab}(x)]^2 + [F_{ab}(x)F^{*ab}(x)]^2\}^{1/2}.$$

3. Suppose that f is an arbitrary holomorphic function of two complex variables ω, ξ in a domain of $\mathbb{C}^2$. Let

$$\mathscr{W}(x) \equiv \mathrm{Re}[f(\omega, \xi)_{|\omega=x^1\cos k+x^2\sin k+ix^3, \xi=x^2+ix^3\sin k+x^4\cos k}],$$

where k is an arbitrary real parameter. Let $C_{ba} = -C_{ab}$ be an arbitrary constant Minkowski tensor field. Then show that

$$A_j(x) \equiv C_j{}^k\mathscr{W}_{,k}$$

solves Maxwell's equations $\Box A_j(x) = 0$, $A^j{}_{,j} = 0$.

4. Consider $T_{(0)}{}^i{}_j(x)$ as given in (6.3.24) and the class of solutions in (6.3.26). Prove that the components of the total three-momentum vector of the electromagnetic field is

$$P_\mu \equiv \int_{\mathbb{R}^3} T_{(0)}{}^4{}_\mu(\mathbf{x}, 0)\, d^3\mathbf{x} = \int_{\mathbb{R}^3} [\alpha^b(\mathbf{k})\bar{\alpha}_b(\mathbf{k})]k_\mu\, d^3\mathbf{k}.$$

6.4. Nonabelian Gauge Fields

Gauge fields are natural generalizations of the electromagnetic field, which is an example of an abelian gauge field. The theory of gauge fields starts with a Lie group G [see equation (3.4.8)]. Suppose that the group G involves r parameters $\alpha_1, \ldots, \alpha_r$. We can associate an r-dimensional *group manifold* with G. The group manifold is a differentiable manifold so that there exists a *tangent space* at every point of the manifold. The tangent space consists of the tangent vectors that are taken to be directional derivatives [see equation (2.1.6)]. There exists a basis $\{\tau_A(p)\colon A = 1,\ldots,r\}$ for the tangent space at a point p. The basis vectors $\tau_A(p)$ are called *generators* of the corresponding Lie *algebra*. The generators satisfy the commutation relations (assuming the summation convention for capital label indices)

$$\begin{aligned}[\tau_A(p), \tau_B(p)] &\equiv \tau_A(p)\tau_B(p) - \tau_B(p)\tau_A(p) \\ &= C^E{}_{AB}\tau_E(p), \end{aligned} \qquad (6.4.1)$$

where the constants $C^E{}_{AB}$ are called *structure constants*. The structure constants can be either real or complex numbers according to the choice of the original Lie group G. From equation (6.4.1) and *Jacobi's identities*

$$[[\tau_A(p), \tau_B(p)], \tau_C(p)] + [[\tau_B(p), \tau_C(p)], \tau_A(p)] + [[\tau_C(p), \tau_A(p)], \tau_B(p)] \equiv 0,$$

it follows that the structure constants must satisfy

$$C^E{}_{BA} = -C^E{}_{AB},$$
$$C^E{}_{AB}C^F{}_{ED} + C^E{}_{DA}C^F{}_{EB} + C^E{}_{BD}C^F{}_{EA} = 0. \tag{6.4.2}$$

It can be proved that for an abelian Lie group G all the structure constants vanish. For a nonabelian Lie group G, at least one of the structure constants is nonzero. We assume that G is nonabelian in the sequel.

Now we define

$$\gamma_{AB} \equiv C^E{}_{AD}C^D{}_{BE}. \tag{6.4.3}$$

It follows that

$$\gamma_{BA} = C^E{}_{BD}C^D{}_{AE} = C^D{}_{AE}C^E{}_{BD} = \gamma_{AB}.$$

For a *semisimple* (the invariant subgroups are nonabelian) Lie group it can be proved that $\det[\gamma_{AB}] \neq 0$. In such a case γ_{AB} can be treated as a *group metric* for which the contravariant tensor γ^{AB} exist to raise capital indices. We *assume* that the Lie group G is semisimple.

Example: Let us consider the semisimple Lie group SU(2), the group of unitary unimodular transformations of two-dimensional complex vectors (see Chapter 4, Sec. 1). This group has three generators τ_μ, which satisfy the commutation relations:

$$[\tau_1, \tau_2] = -\tau_3,$$
$$[\tau_2, \tau_3] = -\tau_1,$$
$$[\tau_3, \tau_1] = -\tau_2.$$

Thus the structure constants are given by

$$C^\mu{}_{\alpha\beta} = -\varepsilon_{\alpha\beta\mu},$$

where $\varepsilon_{\alpha\beta\mu}$ is the three-dimensional totally antisymmetric permutation symbol. The group metric components are given by

$$\gamma_{\rho\sigma} = C^\mu{}_{\rho\delta}C^\delta{}_{\sigma\mu} = \varepsilon_{\rho\delta\mu}\varepsilon_{\sigma\mu\delta} = -2\delta_{\rho\sigma}.$$

(The metric tensor $\gamma_{\rho\sigma}$ is negative definite since the group SU(2) is compact.) The two-dimensional self-representation of SU(2) induces a 2×2 complex matrix representation of the corresponding Lie algebra. This representation is given by the 2 × 2 complex matrix representation of the generators

$$[\tau_\mu] = (i/2)\sigma_\mu,$$

where σ_μ are Pauli matrices. The exponentiation of such a matrix representation is instructive. For example,

$$\begin{aligned} U_1 &\equiv \exp[-\phi\tau_3] = \exp[(-i\phi/2)\sigma_3] \\ &\equiv \sum_{n=0}^{\infty} (n!)^{-1}(-i\phi/2)^n {\sigma_3}^n \\ &= \cos(\phi/2)I - i\sin(\phi/2)\sigma^3 \\ &= \begin{bmatrix} e^{-i\phi/2} & 0 \\ 0 & e^{i\phi/2} \end{bmatrix}. \end{aligned}$$

This unitary unimodular matrix has been considered before in Chapter 4 [preceding equation (4.1.16)]. □

Next we introduce the *gauge potentials* $B^A{}_k(x)$. We assume that the functions $B^A{}_k$ are thrice differentiable in the domain $D \subset \mathbb{R}^4$ of consideration. Nonabelian *gauge field* are defined by

$$F^A{}_{ij}(x) \equiv B^A{}_{j,i} - B^A{}_{i,j} + C^A{}_{BC}B^B{}_iB^C{}_j = -F^A{}_{ji}(x), \tag{6.4.4}$$

where $C^A{}_{BC} \not\equiv 0$.

We note that

$$\begin{aligned} F^A{}_{ij,k} &+ F^A{}_{jk,i} + F^A{}_{ki,j} \\ &= C^A{}_{BC}[[B^B{}_iB^C{}_j]_{,k} + [B^B{}_jB^C{}_k]_{,i} + [B^B{}_kB^C{}_i]_{,j}] \\ &= -C^A{}_{BC}\{B^B{}_iF^C{}_{jk} + B^B{}_jF^C{}_{ki} + B^B{}_kF^C{}_{ij}\}. \end{aligned} \tag{6.4.5}$$

The Lagrangian $\mathscr{L}: D \times \mathbb{R}^{4r} \times \mathbb{R}^{16r} \to \mathbb{R}$ for the gauge field is defined to be

$$\begin{aligned} \mathscr{L}(x, y^A{}_k, y^A{}_{ij}) &\equiv (1/8)\gamma_{AB}d^{ik}d^{jl} \\ &\quad \times [[y^A{}_{ji} - y^A{}_{ij} + C^A{}_{CD}y^C{}_iy^D{}_j] \\ &\qquad \times [y^B{}_{lk} - y^B{}_{kl} + C^B{}_{EF}y^E{}_ky^F{}_l]], \end{aligned} \tag{6.4.6}$$

$$\mathscr{L}(x, y^A{}_k, y^A{}_{ij})|_{y^A{}_k = B^A{}_k(x),\, y^A{}_{ij} = B^A{}_{i,j}(x)} = (1/8)F^A{}_{ij}(x)F_A{}^{ij}(x).$$

From the definition it follows that

$$\begin{aligned} \frac{\partial\mathscr{L}(..)}{\partial y^G{}_s} &= (1/8)\gamma_{AB}d^{ik}d^{jl}[[C^A{}_{GD}\delta^s{}_iy^D{}_j + C^A{}_{CG}\delta^s{}_jy^C{}_i] \\ &\quad \times [y^B{}_{lk} - y^B{}_{kl} + C^B{}_{EF}y^E{}_ky^F{}_l] \\ &\quad + [y^A{}_{ji} - y^A{}_{ij} + C^A{}_{CD}y^C{}_iy^D{}_j] \\ &\quad \times [C^B{}_{GF}\delta^s{}_ky^F{}_l + C^B{}_{EG}\delta^s{}_ly^E{}_k]] \end{aligned}$$

$$= (1/4)\gamma_{AB}[y^A{}_{ji} - y^A{}_{ij} + C^A{}_{CD}y^C{}_i y^D{}_j][C^B{}_{GF}d^{is}y^{Fj} + C^B{}_{EG}d^{js}y^{Ei}]$$
$$= (1/4)C_{AGE}[y^A{}_{ji} - y^A{}_{ij} + C^A{}_{CD}y^C{}_i y^D{}_j][d^{is}y^{Ej} - d^{js}y^{Ei}]$$
$$= -(1/2)C_{AEG}[y^A{}_{ji} - y^A{}_{ij} + C^A{}_{CD}y^C{}_i y^D{}_j]d^{is}y^{Ej},$$

$$\frac{\partial \mathscr{L}(..)}{\partial y^G{}_{sb}} = (1/8)\gamma_{AB}d^{ik}d^{jl}\{\delta^A{}_G[\delta^s{}_j\delta^b{}_i - \delta^s{}_i\delta^b{}_j]$$
$$\times [y^B{}_{lk} - y^B{}_{kl} + C^B{}_{EF}y^E{}_k y^F{}_l]$$
$$+ [y^A{}_{ji} - y^A{}_{ij} + C^A{}_{CD}y^C{}_i y^D{}_j]\delta^B{}_G[\delta^s{}_l\delta^b{}_k - \delta^s{}_k\delta^b{}_l]\}$$
$$= (1/4)(d^{bk}d^{sl} - d^{sk}d^{bl})[y_{Glk} - y_{Gkl} + C_{GED}y^E{}_k y^D{}_l]$$
$$= (1/2)d^{bk}d^{sl}[y_{Glk} - y_{Gkl} + C_{GED}y^E{}_k y^D{}_l]. \tag{6.4.7}$$

Therefore, the corresponding Euler–Lagrange equations are

$$\left[\frac{\partial \mathscr{L}(..)}{\partial y^G{}_s}\right]_{|..} - \frac{d}{dx^b}\left\{\left[\frac{\partial \mathscr{L}(..)}{\partial y^G{}_{sb}}\right]_{|..}\right\}$$
$$= -(1/2)C_{AEG}F^{Asj}(x)B^E{}_j(x) - (1/2)F_G{}^{bs}{}_{,b} = 0,$$

which are equivalent to

$$F_G{}^{ab}{}_{,b} = C_{ADG}F^{Aaj}(x)B^D{}_j(x). \tag{6.4.8}$$

The canonical energy-momentum-stress tensor, as given in (6.1.17), can be computed from (6.4.6) as

$$T^i{}_j(x) = \delta^i{}_j\mathscr{L}(..)_{|..} - \left[\frac{\partial \mathscr{L}(..)}{\partial y^G{}_{si}}\right]_{|..} B^G{}_{s,j}$$
$$= (1/8)\delta^i{}_j F_{Acd}(x)F^{Acd}(x) + (1/2)F_A{}^{si}(x)B^A{}_{s,j}. \tag{6.4.9}$$

The canonical tensor $T_{ij}(x)$ is not symmetric. The symmetrized energy-momentum-stress tensor $\theta_{ij}(x)$ can be constructed by subtracting a "divergence term" from T_{ij}. More precisely,

$$\theta_{ij}(x) \equiv T_{ij}(x) - (1/2)[F_A{}^b{}_i(x)B^A{}_j(x)]_{,b}$$
$$= (1/2)[F_{Abi}(x)B^{Ab}{}_{,j} + (1/4)d_{ij}F_{Akl}(x)F^{Akl}(x)]$$
$$- (1/2)[F_A{}^b{}_i(x)B^A{}_{j,b} + F_A{}^b{}_{i,b}B^A{}_j(x)]$$
$$= (1/2)F_A{}^b{}_i(x)[B^A{}_{b,j} - B^A{}_{j,b}] - (1/2)C_{BDA}F^{Bb}{}_i(x)B^D{}_b(x)B^A{}_j(x)$$
$$+ (1/8)d_{ij}F_{Akl}(x)F^{Akl}(x)$$
$$= (1/2)F_A{}^b{}_i(x)[B^A{}_{b,j} - B^A{}_{j,b} + C^A{}_{ED}B^E{}_j(x)B^D{}_b(x)]$$
$$+ (1/8)d_{ij}F_{Akl}(x)F^{Akl}(x)$$
$$= -(1/2)F_A{}^b{}_i(x)F^A{}_{bj}(x) + (1/8)d_{ij}F_{Akl}(x)F^{Akl}(x)$$
$$= \theta_{ji}(x). \tag{6.4.10}$$

Note that $\theta_{ij}(x)$ is expressible algebraically in terms of the gauge fields rather than the gauge potentials.

Example: We shall obtain the Yasskin class of exact solutions for the gauge field equations (6.4.8). Let $A_i(x)$ be thrice differentiable components of the electromagnetic four-potential, which satisfy Maxwell's equations (6.3.14). We assume for the gauge potential the special form

$$B^A{}_i(x) = k^A A_i(x), \tag{6.4.11}$$

where the k^A are r real constants. Thus the gauge field components become

$$\begin{aligned} F^A{}_{ij} &= B^A{}_{j,i} - B^A{}_{i,j} + C^A{}_{EF} B^E{}_i B^F{}_j \\ &= k^A(A_{j,i} - A_{i,j}) + [C^A{}_{EF} k^E k^F][A_i A_j] \\ &= k^A(A_{j,i} - A_{i,j}) \\ &= k^A F_{ij}. \end{aligned}$$

Therefore, we obtain

$$F_G{}^{sb}{}_{,b} - C_{ADG} F^{Asb} B^D{}_b = k_G F^{sb}{}_{,b} - C_{ADG} k^A k^D F^{sb} A_b \equiv 0.$$

The last term vanishes identically due to the antisymmetry $C_{DAG} = -C_{ADG}$ (see problem 1 of Exercises 6.4). Thus by the *ansatz* (6.4.11), the gauge field equations (6.4.8) are exactly satisfied. □

The gauge Lie group G is assumed to be SU_3 for the strong interactions in quantum chromodynamics, whereas the gauge group is taken to be $SU_2 \times U_1$ for electroweak interactions. The interacting gauge field will be discussed in sec. 6.

EXERCISES 6.4

1. Let $C_{ABD} \equiv \gamma_{AE} C^E{}_{BD}$. Prove that $C_{BAD} = -C_{ABD}$.

2. Let the gauge potentials $B^A{}_i$ be twice differentiable functions in the domain $D \subset \mathbb{R}^4$ of consideration. Prove the differential identities

$$[C^A{}_{DE} B^D{}_j(x) F^{Eij}(x)]_{,i} \equiv 0.$$

3. Consider the symmetrized energy-momentum-stress tensor $\theta_{ij}(x)$ given in equation (6.4.10). Assuming that $B^A{}_i(x)$ are twice differentiable functions in $D \subset \mathbb{R}^4$, prove the differential identities $\theta^{ij}{}_{,j} \equiv 0$ in D by *explicit* computation.

6.5. The Dirac Bispinor Field

In this section we shall use many definitions and equations from Chapter 4. Thus it is advisable to quickly review its contents before proceeding. Let us consider a twice differentiable spinor field $\phi^A(x)$. A logically simple relativistic field equation for such a field is

$$d^{ij}\phi^A{}_{,ij} - m^2\phi^A(x) = \Box\phi^A(x) - m^2\phi^A(x) = 0, \tag{6.5.1}$$

where $m \geq 0$ is the mass parameter. It is known that a class of solutions of the (simple) second-order factorizable ordinary differential equations $y'' - m^2 y = (d/dx + m)(d/dx - m)y = 0$ is provided by the solutions of the first-order ordinary differential equation $y' = my$. Similarly the second-order partial differential equations (6.5.1) allow a class of twice differentiable solution functions ϕ^A which satisfy the system of *first-order* partial differential equations

$$\sigma^{a\bar{B}A}\phi_{\bar{B},a} = m\phi^A(x). \tag{6.5.2}$$

The reason for this assertion is (6.5.2) implies

$$\sigma^b{}_{\bar{C}A}\sigma^{a\bar{B}A}\phi_{\bar{B},ab} = m\sigma^b{}_{\bar{C}A}\phi^A{}_{,b} = m\sigma^b{}_{A\bar{C}}\phi^A{}_{,b}.$$

Thus

$$(1/2)[\sigma^b{}_{\bar{C}A}\sigma^{a\bar{B}A} + \sigma^a{}_{\bar{C}A}\sigma^{b\bar{B}A}]\phi_{\bar{B},ab} = -m\sigma^{bA\bar{C}}\phi_{A,b},$$

or equivalently we get

$$-d^{ab}\delta^{\bar{B}}{}_{\bar{C}}\phi_{\bar{B},ab} = -m^2\phi_{\bar{C}}(x).$$

Hence, we have

$$\Box\phi_{\bar{C}}(x) - m^2\phi_{\bar{C}}(x) = 0,$$

and finally we get

$$\Box\phi^A(x) - m^2\phi^A(x) = 0.$$

Here we have used equation (4.2.18) in the above computations. Note that for the zero rest mass case ($m = 0$), equation (6.5.2) reduces to $\sigma^{a\bar{B}A}\phi_{\bar{B},a} = 0$. This equation is known as the *Weyl equation* of a spinor field.

Equations (6.5.1) and (6.5.2) are mixed spinor-tensor equations and covariant under proper, orthochronous Lorentz transformations. But these equations are *not* covariant under improper Lorentz transformations. To obtain spinor equations that are covariant under general Lorentz transformations we have to consider *two* spinor fields χ_B, and $\phi^{\bar{A}}$. The coupled spinor field equations are taken to be

$$\begin{aligned} \sigma^{k\bar{A}B}\chi_{B,k} &= -m\phi^{\bar{A}}(x), \\ \sigma^k{}_{\bar{B}A}\phi^{\bar{B}}{}_{,k} &= m\chi_A(x). \end{aligned} \tag{6.5.3}$$

We shall now introduce bispinor fields [equation (4.3.2)] and 4 × 4 Dirac matrices [equation (4.3.6)] by

$$\psi(x) = [\psi^u(x)] = \begin{bmatrix} \psi^1(x) \\ \psi^2(x) \\ \psi^3(x) \\ \psi^4(x) \end{bmatrix} \equiv \begin{bmatrix} \chi_1(x) \\ \chi_2(x) \\ \phi^{\bar{1}}(x) \\ \phi^{\bar{2}}(x) \end{bmatrix},$$

$$\gamma^\alpha = [\gamma^{\alpha u}{}_v] \equiv \begin{bmatrix} 0 & \sigma_\alpha \\ \sigma_\alpha & 0 \end{bmatrix}, \tag{6.5.4}$$

$$\gamma^4 = [\gamma^{4u}{}_v] \equiv \begin{bmatrix} 0 & -\sigma^4 \\ \sigma^4 & 0 \end{bmatrix},$$

$$\gamma^k = [\gamma^{ku}{}_v] \equiv \begin{bmatrix} 0 & [\sigma_k{}^{\bar{A}B}] \\ [\sigma^k{}_{\bar{A}B}] & 0 \end{bmatrix},$$

where the bispinor indices u, v take values in $\{1, 2, 3, 4\}$. The coupled spinor field equations (6.5.3) can be cast into the bispinor field equation

$$\gamma^{ku}{}_v \psi^v{}_{,k} + m\psi^u(x) = 0,$$

or more abstractly

$$\gamma^k \psi_{,k} + m\psi(x) = \mathbf{0}. \tag{6.5.5a}$$

The above bispinor field equations for $m > 0$ is the well-known *Dirac equation*, which describes the wave field of a free spin (1/2) massive particle.

The hermitian conjugation of (6.5.5*a*) can be written as

$$\psi^\dagger_{,k}\gamma^{k\dagger} + m\psi(x)^\dagger = \mathbf{0}^\dagger. \tag{6.5.5b}$$

Thus

$$\psi^\dagger_{,\alpha}\gamma^\alpha - \psi^\dagger_{,4}\gamma^4 + m\psi(x)^\dagger = \mathbf{0}^\dagger.$$

Other ways of writing the above equations are

$$i\psi^\dagger_{,\alpha}\gamma^\alpha\gamma^4 - i\psi^\dagger_{,4}[\gamma^4]^2 + im\psi(x)^\dagger\gamma^4 = \mathbf{0}^\dagger,$$

$$\tilde{\psi}_{,k}\gamma^k - m\tilde{\psi}(x) = \mathbf{0}^\dagger, \tag{6.5.5c}$$

$$\tilde{\psi}(x) \equiv i\psi(x)^\dagger\gamma^4.$$

We shall now state and prove the general Lorentz covariance of the Dirac equation (6.5.5*a*).

Theorem (6.5.1): *The Dirac equation (6.5.5a) is covariant under a general Lorentz transformation $\hat{x}^i = a^i{}_j x^j$ and the induced unimodular bispinor transformation $\hat{\psi}(x) = \mathcal{T}\psi(x)$.*

Proof: Suppose that we start with the Dirac equation in the hatted Minkowski coordinates:

$$\gamma^b \frac{\partial\hat{\psi}(\hat{x})}{\partial\hat{x}^b} + m\hat{\psi}(\hat{x}) = \mathbf{0}.$$

By equations (4.3.3) and (4.3.12) we obtain

$$\gamma^b l^k{}_b \mathscr{T} \frac{\partial \psi(x)}{\partial x^k} + m\mathscr{T}\psi(x) = 0.$$

Thus

$$\mathscr{T}\gamma^k\mathscr{T}^{-1}\mathscr{T}\psi_{,k} + m\mathscr{T}\psi(x) = 0,$$

or

$$\mathscr{T}[\gamma^k\psi_{,k} + m\psi(x)] = 0.$$

Since the 4×4 unimodular matrix $\mathscr{T}$ is invertible, we can conclude

$$\gamma^k\psi_{,k} + m\psi(x) = 0.$$

Thus the covariance of the Dirac equation (6.5.5*a*) is proved. ■

The Dirac matrices, satisfying equations (4.3.8*a*) and (4.3.8*b*), are determined only up to a similarity transformation. There exist infinitely many choices of Dirac matrices, and each choice is related to another by a similarity transformation. A popular choice of Dirac matrices (in signature +2) is the following:

$$\begin{aligned} \gamma'^\mu &= \gamma^\mu, \\ \gamma'^4 &= -i\begin{bmatrix} I & O \\ O & -I \end{bmatrix}, \\ \gamma'^5 &\equiv \gamma'^1\gamma'^2\gamma'^3\gamma'^4 = \begin{bmatrix} O & -I \\ I & O \end{bmatrix}, \\ S\gamma^k S^{-1} &= \gamma'^k, \end{aligned} \tag{6.5.6}$$

where $\sqrt{2}S \equiv \begin{bmatrix} iI & I \\ I & iI \end{bmatrix}$.

Writing $\psi'(x) \equiv S\psi(x)$ and multiplying (6.5.5*a*) from the left by S, we can obtain an equivalent form of the Dirac equation as

$$\gamma'^k(x)\psi'_{,k} + m\psi'(x) = 0. \tag{6.5.7}$$

In the sequel we shall *drop the prime* so that the Dirac equation will be formally given by (6.5.7), but the Dirac matrices are given by (6.5.6).

Example: We shall derive the plane-wave solutions of the Dirac equation (6.5.7). For that purpose we shall seek a solution in the form

$$\psi(x) = \xi(p)e^{ip_k x^k},$$

where $\xi(p)$ is the unknown bispinor column vector. The above form implies, by the Dirac equation, the matrix equation

$$[ip_k\gamma^k + mI]\xi(p) = 0. \tag{6.5.8}$$

The above equations yield a system of homogeneous linear equations on the unknown components $\xi^{\mu}(p)$. The necessary condition for the existence of a nontrivial solution is

$$\det[ip_k\gamma^k + mI] = 0. \tag{6.5.9}$$

Thus reducing (6.5.9) to $(d^{kl}p_k p_l + m^2)^2 = 0$, we find that $p_4 = \pm E(\mathbf{p})$, where $E(\mathbf{p}) \equiv [(p_1)^2 + (p_2)^2 + (p_3)^2 + m^2]^{1/2}$. So we have to investigate the following two classes of solutions:

$$\begin{aligned}\psi(x) &= \xi[\mathbf{p}, -E(\mathbf{p})]e^{i[p_\mu x^\mu - E(p)x^4]}\\ &\equiv \mathbf{U}(\mathbf{p})e^{i[p_\mu x^\mu - E(p)x^4]};\end{aligned} \tag{6.5.10a}$$

$$\begin{aligned}\psi(x) &= \xi[-\mathbf{p}, E(\mathbf{p})]e^{-i[p_\mu x^\mu - E(p)x^4]}\\ &\equiv \mathbf{V}(\mathbf{p})e^{-i[p_\mu x^\mu - E(p)x^4]}.\end{aligned} \tag{6.5.10b}$$

These two classes represent solutions of positive and negative energy states respectively. Putting (6.5.10*a*) into the Dirac equation we obtain

$$[ip_\mu\gamma^\mu - iE(\mathbf{p})\gamma^4 + mI]\mathbf{U}(\mathbf{p}) = \mathbf{0}.$$

Defining $\mathbf{U}(\mathbf{p}) \equiv \begin{bmatrix}\chi(\mathbf{p})\\ \phi(\mathbf{p})\end{bmatrix}$, where $\chi(\mathbf{p})$ and $\phi(\mathbf{p})$ are column vectors of size 2×1, the last matrix equation yields the following two matrix equations:

$$[m - E(\mathbf{p})]\chi(\mathbf{p}) + ip_\mu\sigma^\mu\phi(\mathbf{p}) = \mathbf{0}, \tag{6.5.11a}$$

$$ip_\mu\sigma^\mu\chi(\mathbf{p}) + [m + E(\mathbf{p})]\phi(\mathbf{p}) = \mathbf{0}. \tag{6.5.11b}$$

By multiplying with the matrix $i[m + E(\mathbf{p})]^{-1}p_\nu\sigma^\nu$ on the left of (6.5.11*b*) and using (4.1.12) and (6.5.9) we derive (6.5.11*a*). Thus the matrix equations (6.5.11*a*) and (6.5.11*b*) yield a linearly dependent system and we have to solve either (6.5.11*a*) or (6.5.11*b*). Let us solve (6.5.11*b*), which can also be expressed as

$$[I, i(m + E)^{-1}p_\mu\sigma^\mu]\begin{bmatrix}\phi(\mathbf{p})\\ \chi(\mathbf{p})\end{bmatrix} = \mathbf{0}.$$

It is clear that the 2×4 coefficient matrix $[I, i(m + E)^{-1}p_\mu\sigma^\mu]$ is of rank 2. It follows that the original coefficient matrix $[ip_k\gamma^k + mI]$ of the Dirac equation (6.5.8) is of rank 2 for $p_4 = \pm E(\mathbf{p})$, where $\mathbf{p} \in \mathbb{R}^3$ is an arbitrary point. The general solution of (6.5.11*b*) can be obtained by putting

$$\phi(\mathbf{p}) = -i(m + E)^{-1}p_\mu\sigma^\mu\chi(\mathbf{p}),$$

$$\begin{aligned}\mathbf{U}(\mathbf{p}) &= \begin{bmatrix}\chi(\mathbf{p})\\ \phi(\mathbf{p})\end{bmatrix} = \begin{bmatrix}\chi(\mathbf{p})\\ -i(m + E)^{-1}p_\mu\sigma^\mu\chi(\mathbf{p})\end{bmatrix}\\ &= \chi_1(\mathbf{p})\begin{bmatrix}1\\ 0\\ -i(m + E)^{-1}p_3\\ -i(m + E)^{-1}(p_1 + ip_2)\end{bmatrix} + \chi_2(\mathbf{p})\begin{bmatrix}0\\ 1\\ -i(m + E)^{-1}(p_1 - ip_2)\\ -i(m + E)^{-1}p_3\end{bmatrix},\end{aligned}$$

where χ_1, χ_2 are arbitrary complex-valued functions of $\mathbf{p}$. Therefore, a basis for this two-dimensional solution space of positive energy is $\{\mathbf{u}_1(\mathbf{p}), \mathbf{u}_2(\mathbf{p})\}$ where

$$\mathbf{u}_1(\mathbf{p}) = [(m+E)/2m]^{1/2} \begin{bmatrix} 1 \\ 0 \\ -i(m+E)^{-1}p_3 \\ -i(m+E)^{-1}(p_1+ip_2) \end{bmatrix},$$
$$\mathbf{u}_2(\mathbf{p}) = [(m+E)/2m]^{1/2} \begin{bmatrix} 0 \\ 1 \\ -i(m+E)^{-1}(p_1-ip_2) \\ i(m+E)^{-1}p_3 \end{bmatrix}. \tag{6.5.12a}$$

Here we have chosen a suitable normalization. We follow exactly analogous steps for the negative energy $[p^4 = -p_4 = -E(\mathbf{p})]$ solutions to derive the other two basis vectors of the solution space as

$$\mathbf{v}_1(\mathbf{p}) = [(m+E)/2m]^{1/2} \begin{bmatrix} i(m+E)^{-1}p_3 \\ i(m+E)^{-1}(p_1+ip_2) \\ 1 \\ 0 \end{bmatrix},$$
$$\mathbf{v}_2(\mathbf{p}) = [(m+E)/2m]^{1/2} \begin{bmatrix} i(m+E)^{-1}(p_1-ip_2) \\ -i(m+E)^{-1}p_3 \\ 0 \\ 1 \end{bmatrix}. \tag{6.5.12b}$$

By direct computations it can be verified that the above basis vectors satisfy formal "orthogonality" conditions:

$$\begin{aligned} \mathbf{u}_r(\mathbf{p})^\dagger \mathbf{u}_s(\mathbf{p}) &= (1/m)E(\mathbf{p})\delta_{rs}, \\ \mathbf{v}_r(\mathbf{p})^\dagger \mathbf{v}_s(\mathbf{p}) &= (1/m)E(\mathbf{p})\delta_{rs}, \\ \mathbf{u}_r(\mathbf{p})^\dagger \mathbf{v}_s(-\mathbf{p}) &= 0, \\ \mathbf{v}_r(\mathbf{p})^\dagger \mathbf{u}_s(-\mathbf{p}) &= 0. \end{aligned} \tag{6.5.13}$$

We shall now state and prove a theorem regarding the superposition of plane-waves.

Theorem (6.5.2): *Let a Fourier integral be given by*

$$\psi(x) = (2\pi)^{-3/2} \int_{\mathbb{R}^3} \left\{ [m/E(\mathbf{p})]^{1/2} \sum_{r=1}^{2} [\alpha_r(\mathbf{p})\mathbf{u}_r(\mathbf{p}) e^{i[p_\mu x^\mu - E(\mathbf{p})x^4]} + \bar{\beta}_r(\mathbf{p})\mathbf{v}_r(\mathbf{p}) e^{-i[p_\mu x^\mu - E(\mathbf{p})x^4]}] \right\} d^3\mathbf{p}, \tag{6.5.14}$$

such that the above improper integral converges uniformly for otherwise arbitrary choices of the function α_r *and* β_r. *Furthermore, assume that the following improper integrals also converge uniformly*:

$$\int_{\mathbb{R}^3}\left\{[m/E(\mathbf{p})]^{1/2}p_\nu \sum_{r=1}^{2} [\alpha_r(\mathbf{p})\mathbf{u}_r(\mathbf{p})e^{i[p_\mu x^\mu - E(\mathbf{p})x^4]} - \bar{\beta}_r(\mathbf{p})\mathbf{v}_r(\mathbf{p})e^{-i[p_\mu x^\mu - E(\mathbf{p})x^4]}]\right\}d^3\mathbf{p},$$

$$\int_{\mathbb{R}^3}\left\{[mE(\mathbf{p})]^{1/2} \sum_{r=1}^{2} [\alpha_r(\mathbf{p})\mathbf{u}_r(\mathbf{p})e^{i[p_\mu x^\mu - E(\mathbf{p})x^4]} - \bar{\beta}_r(\mathbf{p})\mathbf{v}_r(\mathbf{p})e^{-i[p_\mu x^\mu - E(\mathbf{p})x^4]}]\right\}d^3\mathbf{p}.$$

Then equation (6.5.14) *provides a general class of exact solutions of the Dirac equation* (6.5.5*a*).

Proof: Under the assumption of uniform convergence of the above improper integrals we are allowed to differentiate under the integral sign in (6.5.14). Thus using (6.5.11*a*), (6.5.12*b*), and (6.5.8) we obtain

$$\begin{aligned}
&\gamma^k\psi_{,k} + m\psi(x) \\
&\quad = (2\pi)^{-3/2}\int_{\mathbb{R}^3}\left\{[m/E(\mathbf{p})]^{1/2}\right. \\
&\quad \times \sum_{r=1}^{2} [\alpha_r(\mathbf{p})[ip_\nu\gamma^\nu - iE(\mathbf{p})\gamma^4 + mI]\mathbf{u}_r(\mathbf{p})e^{i[p_\mu x^\mu - E(\mathbf{p})x^4]} \\
&\quad + \beta_r(\mathbf{p})[-ip_\nu\gamma^\nu + iE(\mathbf{p})\gamma^4 + mI]\mathbf{v}_r(\mathbf{p})e^{-i[p_\mu x^\mu - E(\mathbf{p})x^4]}\left.\right\}d^3\mathbf{p} = 0. \quad \blacksquare
\end{aligned}$$

Now we derive the Dirac equation by a variational principle. We introduce the notation:

$$\begin{aligned}
\tilde{\zeta} &= [\tilde{\zeta}_1, \tilde{\zeta}_2, \tilde{\zeta}_3, \tilde{\zeta}_4] \\
&\equiv [\bar{\zeta}^1, \bar{\zeta}^2, \bar{\zeta}^3, \bar{\zeta}^4]\begin{bmatrix} 1 & 0 & 0 & 0 \\ 0 & 1 & 0 & 0 \\ 0 & 0 & -1 & 0 \\ 0 & 0 & 0 & -1 \end{bmatrix} \\
&= i\zeta^\dagger\gamma^4 \\
&= [\bar{\zeta}^1, \bar{\zeta}^2, -\bar{\zeta}^3, -\bar{\zeta}^4].
\end{aligned} \tag{6.5.15}$$

Here $\tilde{\zeta}_u = s(u)\bar{\zeta}^u$, $s(1) = s(2) = -s(3) = -s(4) = 1$. The Lagrangian function is defined by

$$\begin{aligned}
\mathscr{L}(\zeta^v, \tilde{\zeta}_u, \zeta^v_a, \tilde{\zeta}_{ua}) &\equiv (1/2)[\tilde{\zeta}_u\gamma^{au}{}_v\zeta^v{}_a - \tilde{\zeta}_{ua}\gamma^{au}{}_v\zeta^v] + m\tilde{\zeta}_u\zeta^u \\
&= (1/2)\tilde{\zeta}\gamma^a\zeta_a - (1/2)\tilde{\zeta}_a\gamma^a\zeta + m\tilde{\zeta}\zeta, \\
\mathscr{L}(\ldots)|_{\ldots} &= (1/2)\tilde{\psi}(x)\gamma^a\psi_{,a} - (1/2)\tilde{\psi}_{,a}\gamma^a\psi(x) + m\tilde{\psi}(x)\psi(x).
\end{aligned} \tag{6.5.16}$$

Note that $\mathscr{L}$ is a real-valued function from $\mathbb{C}^4 \times \mathbb{C}^4 \times \mathbb{C}^{16} \times \mathbb{C}^{16}$. It follows that

$$\frac{\partial \mathscr{L}(..)}{\partial \zeta^v} = -(1/2)\tilde{\zeta}_{ua}\gamma^{au}{}_v + m\tilde{\zeta}_v,$$

$$\frac{\partial \mathscr{L}(..)}{\partial \zeta^v{}_a} = (1/2)\tilde{\zeta}_u\gamma^{au}{}_v.$$

Thus the Euler–Lagrange equations are

$$\begin{aligned} 0 &= \left[\frac{\partial \mathscr{L}(..)}{\partial \zeta_v}\right]_{|..} - \frac{d}{dx^a}\left\{\left[\frac{\partial \mathscr{L}(..)}{\partial \zeta_{va}}\right]_{|..}\right\} \\ &= -(1/2)\tilde{\psi}_{u,a}\gamma^{au}{}_v + m\tilde{\psi}_v(x) - (1/2)\tilde{\psi}_{u,a}\gamma^{au}{}_v \\ &= -[\tilde{\psi}_{u,a}\gamma^{au}{}_v - m\tilde{\psi}_v(x)]. \end{aligned} \tag{6.5.17}$$

Thus we have

$$\tilde{\psi}_{,a}\gamma^a - m\tilde{\psi}(x) = \mathbf{0}^\dagger,$$

or equivalently

$$\gamma^a\psi_{,a} + m\psi(x) = \mathbf{0}.$$

Example 1: The canonical energy-momentum-stress tensor is computed from (6.5.16) and (6.1.17) as

$$\begin{aligned} T^k{}_j(x) &= \delta^k{}_j\mathscr{L}(..)_{|..} - \left[\frac{\partial \mathscr{L}(..)}{\partial \zeta^v{}_k}\right]_{|..}\psi^v_{,j} - \tilde{\psi}_{u,j}\left[\frac{\partial \mathscr{L}(..)}{\partial \tilde{\zeta}_{uk}}\right]_{|..} \\ &= (1/2)[\tilde{\psi}_{u,j}\gamma^{ku}{}_v\psi^v(x) - \tilde{\psi}_u(x)\gamma^{ku}{}_v\psi^v{}_{,j}] \\ &= (1/2)[\tilde{\psi}_{,j}(x)\gamma^k\psi(x) - \tilde{\psi}(x)\gamma^k\psi_{,j}]. \end{aligned} \tag{6.5.18}$$

The symmetrized energy-momentum stress tensor is

$$\begin{aligned} \theta_{kj}(x) &= T_{kj}(x) + (1/4)\frac{\partial}{\partial x^a}\{\tilde{\psi}(x)[\gamma^a\gamma_j\gamma_k - d_{jk}\gamma^a - \delta^a{}_j\gamma_k + \delta^a{}_k\gamma_j]\psi(x)\} \\ &= (1/4)[\tilde{\psi}_{,j}\gamma_k\psi(x) + \tilde{\psi}_{,k}\gamma_j\psi(x) - \tilde{\psi}(x)\gamma_k\psi_{,j} - \tilde{\psi}(x)\gamma_j\psi_{,k}] \\ &= (1/2)[T_{kj}(x) + T_{jk}(x)]. \quad \square \end{aligned} \tag{6.5.19}$$

Example 2: We can compute the charge-current vector density $j^k(x)$ by equations (6.5.16) and (6.1.2):

$$\begin{aligned} j^k(x) &= -ie\left\{\left[\frac{\partial \mathscr{L}(..)}{\partial \zeta^v{}_k}\right]_{|..}\psi^v(x) - \tilde{\psi}_u(x)\left[\frac{\partial \mathscr{L}_0(..)}{\partial \tilde{\zeta}_{uk}}\right]_{|..}\right\} \\ &= -ie\tilde{\psi}_u(x)\gamma^{ku}{}_v\psi^v(x) \\ &= -ie\tilde{\psi}(x)\gamma^k\psi(x). \quad \square \end{aligned} \tag{6.5.20}$$

Example 3: We shall compute the total electric charge of the Dirac field $\psi(x)$ using (6.1.24) and the class of solutions given in (6.5.14) and the equation (6.5.20).

$$\begin{aligned}
\mathcal{Q} &= \int_{\mathbb{R}^3} j^4(\mathbf{x},0)\,d^3\mathbf{x} = -ie\int_{\mathbb{R}^3} \tilde{\psi}(\mathbf{x},0)\gamma^4\psi(\mathbf{x},0)\,d^3\mathbf{x} \\
&= -(iem/8\pi^3)\sum_{r=1}^{2}\sum_{s=1}^{2}\int_{\mathbb{R}^3}\int_{\mathbb{R}^3}\int_{\mathbb{R}^3}\{-i[E(\mathbf{p})E(\mathbf{p}')]^{-1/2} \\
&\qquad \times [\bar{\alpha}_r(\mathbf{p})\mathbf{u}_r(\mathbf{p})^\dagger e^{-ip_\mu x^\mu} + \beta_r(\mathbf{p})\mathbf{v}_r(\mathbf{p})e^{ip_\mu x^\mu}] \\
&\qquad \times [\alpha_s(\mathbf{p}')u_s(\mathbf{p}')e^{ip'_\nu x^\nu} \\
&\qquad + \bar{\beta}_s(\mathbf{p}')\mathbf{v}_s(\mathbf{p}')^\dagger e^{-ip'_\nu x^\nu}]\}\,d^3\mathbf{x}\,d^3\mathbf{p}'\,d^3\mathbf{p} \\
&= -(em)\sum_{r=1}^{2}\sum_{s=1}^{2}\int_{\mathbb{R}^3}\{[1/E(\mathbf{p})] \\
&\qquad \times [\bar{\alpha}_r(\mathbf{p})\alpha_s(\mathbf{p})u_r(\mathbf{p})^\dagger \mathbf{u}_s(\mathbf{p}) + \beta_r(\mathbf{p})\bar{\beta}_s(\mathbf{p})\mathbf{v}_r(\mathbf{p})^\dagger \mathbf{v}_s(\mathbf{p}) \\
&\qquad + \bar{\alpha}_r(\mathbf{p})\bar{\beta}_s(-\mathbf{p})\mathbf{u}_r(\mathbf{p})^\dagger \mathbf{v}_s(-\mathbf{p}) \\
&\qquad + \beta_r(\mathbf{p})\alpha_s(-\mathbf{p})\mathbf{v}_r(\mathbf{p})^\dagger \mathbf{u}_s(-\mathbf{p})]\}\,d^3\mathbf{p}.
\end{aligned}$$

Using equations (6.5.13) we finally obtain

$$\mathcal{Q} = -e\sum_{r=1}^{2}\int_{\mathbb{R}^3}\{|\alpha_r(\mathbf{p})|^2 + |\beta_r(\mathbf{p})|^2\}\,d^3\mathbf{p}. \tag{6.5.21}$$

The total charge $\mathcal{Q}$ is negative for $e > 0$ and $\alpha_r(\mathbf{p}) \not\equiv 0$, $\beta_r(\mathbf{p}) \not\equiv 0$. This is physically unacceptable since $\psi(x)$ should describe electrons as well as oppositely charged positrons. The situation can be remedied by the process of the *second quantization* whereby $\mathcal{Q}$ can be of both signs. □

EXERCISES 6.5

1. (i) Construct a Lagrangian function $\mathscr{L}$ such that the corresponding Euler–Lagrange equations are

$$\sigma^{k\bar{A}B}\chi_{B,k} = -m\phi^{\bar{A}}(x),$$

$$\sigma^{k}{}_{\bar{B}A}\phi^{\bar{B}}{}_{,k} = m\chi_A(x).$$

(ii) From the above Lagrangian derive the canonical energy-momentum-stress tensor $T^k{}_j(x)$ and the charge-current vector $j^k(x)$.

2. Consider the basis vectors of the solution space of plane-waves in equations (6.5.12*a*) and (6.5.12*b*).

(i) Prove the "orthonormality" conditions

$$\tilde{\mathbf{u}}_r(\mathbf{p})\mathbf{u}_s(\mathbf{p}) = -\tilde{\mathbf{v}}_r(\mathbf{p})\mathbf{v}_s(\mathbf{p}) = \delta_{rs},$$

$$\tilde{\mathbf{u}}_r(\mathbf{p})\mathbf{v}_s(\mathbf{p}) = \tilde{\mathbf{v}}_r(\mathbf{p})\mathbf{u}_s(\mathbf{p}) = 0.$$

(ii) Energy-projection matrices are defined by

$$\Lambda_{\pm}(\mathbf{p}) \equiv (1/2m)[\mp i(p_\mu \gamma^\mu - E(\mathbf{p})\gamma^4) + mI].$$

Show that

$$\Lambda_+(\mathbf{p})\mathbf{u}_r(\mathbf{p}) = \mathbf{u}_r(\mathbf{p}),$$

$$\Lambda_+(\mathbf{p})\mathbf{v}_r(\mathbf{p}) = \mathbf{0},$$

$$\Lambda_-(\mathbf{p})\mathbf{u}_r(\mathbf{p}) = \mathbf{0},$$

$$\Lambda_-(\mathbf{p})\mathbf{v}_r(\mathbf{p}) = \mathbf{v}_r(\mathbf{p}).$$

3. Using the class of solutions in equation (6.5.14), and the equations (6.5.18), (6.1.24), show that the total energy and momentum components of the Dirac field are

$$H = -P_4 = \sum_{r=1}^{2} \int_{\mathbb{R}^3} \{|\alpha_r(\mathbf{p})|^2 - |\beta_r(\mathbf{p})|^2\} E(\mathbf{p})\, d^3\mathbf{p},$$

$$P_\mu = \sum_{r=1}^{2} \int_{\mathbb{R}^3} \{|\alpha_r(\mathbf{p})|^2 - |\beta_r(\mathbf{p})|^2\} p_\mu\, d^3\mathbf{p}.$$

6.6. Interaction of the Dirac Field with Gauge Fields

The electromagnetic field is an abelian gauge field, and it is the simplest case to consider. The Lagrangian for the Dirac field interacting with an electromagnetic field is taken to be [see equations (6.3.18) and (6.5.16)]

$$\begin{aligned}
&\mathscr{L}(\zeta^v, \tilde{\zeta}_u, \zeta^v{}_a, \tilde{\zeta}_{ua}, y_a, y_{ak}) \\
&\quad\equiv (1/2)[\tilde{\zeta}_u \gamma^{au}{}_v(\zeta^v{}_a + iey_a\zeta^v) - (\tilde{\zeta}_{ua} - iey_a\tilde{\zeta}_u)\gamma^{au}{}_v\zeta^v] + m\tilde{\zeta}_u\zeta^u \\
&\qquad - (1/4)d^{ab}d^{kl}(y_{lb} - y_{bl})(y_{ka} - y_{ak}) \\
&\quad= (1/2)[\tilde{\zeta}\gamma^a(\zeta_a + iey_a\zeta) - (\tilde{\zeta}_a - iey_a\zeta)\gamma^a\zeta] \\
&\qquad + m\zeta\tilde{\zeta} - (1/4)d^{ab}d^{kl}(y_{lb} - y_{bl})(y_{ka} - y_{ak}).
\end{aligned} \tag{6.6.1}$$

Thus we may write

$$\begin{aligned}
\mathscr{L}(..)|_{..} &= (1/2)\{\tilde{\psi}(x)\gamma^a[\psi_{,a} + ieA_a(x)\psi(x)] - [\tilde{\psi}_{,a} - ieA_a(x)\psi(x)]\gamma^a\psi(x)\} \\
&\quad + m\tilde{\psi}(x)\psi(x) - (1/4)F^{ab}(x)F_{ab}(x).
\end{aligned}$$

Note that $\mathscr{L}(..)$ is a real-valued relativistic invariant. Furthermore, $\mathscr{L}(..)$ is invariant under the combined local gauge transformations:

$$\begin{aligned}
\psi'(x) &= \psi(x)e^{ie\lambda(x)}, \\
A'_k(x) &= A_k(x) - \lambda_{,k}.
\end{aligned} \tag{6.6.2}$$

From (6.6.1) it follows that

$$\frac{\partial \mathcal{L}(..)}{\partial \zeta^v} = (ie/2)\tilde{\zeta}_u \gamma^{au}{}_v y_a - (1/2)(\tilde{\zeta}_{ua} - iey_a\tilde{\zeta}_u)\gamma^{au}{}_v + m\tilde{\zeta}_v,$$

$$= [-(1/2)\tilde{\zeta}_{ua} + iey_a\tilde{\zeta}_u]\gamma^{au}{}_v + m\tilde{\zeta}_v,$$

$$\frac{\partial \mathcal{L}(..)}{\partial \zeta^v_a} = (1/2)\tilde{\zeta}_u\gamma^{au}{}_v,$$

$$\frac{\partial \mathcal{L}(..)}{\partial y_a} = (ie)\tilde{\zeta}_u\gamma^{au}{}_v\zeta^v, \tag{6.6.3}$$

$$\frac{\partial \mathcal{L}(..)}{\partial y_{ab}} = (y^{ba} - y^{ab}).$$

Thus the Euler–Lagrange equations are

$$0 = \left[\frac{\partial \mathcal{L}(..)}{\partial \zeta^v}\right]_{|..} - \frac{d}{dx^a}\left\{\left[\frac{\partial \mathcal{L}(..)}{\partial \zeta^v{}_a}\right]_{|..}\right\}$$

$$= [-(1/2)\tilde{\psi}_{u,a} + ie\tilde{\psi}_u(x)A_a(x)]\gamma^{au}{}_v + m\tilde{\psi}_v(x) - (1/2)\tilde{\psi}_{u,a}\gamma^{au}{}_v$$

$$= -[\tilde{\psi}_{u,a} - ieA_a(x)\tilde{\psi}_u(x)]\gamma^{au}{}_v + m\tilde{\psi}_v(x), \tag{6.6.4a}$$

and

$$0 = \left[\frac{\partial \mathcal{L}(..)}{\partial y_a}\right]_{|..} - \frac{d}{dx^b}\left\{\left[\frac{\partial \mathcal{L}(..)}{\partial y_{ab}}\right]_{|..}\right\}$$

$$= ie\tilde{\psi}_u(x)\gamma^{au}{}_v\psi^v(x) - F^{ab}{}_{,b}. \tag{6.6.4b}$$

These equations may be written in matrix form as

$$[\tilde{\psi}_{,a} - ieA_a(x)\tilde{\psi}(x)]\gamma^a - m\tilde{\psi}(x) = \mathbf{0}^\dagger,$$

or

$$\gamma^a[\psi_{,a} + ieA_a(x)\psi(x)] + m\psi(x) = \mathbf{0},$$

and

$$F^{ab}{}_{,b} = ie\tilde{\psi}(x)\gamma^a\psi(x).$$

If we define the *gauge-covariant derivative* as

$$D_a\psi(x) \equiv \psi_{,a} + ieA_a(x)\psi(x), \tag{6.6.5}$$

then we can write the combined Maxwell-Dirac equations (6.6.4*a*) and (6.6.4*b*) as

$$\gamma^a D_a\psi(x) + m\psi(x) = \mathbf{0},$$

$$\bar{D}_a\tilde{\psi}(x)\gamma^a - m\tilde{\psi}(x) = \mathbf{0}^\dagger, \tag{6.6.6}$$

$$F^{ab}{}_{,b} = ie\tilde{\psi}(x)\gamma^a\psi(x).$$

Example 1: Suppose that the Dirac wave functions ψ^u are twice differentiable. In that case from (6.6.4*a*) and (4.3.8*a*) we obtain

$$
\begin{aligned}
0 &= [\gamma^b(\partial_b + ieA_b(x)) - m\mathbf{I}][\gamma^a(\psi_{,a} + ieA_a(x)\psi(x)) + m\psi(x)] \\
&= \gamma^b\gamma^a\{\psi_{,ab} - e^2A_a(x)A_b(x)\psi(x) + ie[A_{a,b}\psi(x) + A_a(x)\psi_{,b} + A_b(x)\psi_{,a}]\} \\
&\quad - m^2\psi(x) \\
&= [d^{ab}\mathbf{I} + (1/2)(\gamma^b\gamma^a - \gamma^a\gamma^b)] \\
&\quad \times [\psi_{,ab} - e^2A_a(x)A_b(x)\psi(x) + (ie/2)(A_{a,b} + A_{b,a})\psi(x) \\
&\qquad + (ie/2)F_{ab}(x)\psi(x) + ie(A_a(x)\psi_{,b} + A_b(x)\psi_{,a})] - m^2\psi(x) \\
&= D^aD_a\psi(x) + [(ie/4)(\gamma^a\gamma^b - \gamma^b\gamma^a)F_{ab}(x) - m^2\mathbf{I}]\psi(x).
\end{aligned} \tag{6.6.7}
$$

The term $(ie/4)(\gamma^a\gamma^b - \gamma^b\gamma^a)F_{ab}(x)$ represents the magnetic moment of the Dirac particle (an electron or μ-meson) and the corresponding gyromagnetic ratio 2 agrees well with experimental results. □

Example 2: Let us consider the coupled Maxwell-Dirac field equations (6.6.4*a*), (6.6.4*b*) for the zero rest mass case ($m = 0$). We pose the question whether or not the spin (1/2) particle can acquire a nonzero mass due to its own electromagnetic self-interaction. We will try to obtain such a solution with the following assumptions:

$$
\begin{aligned}
&A_1(x) = A_2(x) \equiv 0, \\
&A_3(x) = a(x^1, x^2), \\
&A_4(x) = W(x^1, x^2), \\
&A^b{}_{,b} \equiv 0, \qquad \psi(x) = \begin{bmatrix} \xi(x^4) \\ 0 \\ -i\xi(x^4) \\ 0 \end{bmatrix}, \qquad \xi(x^4) \neq 0.
\end{aligned} \tag{6.6.8}
$$

The Dirac equation (6.6.4*a*) for the case $m = 0$ reduces to *one* independent equation:

$$
e[a(x^1, x^2) + W(x^1, x^2)] = i\frac{d}{dx^4}\log[\xi(x^4)].
$$

By separation of variables we obtain

$$
\begin{aligned}
a(x^1, x^2) + W(x^1, x^2) &= M/e, \\
\xi(x^4) &= \alpha e^{-iMx^4},
\end{aligned} \tag{6.6.9}
$$

where $M \in \mathbb{R}$ and $\alpha \in \mathbb{C}$ are arbitrary constants. The electromagnetic equations (6.6.4*b*) reduce to two independent equations:

$$
\begin{aligned}
a_{,11} + a_{,22} &= -2e|\alpha|^2, \\
W_{,11} + W_{,22} &= 2e|\alpha|^2.
\end{aligned}
$$

The general solutions of the above equations, consistent with (6.6.9), are

$$
\begin{aligned}
a(x^1, x^2) &= -(e/2)|\alpha|^2[(x^1)^2 + (x^2)^2] - h(x^1, x^2), \\
W(x^1, x^2) &= (e/2)|\alpha|^2[(x^1)^2 + (x^2)^2] + h(x^1, x^2) + (M/e),
\end{aligned}
\tag{6.6.10}
$$

where $h(x^1, x^2)$ is an arbitrary real harmonic functon. Equations (6.6.8), (6.6.9), and (6.6.10) constitute a special class of local solutions of (6.6.4*a*) and (6.6.4*b*) with $m = 0$. By applying a gauge transformation, the solution can be simplified into the case where $M = 0$. □

We shall now formulate the Lagrangian for a Dirac field in interaction with a general nonabelian gauge field. Recall that there exist a gauge group G and the generators τ_A of the corresponding Lie algebra that satisfy the commutation relations (6.4.1). An n-dimensional representation [see equation (3.3.1)] of the group G induces $n \times n$ matrix representations $[\tau_A{}^K{}_J]$ of the generators. Here the indices $A, B, \ldots, F$ takes values in $\{1, \ldots, r\}$ and indices J, K take values in $\{1, \ldots, n\}$. We will apply the summation convention to both types of indices. The matrices

$$
[T_A{}^K{}_J] \equiv -i[\tau_A{}^K{}_J], \tag{6.6.11}
$$

can be chosen to be hermitian whenever $G = SU(N)$ for some positive integer N.

The Lagrangian for the interacting fields is defined by

$$
\begin{aligned}
&\mathscr{L}(\zeta^{Kv}, \tilde{\zeta}_{Ku}, \zeta^{Kv}{}_a, \tilde{\zeta}_{Kua}, y^A{}_j, y^A{}_{lk}) \\
&\quad \equiv \tilde{\zeta}_{Ku}\gamma^{au}{}_v[\zeta^{Kv}{}_a + igT_A{}^K{}_J y^A{}_a \zeta^{Jv}] \\
&\qquad + (1/8)\gamma_{AB} d^{bk} d^{jl}[y^A{}_{jb} - y^A{}_{bj} + gC^A{}_{CD} y^C{}_b y^D{}_j] \\
&\qquad \times [y^B{}_{lk} - y^B{}_{kl} + gC^B{}_{EF} y^E{}_k y^F{}_l],
\end{aligned}
\tag{6.6.12}
$$

$$
\tilde{\zeta}_K \equiv i(\zeta^K)^\dagger \gamma^4,
$$

$$
F^A{}_{kl}(x) \equiv [y^A{}_{lk} - y^A{}_{kl} + gC^A{}_{EF} y^E{}_k y^F{}_l]|_{y^A{}_l = B^A{}_l(x),\, y = B^A{}_{l,k}},
$$

$$
\mathscr{L}(\ldots)|_{\ldots} = \tilde{\psi}_K(x)\gamma^a[\psi^K{}_{,a} + igT_A{}^K{}_J B^A{}_a(x)\psi^J(x)] + (1/8)\gamma_{AB} F^A{}_{kl}(x) F^{Bkl}(x),
$$

where $g \neq 0$ is the coupling constant (or charge) of the gauge field interaction.

It follows from (6.6.12) that

$$
\frac{\partial \mathscr{L}(\ldots)}{\partial \zeta^{Jv}} = ig\tilde{\zeta}_{Ku}\gamma^{au}{}_v T_A{}^K{}_J y^A{}_a,
$$

$$
\frac{\partial \mathscr{L}(\ldots)}{\partial \zeta^{Jv}{}_a} = \tilde{\zeta}_{Ju}\gamma^{au}{}_v,
$$

$$
\begin{aligned}
\frac{\partial \mathscr{L}(\ldots)}{\partial \zeta^A{}_s} &= ig\tilde{\zeta}_{Ku}\gamma^{su}{}_v T_A{}^K{}_J \zeta^{Jv} \\
&\quad - (g/2)C_{EDA} d^{sb}[y^E{}_{jb} - y^E{}_{bj} + gC^E{}_{CF} y^C{}_b y^F{}_j] y^{Dj},
\end{aligned}
$$

$$
\frac{\partial \mathscr{L}(\ldots)}{\partial y^A{}_{sr}} = -(1/2)d^{sa}d^{rb}[y_{Aba} - y_{Aab} + gC_{ADE} y^D{}_a y^E{}_b].
$$

Thus the Euler–Lagrange equations are

$$0 = \left[\frac{\partial \mathscr{L}(..)}{\partial \zeta^{Jv}}\right]_{|..} - \frac{d}{dx^a}\left\{\left[\frac{\partial \mathscr{L}(..)}{\partial \zeta^{Jv}{}_a}\right]_{|..}\right\}$$

$$= ig\tilde{\psi}_{Ku}(x)\gamma^{au}{}_v T_A{}^K{}_J B^A{}_a(x) - \tilde{\psi}_{Ju,a}\gamma^{au}{}_v.$$

Hence we have

$$[\tilde{\psi}_{J,a} - ig\tilde{\psi}_K(x)T_A{}^K{}_J B^A{}_a(x)]\gamma^a = 0^\dagger,$$

or equivalently

$$\gamma^a[\psi^K{}_{,a} + igT_A{}^K{}_J B^A{}_a(x)\psi^J(x)] = \mathbf{0}. \tag{6.6.13a}$$

Furthermore, the Euler–Lagrange equations for the gauge fields are

$$0 = \left[\frac{\partial \mathscr{L}(..)}{\partial y^A{}_s}\right]_{|..} - \frac{d}{dx^r}\left\{\left[\frac{\partial \mathscr{L}(..)}{\partial y^A{}_{sr}}\right]_{|..}\right\}$$

$$= ig\tilde{\psi}_{Ku}(x)\gamma^{su}{}_v T_A{}^K{}_J\psi^{Jv}(x) - (g/2)C_{EDA}F^{Es}{}_j(x)B^{Dj}(x) + (1/2)F_A{}^{sr}{}_{,r}.$$

The above equations are equivalent to

$$F_A{}^{ar}{}_{,r} = gC_{EDA}F^{Ea}{}_j(x)B^{Dj}(x) - 2ig\tilde{\psi}_K(u)(x)\gamma^a T_A{}^K{}_J\psi^J(x). \tag{6.6.13b}$$

The Lagrangian in (6.6.12) is not only invariant under Poincaré transformations, it is also invariant under some gauge transformations. To investigate the local gauge invariance, we define the following *local gauge transformations*:

$$\exp[M] \equiv \sum_{n=0}^{\infty} [M]^n/n!,$$

$$\mathbf{U}(x) \equiv \exp[i\theta^A(x)T_A],$$

$$\psi^v(x) \equiv \begin{bmatrix} \psi_1^v(x) \\ \vdots \\ \psi_n^v(x) \end{bmatrix}, \tag{6.6.14a}$$

$$\tilde{\psi}^u(x) \equiv s(u)\psi^u(x)^\dagger.$$

$$\psi^v(x) = \mathbf{U}(x)^{-1}\psi'^v(x),$$

$$\tilde{\psi}_u(x) = \tilde{\psi}'_u(x)\mathbf{U}(x),$$

$$T_A B'^A{}_a(x) = \mathbf{U}(x)T_A\mathbf{U}(x)^{-1}B^A{}_a(x) - (i/g)\mathbf{U}(x)[\mathbf{U}(x)^{-1}]_{,a}, \tag{6.6.14b}$$

where we have $\mathbf{U}(x)^\dagger = \exp[-i\theta^A(x)T_A] = \mathbf{U}(x)^{-1}$ and $s(1) = s(2) = -s(3) = -s(4) = 1$.

The transformation of gauge field components is described in the following lemma.

Lemma (6.6.1): *Under the local gauge transformation* (6.6.14*b*) *the gauge field components transform by the rule*

$$T_A F'^A{}_{kl}(x) = \mathbf{U}(x) T_A \mathbf{U}(x)^{-1} F^A{}_{kl}(x). \qquad (6.6.14c)$$

Proof: From equation (6.6.14*b*) it follows that

$$\begin{aligned} T_A B'^A{}_{k,l} &= \mathbf{U}(x) T_A \mathbf{U}(x)^{-1} B^A{}_{k,l} + \mathbf{U}_{,l} T_A \mathbf{U}(x)^{-1} B^A{}_k(x) \\ &\quad + \mathbf{U}(x) T_A [\mathbf{U}(x)^{-1}]_{,l} B^A{}_k(x) \\ &\quad - (i/g)\{\mathbf{U}_{,l}[\mathbf{U}(x)^{-1}]_{,k} + \mathbf{U}(x)[\mathbf{U}(x)^{-1}]_{,kl}\}. \end{aligned}$$

From the definition of the gauge field components in (6.6.12) we obtain

$$\begin{aligned} T_A F'^A{}_{kl} &= T_A B'^A{}_{l,k} - T_A B'^A{}_{k,l} + g C^A{}_{EF} T_A B'^E{}_k(x) B'^F{}_l(x) \\ &= T_A B'^A{}_{l,k} - T_A B'^A{}_{k,l} + ig[T_E B'^E{}_k(x), T_F B'^F{}_l(x)] \\ &= \mathbf{U}(x) T_A \mathbf{U}(x)^{-1} F^A{}_{kl}(x) \\ &\quad + \{\mathbf{U}_{,k} T_A \mathbf{U}(x)^{-1} B^A{}_l(x) - \mathbf{U}_{,l} T_A \mathbf{U}(x)^{-1} B^A{}_k(x) + \mathbf{U}(x)[\mathbf{U}(x)^{-1}]_{,k} \\ &\qquad \times \mathbf{U}(x) T_A \mathbf{U}(x)^{-1} B^A{}_l(x) - \mathbf{U}(x)[\mathbf{U}(x)^{-1}]_{,l} \mathbf{U}(x) T_A \mathbf{U}(x)^{-1} B^A{}_k(x)\} \\ &\quad + (i/g)\{\mathbf{U}_{,l}[\mathbf{U}(x)^{-1}]_{,k} - \mathbf{U}_{,k}[\mathbf{U}(x)^{-1}]_{,l} \\ &\qquad - \mathbf{U}(x)\mathbf{U}^{-1}{}_{,k}\mathbf{U}(x)[\mathbf{U}(x)^{-1}]_{,l} \\ &\qquad + \mathbf{U}(x)\mathbf{U}^{-1}{}_{,l}\mathbf{U}(x)[\mathbf{U}(x)^{-1}]_{,k}\}. \end{aligned}$$

Using $[\mathbf{U}(x)\mathbf{U}(x)^{-1}]_{,a} = \mathbf{0}$, so that $[\mathbf{U}(x)^{-1}]_{,a} = -\mathbf{U}(x)^{-1}\mathbf{U}_{,a}\mathbf{U}(x)^{-1}$, we see that both of the terms enclosed in braces vanish and equation (6.6.14*c*) holds. ■

Theorem (6.6.1): *The Lagrangian in* (6.6.12) *is invariant under the gauge transformations in* (6.6.14*a*) *and* (6.6.14*b*).

Proof: We use equations (6.6.14*a*)–(6.6.14*c*) and assume that $\gamma_{AB} = -c\,\mathrm{Trace}[T_A T_B]$ where c is a suitable real constant. Then the Lagrangian can be expressed as

$$\begin{aligned} \mathcal{L}(..)|_{..} &= \tilde{\psi}_u(x)\gamma^{au}{}_v[\psi^v{}_{,a} + igT_A B^A{}_a(x)\psi^v(x)] + (1/8)\gamma_{AB} F^A{}_{kl}(x) F^{Bkl}(x) \\ &= \tilde{\psi}'_u(x)\mathbf{U}(x)\gamma^{au}{}_v[\mathbf{U}(x)^{-1}\psi'^v{}_{,a} + (\mathbf{U}(x)^{-1})_{,a}\psi'^v(x) \\ &\qquad + igT_A B^A{}_a(x)\mathbf{U}(x)^{-1}\psi'^v(x)] \\ &\quad - (c/8)\,\mathrm{Trace}[T_A T_B] F^A{}_{kl}(x) F^{Bkl}(x) \\ &= \tilde{\psi}'_u(x)\gamma^{au}{}_v[\psi'^v{}_{,a} + ig(\mathbf{U}(x) T_A \mathbf{U}(x)^{-1} B^A{}_a(x) \\ &\qquad - (i/g)\mathbf{U}(x)(\mathbf{U}(x)^{-1})_{,a})\psi'^v(x)] \\ &\quad - (c/8)\,\mathrm{Trace}\{[\mathbf{U}(x)^{-1} T_A F'^A{}_{kl}(x)\mathbf{U}(x)\mathbf{U}(x)^{-1} T_B F'^{Bkl}(x)\mathbf{U}(x)]\} \\ &= \tilde{\psi}'_u(x)\gamma^{au}{}_v[\psi'^v{}_{,a} + igT_A B'^A{}_a(x)] - (c/8)\,\mathrm{Trace}[T_A T_B] F'^A{}_{kl}(x) F'^{Bkl}(x) \\ &= \tilde{\psi}'_u(x)\gamma^{au}{}_v[\psi'^v{}_{,a} + igT_A B'^A{}_a(x)] + (1/8)\gamma_{AB} F'^A{}_{kl}(x) F'^{Bkl}(x). \quad ■ \end{aligned}$$

We can define a *gauge-covariant derivative* by

$$D_a\psi^K(x) \equiv \psi^K{}_{,a} + igT_A{}^K{}_J B^A{}_a(x)\psi^J(x). \tag{6.6.15}$$

With the above definition, the Lagrangian and the Euler–Lagrange equations can be written as

$$\begin{aligned}
&\mathscr{L}(\ldots)|_{\ldots} = \tilde{\psi}_K(x)\gamma^a D_a\psi^K(x) + (1/8)\gamma_{AB}F^A{}_{kl}(x)F^{Bkl}(x),\\
&\gamma^a D_a\psi^K(x) = 0, \qquad\qquad (6.6.16)\\
&F_A{}^{ab}{}_{,b} = gC_{EDA}F^{Ea}{}_j(x)B^{Dj}(x) - 2ig\tilde{\psi}_K(x)\gamma^a T_A{}^K{}_J\psi^J(x).
\end{aligned}$$

Now we shall discuss a special example of the interacting gauge field and the Dirac field.

Example: Let us choose the semisimple gauge group $G = \mathrm{SU}(2)$. It is the group of unimodular unitary group of transformations of the complex space $\mathscr{V}_2$. The group has three generators $\tau_\mu = iT_\mu$ such that $[T_\alpha, T_\beta] = i\varepsilon_{\alpha\beta\gamma}T_\gamma$. The group metric is $\gamma_{\rho\sigma} = -2\delta_{\rho\sigma}$. The 2×2 complex matrix representation of T_μ is given by $T_\mu = (1/2)\sigma_\mu$, where σ_μ is a Pauli matrix. The Lagrangian (6.6.12) is

$$\begin{aligned}
\mathscr{L}(\ldots)|_{\ldots} &= \tilde{\psi}_K(x)\gamma^a[\psi^K{}_{,a} + (ig/2)\sigma_\alpha{}^K{}_J B^\alpha{}_a(x)\psi^J(x)]\\
&\quad - (1/4)\delta_{\alpha\beta}F^\alpha{}_{kl}(x)F^{\beta kl}(x)\\
&= \tilde{\psi}_u(x)\gamma^{au}{}_v[\psi^v{}_{,a} + (ig/2)\sigma_\alpha B^\alpha{}_a(x)\psi^v(x)]\\
&\quad - (1/4)\delta_{\alpha\beta}F^\alpha{}_{kl}(x)F^{\beta kl}(x).
\end{aligned}$$

The field equations (6.6.13*a*) and (6.6.13*b*) yield in this case

$$\begin{aligned}
&\gamma^a[\psi^K{}_{,a} + (ig/2)\sigma_\alpha{}^K{}_J B^\alpha{}_a(x)\psi^J(x)] = 0,\\
&F_\alpha{}^{ab}{}_{,b} = -g\varepsilon_{\mu\nu\alpha}F^{\mu a}{}_j(x)B^{\nu j}(x) - ig\tilde{\psi}_K(x)\gamma^a\sigma_\alpha{}^K{}_J\psi^J(x). \quad \square
\end{aligned}$$

In the *electroweak* field theory the gauge group is taken to be $G = \mathrm{SU}(2) \times \mathrm{U}(1)$. Furthermore in *quantum chromodynamics* (the theory of the strong interaction of quarks and gluons) the gauge group is chosen to be SU(3).

EXERCISES 6.6

1. Consider the Lagrangian function given in equation (6.6.1). Prove that

$$\text{(i)} \quad \frac{\partial\mathscr{L}(\ldots)}{\partial y_{ak}} + \frac{\partial\mathscr{L}(\ldots)}{\partial y_{ka}} \equiv 0,$$

$$\text{(ii)} \quad \frac{\partial\mathscr{L}(\ldots)}{\partial y_a} - ie\left[\frac{\partial\mathscr{L}(\ldots)}{\partial\zeta^v{}_a}\zeta^v - \frac{\partial\mathscr{L}_0(\ldots)}{\partial\tilde{\zeta}_{ua}}\tilde{\zeta}_u\right] \equiv 0,$$

$$\text{(iii)} \quad \frac{\partial\mathscr{L}(\ldots)}{\partial\zeta^v{}_a}\zeta^v - \frac{\partial\mathscr{L}(\ldots)}{\partial\tilde{\zeta}_u}\tilde{\zeta}_u + \frac{\partial\mathscr{L}(\ldots)}{\partial\zeta^v{}_a}\zeta^v{}_a - \frac{\partial\mathscr{L}(\ldots)}{\partial\tilde{\zeta}_{ua}}\tilde{\zeta}_{ua} \equiv 0.$$

2. Obtain the symmetrized energy-momentum-stress tensor $\theta_{ij}(x)$ from the Lagrangian (6.6.12) such that

$$\theta_{ij}(x) - T_{ij}(x) = \Sigma_{ij}{}^{a}{}_{,a},$$

where $\Sigma_{ij}{}^{a}(x)$ is a $\mathscr{C}^2$-tensor field satisfying

$$\Sigma^{ija}{}_{,ai}(x) \equiv 0.$$

3. Using Euler–Lagrange equations (6.6.13*a*), (6.6.13*b*), and appropriate conditions on differentiability, prove the differential identity

$$[C_{EDA}F^{Ea}{}_{j}(x)B^{Dj}(x) - 2i\tilde{\psi}_K(x)\gamma^a T_A{}^K{}_J\psi^J(x)]_{,a} \equiv 0.$$

References

1. Y. Choquet-Bruhat, C. Dewitt-Morette, and M. Dillard-Bleick, *Analysis, manifolds, and physics*, North-Holland Publ. Co., Amsterdam, 1977. [pp. 154, 168, 174]
2. E. M. Corson, *Introduction to tensors, spinors, and relativistic wave-equations*, Blackie and Son Ltd., London, 1955. [pp. 72, 176, 179, 213]
3. K. Huang, *Quarks, leptons, and gauge fields*, World Scientific Publ. Co., Singapore, 1982. [pp. 204, 227, 231]
4. D. Lovelock and H. Rund, *Tensors, differential forms, and variational principles*, John Wiley and Sons, New York, 1975. [p. 176]
5. Y. Takahashi, *An introduction of field quantization*, Pergamon Press Ltd., Toronto, 1969. [pp. 183, 186, 195, 213, 217]
6. P. B. Yasskin, Phys. Rev. *D* **12** (1975), 2212–2217.

7 The Extended (or Covariant) Phase Space and Classical Fields*

7.1. Classical Fields

In Hamiltonian mechanics and the subsequent quantum (particle) mechanics the basic equations remain covariant under canonical transformations (see Chapter 5). A particular example of canonical transformation is given by

$$\hat{q}^a = -p^a, \quad \hat{p}^a = q^a. \tag{7.1.1}$$

Born and Landé called this transformation the *reciprocity* transformation. The usual quantum field theories do not allow covariance under reciprocity. Yukawa and his students incorporated the idea of reciprocity in their formulations of nonlocal fields. In such formulations, fields had to be defined on the eight-dimensional extended phase space (or covariant phase space) M_8 (see Sect. 5.4).

The present author constructed field theories in the four-dimensional complex space–time in order to quantize space–time and at the same time to utilize the larger group of covariance to interpret internal groups.

However, if complex coordinates are interpreted as $z^a = q^a + ip^a$ according to Bargman's representation of quantum mechanics, then fields are automatically defined over the extended phase space (or the covariant phase space) M_8. In this space a subgroup of transformations can be defined as

$$\begin{aligned} \hat{q}^a &= (\cos\phi)q^a - (\sin\phi)p^a, \\ \hat{p}^a &= (\sin\phi)q^a + (\cos\phi)p^a, \end{aligned} \tag{7.1.2}$$

where $\phi \in (-\pi, \pi)$ is the group parameter. For the special value $\phi = \pi/2$, (7.1.2) reduces to (7.1.1). This transformation is called a *generalized reciprocity*. It is a subgroup of canonical transformations (problem 3 of Exercises 5.4) as well as that of the 36-parameter extended Poincaré group. This subgroup naturally leads to a flat extended phase space M_8 with the "line element"

$$(d\ell)^2 = d_{ij}(dq^i dq^j + dp^i dp^j). \tag{7.1.3}$$

Thus M_8 is of signature + 4.

---

* This chapter is a review of a research topic pursued by several scientists.

The special theory of relativity presupposes that the maximum speed of propagation of action is the speed of light. Similarly, the metric in (7.1.3) implies a "maximal proper acceleration." This point of view has been explored by Caianiello and others.

Notation will be elaborated now. The physical units are so chosen that $a = b = c = 1$, where a, b are the fundamental length and momentum respectively ($ab = \hbar$) and c is the speed of light. All physical quantities are expressed as (dimensionless) numbers. A point in the extended phase space is denoted by

$$\begin{aligned}\xi &\equiv (\xi^1, \xi^2, \xi^3, \xi^4, \xi^5, \xi^6, \xi^7, \xi^8)\\ &\equiv (q^1, q^2, q^3, q^4, p^1, p^2, p^3, p^4) \equiv (q, p).\end{aligned}$$

Greek indices take values from $\{1, 2, 3\}$, lower case Roman indices take from $\{1, 2, 3, 4\}$, and capital Roman indices take from $\{1, 2, 3, 4, 5, 6, 7, 8\}$. (*The last convention differs from those of the preceding chapters.*) In all cases summation convention is followed. Furthermore, in expressions like $a^i_j A^{q^j}$, $q^j\phi_{,p^j}$, and $\xi^j T^4_{4+j}$, the index j is automatically summed from 1 to 4. Partial derivatives are denoted by

$$\phi^{\cdot\cdot}_{,A} \equiv \frac{\partial \phi^{\cdot\cdot}(\xi)}{\partial \xi^A}, \qquad \phi^{\cdot\cdot}_{,q^j} \equiv \phi^{\cdot\cdot}_{,j} \equiv \frac{\partial \phi^{\cdot\cdot}(q,p)}{\partial q^j},$$

$$\phi^{\cdot\cdot}_{,p^j} \equiv \phi^{\cdot\cdot}_{,j+4} \equiv \frac{\partial \phi^{\cdot\cdot}(q,p)}{\partial p^j}.$$

In the eight-dimensional notation, the "line element" (7.1.3) can be expressed as

$$\begin{aligned}(d\ell)^2 &= D_{AB}\, d\xi^A d\xi^B,\\ [D_{AB}] &\equiv \left[\begin{array}{c|c} d_{ij} & 0 \\ \hline 0 & d_{ij}\end{array}\right].\end{aligned} \tag{7.1.4}$$

The nonhomogeneous linear transformations that leave the above line element invariant is given by

$$\begin{aligned}\hat{\xi}^A &= \ell^A + \ell^A_B \xi^B,\\ [\ell^A] &\equiv [b^i, d^i], \qquad [\ell^A_B] \equiv \left[\begin{array}{c|c} a^i_j & b^i_j \\ \hline c^i_j & e^i_j\end{array}\right],\\ \hat{q}^i &= b^i + a^i{}_j q^j + b^i{}_j p^j,\\ \hat{p}^i &= d^i + c^i{}_j q^j + e^i{}_j p^j,\\ D_{AB}\ell^A_C \ell^B_C &= D_{CD},\\ d_{ij}(a^i{}_k a^j{}_l + c^i{}_k c^j{}_l) &= d_{ij}(b^i{}_k b^j{}_l + e^i{}_k e^j{}_l) = d_{kl},\\ d_{ij}(a^i{}_k b^j{}_l + c^i{}_k e^j{}_l) &= 0.\end{aligned} \tag{7.1.5}$$

The above equations yield the 36-parameter extended Poincaré group. In case $d^i \equiv 0$, $b^i{}_j = c^i{}_j \equiv 0$, $a^i{}_j \equiv e^i{}_j$, the usual ten-parameter Poincaré group is recovered. Furthermore, in case $b^i = d^i \equiv 0$, $a^i{}_j = e^i{}_j = (\cos\phi)\delta^i{}_j$, $c^i{}_j = -b^i{}_j = (\sin\phi)\delta^i{}_j$, the generalized reciprocity (7.1.2) is obtained.

Now the tensor transformation rules under an extended Poincaré transformation will be discussed [compare the equation (2.4.11)].

$$\begin{gathered}\hat{T}^{A_1\cdots A_r}{}_{B_1\cdots B_s}(\hat{\xi}) = \ell^{A_1}{}_{C_1}\cdots\ell^{A_r}{}_{C_r}a^{D_1}{}_{B_1}\cdots a^{D_s}{}_{B_s}T^{C_1\cdots C_r}{}_{D_1\cdots D_s}(\xi),\\ [a^A{}_B] \equiv [\ell^A{}_B]^{-1}.\end{gathered} \tag{7.1.6}$$

Two simple examples will be dealt with. Consider a scalar field $\phi(\xi)$. The transformation is given by simply

$$\begin{gathered}\hat{\phi}(\hat{\xi}) = \phi(\xi),\\ \hat{\phi}(\hat{q},\hat{p}) = \phi(q,p).\end{gathered} \tag{7.1.7a}$$

Now consider a vector field $A^B(\xi)$. The transformation rule is the following equations:

$$\begin{aligned}&\hat{A}^B(\hat{\xi}) = \ell^B{}_C A^C(\xi),\\ &A^{q^i}(q,p) \equiv A^i(\xi),\\ &A^{p^i}(q,p) \equiv A^{i+4}(\xi),\\ &\hat{A}^{\hat{q}^i}(\hat{q},\hat{p}) = a^i{}_j A^{q^j}(q,p) + b^i{}_j A^{p^j}(q,p),\\ &\hat{A}^{p^i}(\hat{q},\hat{p}) = c^i{}_j A^{q^j}(q,p) + e^i{}_j A^{p^j}(q,p).\end{aligned} \tag{7.1.7b}$$

Note that the summation convention is operating on the index j above.

The contravariant metric tensor components are given by

$$[D^{AB}] \equiv [D_{AB}]^{-1} = [D_{AB}]. \tag{7.1.8}$$

The raising and lowering of indices can be performed as follows:

$$\begin{aligned}T_A(\xi) &\equiv D_{AB}T^B(\xi), & T^A(\xi) &= D^{AB}T_B(\xi)\\ T_{q^i}(q,p) &\equiv d_{ij}T^{q^j}(q,p), & T^{q^i}(q,p) &= d^{ij}T_{q^j}(q,p),\\ T_{p^i}(q,p) &\equiv d_{ij}T^{p^j}(q,p), & T^{p^i}(q,p) &= d^{ij}T_{p^j}(q,p).\end{aligned} \tag{7.1.9}$$

Expressions like

$$\begin{aligned}T_A(\xi)T^A(\xi) &= d_{ij}[T^{q^i}(q,p)T^{q^j}(q,p) + T^{p^i}(q,p)T^{p^j}(q,p)],\\ U_{AB}(\xi)V^{AB}(\xi) &= U_{q^iq^j}(q,p)V^{q^iq^j}(q,p) + U_{q^ip^j}(q,p)V^{q^ip^j}(q,p)\\ &\quad + U_{p^iq^j}(q,p)V^{p^iq^j}(q,p) + U_{p^ip^j}(q,p)V^{p^ip^j}(q,p)\end{aligned} \tag{7.1.10}$$

are invariant under the extended Poincaré group.

Now Sect. 6.1 will be generalized for the extended phase space M_8. The invariant action integral for a complex-valued tensor field

$\phi^{A\cdots} \in C^2(D \subset M_8; \mathbb{C})$ can be defined as

$$\mathscr{A}(\phi^{\cdot\cdot}) \equiv \int_{\bar{D}} [\mathscr{L}(\xi, \zeta^{\cdot\cdot}, \bar{\zeta}^{\cdot\cdot}; \zeta^{\cdot\cdot}_A, \bar{\zeta}^{\cdot\cdot}_A)]_{|\zeta^{\cdot\cdot}=\phi^{\cdot\cdot}(\xi),\ldots,\bar{\zeta}^{\cdot\cdot}_A=\bar{\phi}^{\cdot\cdot}_{,A}} d^8\xi, \tag{7.1.11}$$

where D is a bounded, simply connected domain in M_8 with a piecewise smooth, orientable, closed boundary ∂D.

The variational principle

$$\delta\mathscr{A}(\phi^{\cdot\cdot}) = 0, \qquad \delta\phi^{\cdot\cdot}{}_{|\partial D} = 0$$

yields the Euler–Lagrange equation [compare the equation (6.1.7)]

$$\frac{\partial\mathscr{L}(..)}{\partial\zeta^{\cdot\cdot}}_{|..} - \frac{d}{d\xi^A}\left\{\left[\frac{\partial\mathscr{L}(..)}{\partial\zeta^{\cdot\cdot}_A}\right]_{|..}\right\} = 0. \tag{7.1.12}$$

The infinitesimal version of the extended Poincaré transformation (7.1.5) can be summed up as

$$\begin{aligned} \delta\xi^A &\equiv \hat{\xi}^A - \xi^A = \varepsilon^A + \varepsilon^A{}_B\xi^B, \\ \varepsilon_{BA} &= -\varepsilon_{AB} + r_{AB}(\varepsilon_{CD}), \end{aligned} \tag{7.1.13}$$

where $|\varepsilon^A|$, $|\varepsilon^A{}_B|$ are arbitrarily small positive numbers. Variations and the Lie variations of a complex tensor field $\phi^{A\cdot\cdot}(\xi)$ are furnished by

$$\begin{aligned} \delta\phi^{A\cdot\cdot} &\equiv \hat{\phi}^{A\cdot\cdot}(\hat{\xi}) - \phi^{A\cdot\cdot}(\xi) \\ &= (1/2)\varepsilon^{CD}S_{CD}{}^{A\cdot\cdot}{}_{B\cdot\cdot}\phi^{B\cdot\cdot}(\xi) + r^{A\cdot\cdot}(\varepsilon^{\cdot\cdot}), \\ \delta_L\phi^{A\cdot\cdot} &\equiv \hat{\phi}^{A\cdot\cdot}(\xi) - \phi^{A\cdot\cdot}(\xi) \\ &= -\varepsilon^B\phi^{A\cdot\cdot}{}_{,B} + (1/2)\varepsilon^{CD}[(\xi_C\phi^{A\cdot\cdot}{}_{,D} - \xi_D\phi^{A\cdot\cdot}{}_{,C}) \\ &\quad + S_{CD}{}^{A\cdot\cdot}{}_{B\cdot\cdot}\phi^{B\cdot\cdot}(\xi)] + \hat{r}^{A\cdot\cdot}(\varepsilon^{\cdot\cdot}), \end{aligned} \tag{7.1.14}$$

where $S_{CD}{}^{A\cdot\cdot}{}_{B\cdot\cdot}$ is a numerical tensor representing the "extended spin" and remainder terms $r^{A\cdot\cdot}$, $\hat{r}^{A\cdot\cdot}$ are of order $O(\varepsilon^2)$.

The invariance of the action integral with respect to transformations (7.1.13) leads to Noether's theorem (compare Theorem 6.1.2):

$$\frac{d}{d\xi^A}\left\{[\mathscr{L}(..)]_{|..}\delta\xi^A + \left[\frac{\partial\mathscr{L}(..)}{\partial\zeta^{\cdot\cdot}{}_A}\right]_{|..}\delta_L\phi^{\cdot\cdot} + \left[\frac{\partial\mathscr{L}(..)}{\partial\bar{\zeta}^{\cdot\cdot}_A}\right]_{|..}\delta_L\bar{\phi}^{\cdot\cdot}\right\} = 0. \tag{7.1.15}$$

By considering different possible cases for $\delta_L\phi^{\cdot\cdot}$, one obtains various differential conservation laws. For example, in the case $\varepsilon^A \neq 0$, $\varepsilon_{AB} \equiv 0$, using (7.1.13) and (7.1.14), the equation (7.1.15) yields

$$\begin{aligned} &T^A{}_{B,A} = 0, \\ &T^A{}_B(\xi) \equiv \delta^A{}_B[\mathscr{L}(..)]_{|..} - \frac{\partial\mathscr{L}(..)}{\partial\zeta^{\cdot\cdot}_A}\phi^{\cdot\cdot}{}_{,B} - \left[\frac{\partial\mathscr{L}(..)}{\partial\bar{\zeta}^{\cdot\cdot}_A}\right]_{|..}\bar{\phi}^{\cdot\cdot}{}_{,B}, \\ &T^4{}_j(\xi) \equiv T^{q^4}{}_{q^j}(q,p) \\ &\quad = \delta^4{}_j[\mathscr{L}(..)]_{|..} - \left[\frac{\partial\mathscr{L}(..)}{\partial\zeta^{\cdot\cdot}_4}\right]_{|..}\phi^{\cdot\cdot}{}_{,q^j} - \left[\frac{\partial\mathscr{L}(..)}{\partial\bar{\zeta}^{\cdot\cdot}_4}\right]_{|..}\bar{\phi}^{\cdot\cdot}{}_{,q^j}. \end{aligned} \tag{7.1.16}$$

In case $\varepsilon^A \equiv 0$, $\varepsilon_{AB} \neq 0$, the equation (7.1.15) provides another conservation law, namely,

$$
\begin{aligned}
j^A{}_{BC,A} = 0, \qquad & j^A{}_{BC}(\xi) \equiv M^A{}_{BC}(\xi) + \mathscr{S}^A{}_{BC}(\xi) = -j^A{}_{CB}(\xi), \\
& M^A{}_{BC}(\xi) \equiv \xi_C T^A{}_B(\xi) - \xi_B T^A{}_C(\xi), \\
& \mathscr{S}^A{}_{BC}(\xi) \equiv \left[\frac{\partial \mathscr{L}(..)}{\partial \zeta_A^{\cdot\cdot}}\right]_{|..} S_{BC}{}^{\cdot\cdot}_{\cdot\cdot}\phi^{\cdot\cdot}(\xi) + \text{(c.c.)}.
\end{aligned}
\tag{7.1.17}
$$

The infinitesimal version of the generalized reciprocity transformation (7.1.2) is given by

$$
\begin{aligned}
&\varepsilon^A \equiv 0, \qquad \varepsilon_{ij} = \varepsilon_{q^i q^j} \equiv 0, \qquad \varepsilon_{i+4,\, j+4} = \varepsilon_{p^i p^j} \equiv 0, \\
&\varepsilon_{i,\, j+4} = \varepsilon_{q^i p^j} = -\varepsilon_{p^j q^i} \equiv \varepsilon d_{ij}, \qquad \varepsilon \neq 0.
\end{aligned}
\tag{7.1.18}
$$

As a consequence of this invariance the following conservation law emerges:

$$
\begin{aligned}
& j^{(A)}{}_{,A} = 0, \qquad j^{(A)}(\xi) \equiv T^{(A)}(\xi) + (1/2)B^{(A)}(\xi), \\
& T^{(A)}(\xi) \equiv \xi^j T^A{}_{j+4}(\xi) - \xi^{j+4} T^A{}_j(\xi), \\
& B^{(A)}(\xi) \equiv \left[\frac{\partial \mathscr{L}(..)}{\partial \zeta_A^{\cdot\cdot}}\right]_{|..} d^{ij}(S_{i+4,j}{}^{\cdot\cdot}_{\cdot\cdot} - S_{i,j+4}{}^{\cdot\cdot}_{\cdot\cdot})\phi^{\cdot\cdot}(\xi) + \text{(c.c.)}, \\
& T^{(4)}(q,p) = q^j T^4{}_{p^j}(q,p) - p^j T^4{}_{q^j}(q,p), \\
& B^{(4)}(q,p) = \left[\frac{\partial \mathscr{L}(..)}{\partial \zeta_4^{\cdot\cdot}}\right]_{|..} d^{ij}(S_{p^i q^j}{}^{\cdot\cdot}_{\cdot\cdot} - S_{q^i p^j}{}^{\cdot\cdot}_{\cdot\cdot})\phi^{\cdot\cdot}(q,p) + \text{(c.c.)}.
\end{aligned}
\tag{7.1.19}
$$

In case the action integral (7.1.11) is invariant under an infinitesimal phase transformation (nongeometrical!)

$$
\begin{aligned}
\hat{\phi}(\xi) &= \phi(\xi)\exp(i\varepsilon), \\
\hat{\bar{\phi}}(\xi) &= \bar{\phi}(\xi)\exp(-i\varepsilon),
\end{aligned}
\tag{7.1.20}
$$

where $|\varepsilon| > 0$, a conservation rule follows, namely,

$$
\begin{aligned}
& v^A{}_{,A} = 0, \\
& v^A(\xi) \equiv i\left\{\left[\frac{\partial \mathscr{L}(..)}{\partial \bar{\zeta}_A^{\cdot\cdot}}\right]_{|..} \bar{\phi}^{\cdot\cdot}(\xi) - \text{(c.c.)}\right\}.
\end{aligned}
\tag{7.1.21}
$$

One would consider $\xi^4 = q^4$, the time coordinate as the preferred one to obtain integral conservation laws. Furthermore, a hypercylindrical, bounded domain $D \subset M_8$ will be picked up such that D lies between two distinct $\xi^4 =$ constant hyperplanes and surrounded by a wall σ (see Figs. 26, 27). It is assumed that $T^A{}_B(\xi) = v^A(\xi) \equiv 0$ outside the phase space tube containing D. Across the wall σ, the usual jump conditions

$$
T^A{}_B n_{A|\sigma} = j^A{}_{BC} n_{A|\sigma} = v^A n_{A|\sigma} = 0
$$

are imposed, where n^A is the unit outward normal to σ. Applying the eight-

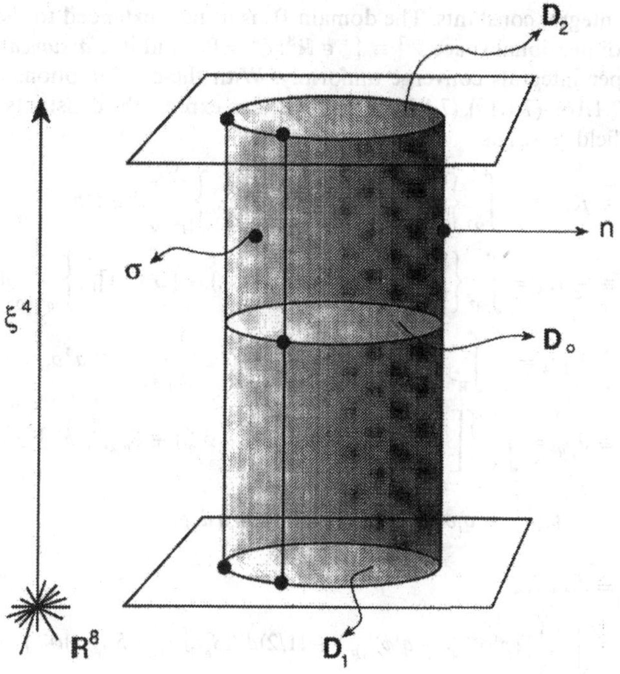

FIGURE 27. An extended world-tube in $\mathbb{R}^8$.

dimensional Gauss theorem to the differential conservation laws, we derive the following integral constants:

$$K_A \equiv \int_{\bar{D}_0} T^4{}_A(\xi)\, d^7\xi,$$

$$J_{BC} \equiv \int_{\bar{D}_0} j^4{}_{BC}(\xi)\, d^7\xi = -J_{CB},$$

$$J \equiv \int_{\bar{D}_0} [T^{(4)}(\xi) + (1/2)B^{(4)}(\xi)]\, d^7\xi, \qquad (7.1.22)$$

$$N \equiv \int_{\bar{D}_0} v^4(\xi)\, d^7\xi,$$

$$d^7\xi \equiv d\xi^1\, d\xi^2\, d\xi^3\, d\xi^5\, d\xi^6\, d\xi^7\, d\xi^8,$$

where D_0 is a seven-dimensional bounded domain for which $\xi^4 = q^4 = \text{const.}$ (see Figure 27).

These integrals are independent of the time coordinate $\xi^4 \equiv q^4$. Therefore, K_A, J_{BC}, J, N are called constants of motion (or evolution) of the field, or, in short, integral constants. The domain D_0 is usually extended to the whole of seven-dimensional space $\mathbb{R}^7 \equiv \{\xi \in \mathbb{R}^8 : \xi^4 = 0\}$, and it is *assumed* that these improper integrals converge uniformly. With these assumptions and equations (7.1.16)–(7.1.19), (7.1.21), (7.1.22) we can express the constants of motion of the field $\varphi^{\cdot\cdot}(\xi)$ as

$$K_\alpha \equiv K_{q^\alpha} = -\int_{\mathbb{R}^7} \left\{ \left[\frac{\partial \mathscr{L}(..)}{\partial \zeta^{\cdot\cdot}_4} \right]_{|..} \phi^{\cdot\cdot}_{,q^\alpha} + (\text{c.c.}) \right\}_{|q^4=0} d^3q\, d^4p,$$

$$H \equiv -K_4 = \int_{\mathbb{R}^7} \left\{ \left[\frac{\partial \mathscr{L}(..)}{\partial \zeta^{\cdot\cdot}{}_4} \right]_{|..} \phi^{\cdot\cdot}_{,q^4} + (\text{c.c.}) - [\mathscr{L}(..)]_{|..} \right\}_{|q^4=0} d^3q\, d^4p,$$

$$X_j \equiv K_{j+4} = -\int_{\mathbb{R}^7} \left\{ \left[\frac{\partial \mathscr{L}(..)}{\partial \zeta^{\cdot\cdot}_4} \right]_{|..} \phi^{\cdot\cdot}_{,p^j} + (\text{c.c.}) \right\}_{|q^4=0} d^3q\, d^4p,$$

$$J_{ij} \equiv J_{q^iq^j} = \int_{\mathbb{R}^7} \left\{ \left[\frac{\partial \mathscr{L}(..)}{\partial \zeta^{\cdot\cdot}_4} \right]_{|..} [(q_i\phi^{\cdot\cdot}_{,q^j} - q_j\phi^{\cdot\cdot}_{,q^i}) + S_{q^iq^j}{}^{\cdot\cdot}{}_{..}\phi^{\cdot\cdot}] + (\text{c.c.}) \right.$$

$$\left. + (q_j\delta^4{}_i - q_i\delta^4{}_j)[\mathscr{L}(..)]_{|..} \right\}_{|q^4=0} d^3q\, d^4p, \tag{7.1.23}$$

$$L_{ij} \equiv J_{i+4,\, j+4},$$

$$J \equiv \int_{\mathbb{R}^7} \left\{ [(p^i\phi^{\cdot\cdot}{}_{,q^i} - q^i\phi^{\cdot\cdot}{}_{,p^i}) + (1/2)d^{ij}(S_{p^iq^j}{}^{\cdot\cdot}{}_{..} - S_{q^ip^j}{}_{..})\phi^{\cdot\cdot}] \right.$$

$$\left. \times \left[\frac{\partial \mathscr{L}(..)}{\partial \zeta^{\cdot\cdot}{}_4} \right]_{|..} + (\text{c.c.}) - p^4[\mathscr{L}(..)]_{|..} \right\}_{|q^4=0} d^3q\, d^4p,$$

$$Q \equiv e_0 J,$$

$$N \equiv i\int_{\mathbb{R}^7} \left\{ \left[\frac{\partial \mathscr{L}(..)}{\partial \bar{\zeta}^{\cdot\cdot}_4} \right]_{|..} \bar{\phi}^{\cdot\cdot} - (\text{c.c.}) \right\}_{|q^4=0} d^3q\, d^4p,$$

where e_0 is the charge parameter.

Now, physical interpretations of various constants in (7.1.23) are in order. With regards to the $\phi^{\cdot\cdot}$-field distribution in the extended phase space M_8:

(i) K_α are components of the total three-momentum;
(ii) H is the total energy;
(iii) X_α are components of spatial center of the distribution;
(iv) X_4 is the time center of the distribution;
(v) J_{ij} are components of the total "relativistic" angular momentum;
(vi) L_{ij} are components of the total "angular momentum" in the four-momentum space;
(vii) Q, which arises from the sum of *isotropic* contribution $e_0 T^{(4)}(\xi)$ and *baryonic* contributions $(e_0/2)B^4(\xi)$, is the total charge;
(viii) N is the total content of the field (which can represent the total number of quanta after the second quantization).

To contrast with the usual relativistic field theory, we may notice that only integral constants that emerge in that arena are K_α, H, J_{ij}, $Q \equiv eN$ [see equation (6.1.24)]. Therefore, in the present formulation, we have many more integral constants.

7.2. The Generalized Klein–Gordon Equation

In this section the generalized Klein–Gordon equation for a scalar field defined over M_8 will be discussed. Yukawa pursued Born's reciprocity covariance in his theory of the nonlocal scalar field. He advocated *two* equations (one for the space–time variables, another for the four-momentum variables) for this scalar field. In this section, the generalized Klein–Gordon equation is obtained by "gluing" his two equations together, so to say. The first advantage of this procedure is that group of covariance is enlarged to the extended Poincaré group, and, secondly, the scalar field thus obtained can describe an arbitrary number of mesons in a unified manner.

The Lagrangian function in this case is chosen to be $\mathscr{L}: \mathbb{C} \times \mathbb{C} \times \mathbb{C}^8 \times \mathbb{C}^8 \to \mathbb{R}$ such that

$$\mathscr{L}(\zeta, \bar{\zeta}, \zeta_A, \bar{\zeta}_A) \equiv -[D^{AB}\bar{\zeta}_A\zeta_B + \mu^2\bar{\zeta}\zeta],$$
$$\frac{\partial\mathscr{L}(..)}{\partial\bar{\zeta}} = -\mu^2\zeta, \qquad \frac{\partial\mathscr{L}(..)}{\partial\bar{\zeta}_A} = -D^{AB}\zeta_B, \tag{7.2.1}$$

where μ is the (real) mass parameter.

The Euler–Lagrange equations (7.1.12) [compare equation (6.2.3)] yield in this case

$$\left[\frac{\partial\mathscr{L}(..)}{\partial\bar{\zeta}}\right]_{|..} - \frac{d}{d\xi^A}\left\{\left[\frac{\partial\mathscr{L}(..)}{\partial\bar{\zeta}_A}\right]_{|..}\right\}$$
$$= -\mu^2\phi(\xi) + D^{AB}\phi_{,AB} = d^{ij}(\phi_{,q^iq^j} + \phi_{,p^ip^j}) - \mu^2\phi(q,p) = 0. \tag{7.2.2}$$

The above equation is obviously invariant under the extended Poincaré group. From the transformation rule $\hat{\phi}(\hat{\xi}) = \phi(\xi)$ and the equation (7.1.14), it is clear that

$$S_{CD} \equiv 0, \qquad r(\varepsilon) = \hat{r}(\varepsilon^{\cdot\cdot}) \equiv 0.$$

The various tensor densities obeying differential conservation rules can be computed from (7.2.1), (7.1.16), (7.1.17), (7.1.19), (7.1.21). These are furnished by

$$T^A{}_B(\xi) = \delta^A{}_B[\mathscr{L}(..)]_{|..} - \left[\frac{\partial\mathscr{L}(..)}{\partial\zeta_A}\right]_{|..}\phi_{,B} - \left[\frac{\partial\mathscr{L}(..)}{\partial\bar{\zeta}_A}\right]_{|..}\bar{\phi}_{,B}$$
$$= D^{AC}[\bar{\phi}_{,C}\phi_{,B} + \bar{\phi}_{,B}\phi_{,C}] - \delta^A{}_B[D^{CE}\bar{\phi}_{,C}\phi_{,E} + \mu^2|\phi(\xi)|^2],$$

$$
\begin{aligned}
j^{A}{}_{BC}(\xi) &= M^{A}{}_{BC}(\xi) \\
&= \xi_C\{D^{AE}[\bar{\phi}_{,E}\phi_{,B} + \bar{\phi}_{,B}\phi_{,E}] + \delta^{A}{}_{B}[\mathscr{L}(..)]_{|..}\} \\
&\quad - \xi_B\{D^{AE}[\bar{\phi}_{,E}\phi_{,C} + \bar{\phi}_{,C}\phi_{,E}] + \delta^{A}{}_{C}[\mathscr{L}(..)_{|..}\},
\end{aligned}
$$

$$
\mathscr{S}^{A}{}_{BC}(\xi) \equiv 0, \tag{7.2.3}
$$

$$
\begin{aligned}
j^{(A)}(\xi) &= T^{(A)}(\xi) \\
&= \xi^{j}\{D^{AC}(\bar{\phi}_{,C}\phi_{,j+4} + \bar{\phi}_{,j+4}\phi_{,C}) + \delta^{A}{}_{j+4}[\mathscr{L}(..)_{|..}\} \\
&\quad - \xi^{j+4}\{D^{AC}(\bar{\phi}_{,C}\phi_{,j} + \bar{\phi}_{,j}\phi_{,C}) + \delta^{A}{}_{j}[\mathscr{L}(..)]_{|..}\},
\end{aligned}
$$

$$
B^{A}(\xi) \equiv 0,
$$

$$
v^{A}(\xi) = iD^{AB}[\bar{\phi}_{,B}\phi(\xi) - \bar{\phi}(\xi)\phi_{,B}].
$$

The constants of motion (or integral constants) for the scalar field follow from (7.2.1) and (7.1.23). These are given by

$$
K_\alpha \equiv K_{q^\alpha} = -\int_{\mathbb{R}^7} [\bar{\phi}_{,q^4}\phi_{,q^\alpha} + (\text{c.c.})]_{|q^4=0}\, d^3q\, d^4p,
$$

$$
\begin{aligned}
H \equiv -K_4 = \int_{\mathbb{R}^7} [&\bar{\phi}_{,q^\alpha}\phi_{,q^\alpha} + \bar{\phi}_{,q^4}\phi_{,q^4} \\
&+ \bar{\phi}_{,p^\alpha}\phi_{,p^\alpha} - \bar{\phi}_{,p^4}\phi_{,p^4} + \mu^2|\phi|^2]_{|q^4=0}\, d^3q\, d^4p,
\end{aligned}
$$

$$
X_i \equiv K_{i+4} \equiv K_{p^i} = -\int_{\mathbb{R}^7} [\bar{\phi}_{,q^4}\phi_{,p^i} + (\text{c.c.})]_{|..}\, d^3q\, d^4p,
$$

$$
J_{ij} = \int_{\mathbb{R}^7} \{(q_i\bar{\phi}_{,q^j} - q_j\bar{\phi}_{,q^i})\phi_{,q^4} + (\text{c.c.}) + (q_j\delta^4{}_i - q_i\delta^4{}_j)[\mathscr{L}(..)]_{|..}\}_{|..}\, d^3q\, d^4p, \tag{7.2.4}
$$

$$
L_{ij} = \int_{\mathbb{R}^7} [(p_i\bar{\phi}_{,p^j} - p_j\bar{\phi}_{,p^i})\phi_{,q^4} + (\text{c.c.})]_{|..}\, d^3q\, d^4p,
$$

$$
Q = e_0 J = e_0 \int_{\mathbb{R}^7} \{(p^i\bar{\phi}_{,q^i} - q^i\bar{\phi}_{,p^i})\phi_{,q^4} + (\text{c.c.}) - p^4[\mathscr{L}(..)]_{|..}\}_{|..}\, d^3q\, d^4p,
$$

$$
B = 0,
$$

$$
N = i\int_{\mathbb{R}^7} [\bar{\phi}(q,p)\cdot\phi_{,q^4} - \bar{\phi}_{,q^4}\cdot\phi(q,p)]_{|..}\, d^3q\, d^4p.
$$

Now we shall introduce the Fourier dual of the seven-dimensional space of $(q^1, q^2, q^3, p^1, p^2, p^3, p^4)$. A subset D_7 of this dual space will be essential in subsequent computations. It is defined by

$$
D_7 \equiv \{(\mathbf{k}, x) \in \mathbb{R}^7\colon k_\alpha k_\alpha + x_\alpha x_\alpha + \mu^2 > (x_4)^2\}. \tag{7.2.5}
$$

In this domain the following function satisfies the inequality

$$\begin{aligned} \Omega(\mathbf{k}, x) &\equiv \sqrt{k_\alpha k_\alpha + x_\alpha x_\alpha + \mu^2 - (x_4)^2} > 0, \\ \omega = W(\mathbf{k}, \mathbf{x}) &\equiv \sqrt{k_\alpha k_\alpha + x_\alpha x_\alpha + \mu^2} > 0, \\ \omega' = W(\mathbf{k}', \mathbf{x}') &\equiv \sqrt{k'_\alpha k'_\alpha + x'_\alpha x'_\alpha + \mu^2} > 0. \end{aligned} \tag{7.2.6}$$

A class of solutions of the generalized Klein–Gordon equation (7.2.2) can be furnished by [compare equation (6.2.7)]

$$\begin{aligned} \phi(q, p) &= (2\pi)^{-7/2} \int_{D_7} [2\Omega(\mathbf{k}, x)]^{-1/2} \{\alpha(\mathbf{k}, x) \exp[i(k_a q^a + x_a p^a)] \\ &\quad + \overline{\beta(\mathbf{k}, x)} \exp[-i(k_a q^a + x_a p^a)]\} \, d^3\mathbf{k} \, d^4x, \\ \bar{\phi}(q, p) &= (2\pi)^{-7/2} \int_{D_7} [2\Omega(\mathbf{x}, x)]^{-1/2} \{\overline{\alpha(\mathbf{k}, x)} \exp[-i(k_a q^a + x_a p^a)] \\ &\quad + \beta(\mathbf{k}, x) \exp[i(k_a q^a + x_a p^a)]\} \, d^3\mathbf{k} \, d^4x, \\ k_4 &\equiv -\Omega(\mathbf{k}, x) = -\sqrt{k_\alpha k_\alpha + x_\alpha x_\alpha + \mu^2 - x_4^2}, \\ \phi_{,q^4} &= -i(2\pi)^{-7/2} \int_{D_7} (\Omega/2)^{1/2} \{\alpha \exp[i(k_a q^a + x_a p^a)] \\ &\quad - \bar{\beta} \exp[-i(k_a q^a + x_a p^a)]\} \, d^3\mathbf{k} \, d^4x, \\ \bar{\phi}_{,q^\alpha} &= -i(2\pi)^{-7/2} \int_{D_7} (2\Omega)^{-1/2} k_\alpha \{\bar{\alpha} \exp[-i(k_a q^a + x_a p^a)] \\ &\quad - \beta \exp[i(k_a q^a + x_a p^a)]\} \, d^3\mathbf{k} \, d^4x, \\ \bar{\phi}_{,p^a} &= -i(2\pi)^{-7/2} \int_{D_7} x_a \{\bar{\alpha} \exp[-i(k_a q^a + x_a p^a)] \\ &\quad - \beta \exp[i(k_a q^a + x_a p^a)]\} \, d^3\mathbf{k} \, d^4x. \end{aligned} \tag{7.2.7}$$

Here we have assumed that the above seven-dimensional improper integrals converge uniformly and that differentiations under the integral signs are permitted. The complex-valued functions α, β are otherwise arbitrary. We shall compute the constants of motion (7.2.4) using this class of solutions (which may be considered as plane wave superposition). The easiest one to calculate is N, and it will be accomplished presently. Using (7.2.4), (7.2.7) we have

$$\begin{aligned} N &= i \int_{\mathbb{R}^7} [\bar{\phi}(q, p) \phi_{,q^4}]_{|q^4=0} \, d^3q \, d^4p + (\text{c.c.}) \\ &= (2)^{-1} \int_{D_7} \left\{ \int_{D_7} (\Omega/\Omega')^{1/2} [(\bar{\alpha}'\alpha - \beta'\bar{\beta}) \delta^3(\mathbf{k} - \mathbf{k}') \delta^4(x - x') \right. \\ &\quad \left. + (\beta'\alpha - \bar{\alpha}'\bar{\beta}) \delta^3(\mathbf{k} + \mathbf{k}') \delta^3(\mathbf{x} + \mathbf{x}')] \, d^3\mathbf{k}' \, d^4x' \right\} d^3\mathbf{k} \, d^4x + (\text{c.c.}) \end{aligned}$$

$$= (2)^{-1} \int_{D_7} \left\{ \int_{\mathbb{R}^6} \int_{-\omega'}^{\omega'} [(\bar{\alpha}'\alpha - \beta'\bar{\beta})\delta^3(\mathbf{k} - \mathbf{k}') \right.$$
$$\times \delta^3(\mathbf{x} - \mathbf{x}')\delta(x_4 - x_4') + (\beta'\alpha - \bar{\alpha}'\bar{\beta})\delta^3(\mathbf{k} + \mathbf{k}')$$
$$\left. \times \delta^3(\mathbf{x} + \mathbf{x}')\delta(x_4 + x_4')] \, d^3\mathbf{k}' \, d^3\mathbf{x}' \, dx_4' \right\} d^3\mathbf{k} \, d^4x + (\text{c.c.})$$
$$= (2)^{-1} \int_{D_7} \{(|\alpha|^2 - |\beta|^2)$$
$$+ [\beta(-\mathbf{k}, -x)\alpha(\mathbf{k}, x) - \bar{\alpha}(-\mathbf{k}, -x)\bar{\beta}(\mathbf{k}, x)]\} \, d^3\mathbf{k} \, d^4x + (\text{c.c.}). \tag{7.2.8}$$

Now let us consider the second part of the integral (with a substitution $k' = -k, x' = -x$)

$$I \equiv \int_{-\infty}^{\infty} \cdots\cdots \int_{-\infty}^{\infty} \int_{-\omega}^{\omega} [\beta(-\mathbf{k}, -\mathbf{x}, -\mathbf{x}_4)\alpha(\mathbf{k}, \mathbf{x}, x_4)$$
$$- \bar{\alpha}(-\mathbf{k}, -\mathbf{x}, -\mathbf{x}_4)\bar{\beta}(\mathbf{k}, \mathbf{x}, x_4)] \, d^3\mathbf{k} \, d^4x + (\text{c.c.})$$
$$= -\int_{\infty}^{-\infty} \cdots\cdots \int_{\infty}^{-\infty} \int_{\omega}^{-\omega} [\beta(\mathbf{k}', \mathbf{x}', x_4')\alpha(-\mathbf{k}', -\mathbf{x}', -x_4')$$
$$- \bar{\alpha}(\mathbf{k}, \mathbf{x}', x_4')\bar{\beta}(-\mathbf{k}', -\mathbf{x}', -x_4')] \, d^3k' \, d^4x' + (\text{c.c.})$$
$$= -I = 0. \tag{7.2.9}$$

Therefore, from (7.2.8) we obtain

$$N = \int_{D_7} [|\alpha(\mathbf{k}, x)|^2 - |\beta(\mathbf{k}, x)|^2] \, d^3\mathbf{k} \, d^4x. \tag{7.2.10}$$

Now let us compute another important integral constant, namely, the total energy H from equations (7.2.4), (7.2.7):

$$H = (2)^{-1}(2\pi)^{-7} \int_{\mathbb{R}^7} \int_{D_7} \int_{D_7} (\Omega\Omega')^{-1/2}$$
$$\times \{(\Omega'\Omega + k_\alpha' k_\alpha + x_\alpha' x_\alpha - x_4' x_4)$$
$$\times [\bar{\alpha}' \exp[-i(..)'] - \beta' \exp[i(..)']]$$
$$\times [\alpha \exp[i(..)] - \bar{\beta} \exp[-i(..)]]$$
$$+ \mu^2 [\bar{\alpha}' \exp[-i(..)'] + \beta' \exp[i(..)']]$$
$$\times [\alpha \exp[i(..)] + \bar{\beta} \exp[-i(..)]]\} \, d^3q \, d^4p \, d^3\mathbf{k}' \, d^4x' \, d^3\mathbf{k} \, d^4x$$

$$= (2)^{-1} \int_{D_7} \int_{D_7} (\Omega\Omega')^{1/2}$$
$$\times [(\Omega'\Omega + k_\alpha' k_\alpha + x_\alpha' x_\alpha - x_4' x_4 + \mu^2)$$
$$\times (\bar{\alpha}'\alpha + \beta'\bar{\beta})\delta^3(\mathbf{k} - \mathbf{k}')\delta^4(x - x')$$
$$- (\Omega'\Omega + k_\alpha' k_\alpha + x_\alpha' x_\alpha - x_4' x_4 - \mu^2)$$
$$\times (\bar{\alpha}'\bar{\beta} + \beta'\alpha)\delta^3(\mathbf{k} + \mathbf{k}')\delta^4(x + x')] \, d^3\mathbf{k}' \, d^4x' \, d^3\mathbf{k} \, d^4x. \tag{7.2.11}$$

Now let us digress slightly to consider an integral of the type

$$\int_{R^6}\left[\int_{-\omega'}^{\omega'} f(\mathbf{k}',\mathbf{k},\mathbf{x}',\mathbf{x},x_4',x_4)\delta(x_4-x_4')\,dx_4'\right]\delta^3(\mathbf{k}-\mathbf{k}')\delta^3(\mathbf{x}-\mathbf{x}')\,d^3\mathbf{k}'\,d^3\mathbf{x}'$$

$$=\begin{cases} f(\mathbf{k},\mathbf{k},\mathbf{x},\mathbf{x},x_4,x_4) & \text{for } x_4\in(-\omega,\omega),\\ 0 & \text{for } x_4\notin(-\omega,\omega).\end{cases}$$

Substituting the above result into the equation (7.2.11) we obtain

$$\begin{aligned} H &= (2)^{-1}\int_{D_7}(\Omega)^{-1}\{[\Omega^2+k_\alpha k_\alpha+x_\alpha x_\alpha-x_4^2+\mu^2][|\alpha|^2+|\beta|^2] \\ &\quad -[\Omega^2-k_\alpha k_\alpha-x_\alpha x_\alpha+x_4^2-\mu^2] \\ &\quad \times[\bar{\alpha}(-\mathbf{k},-x)\bar{\beta}(\mathbf{k},x)+\alpha(\mathbf{k},x)\beta(-\mathbf{k},-x)]\}\,d^3\mathbf{k}\,d^4x \\ &= \int_{D_7}[|\alpha(\mathbf{k},x)|^2+|\beta(\mathbf{k},x)|^2]\Omega(\mathbf{k},x)\,d^3\mathbf{k}\,d^4x. \end{aligned} \tag{7.2.12}$$

Now we shall summarize computations of many constants of motion:

$$\begin{aligned} K_\alpha &= \int_{D_7}[f_+(\mathbf{k},x)+f_-(\mathbf{k},x)]k_\alpha\,d^3\mathbf{k}\,d^4x,\\ H &= -K_4 = \int_{D_7}[f_+(\mathbf{k},x)+f_-(\mathbf{k},x)]\Omega(\mathbf{k},x)\,d^3\mathbf{k}\,d^4x,\\ X_\alpha &= \int_{D_7}[f_+(\mathbf{k},x)+f_-(\mathbf{k},x)]x_\alpha\,d^3\mathbf{k}\,d^4x,\\ B &= 0,\\ N &= \int_{D_7}[f_+(\mathbf{k},x)-f_-(\mathbf{k},x)]\,d^3\mathbf{k}\,d^4x, \end{aligned} \tag{7.2.13}$$

$$f_+(\mathbf{k},x)\equiv|\alpha(\mathbf{k},x)|^2\geq 0,\qquad f_-(\mathbf{k},x)\equiv|\beta(\mathbf{k},x)|^2\geq 0.$$

Some physical interpretations can be reached now. The functions $f_+(\mathbf{k},x)$, $f_-(\mathbf{k},x)$ can be naturally interpreted as a Boltzmann type of distribution function for the similarly created particles and antiparticles. This view point supports the statistical interpretation of quantum mechanics for Bosons. All the integral constants belong to the scalar field representing the unified meson field ensemble. In the equation (7.2.13),

(i) K_α represent the total three-momentum components;
(ii) H represents the total energy;
(iii) X_α are components of the first moment for the distribution;
(iv) X_4 is the center of time for the distribution;
(v) $B = 0$ is the baryonic content;
(vi) N is particles minus antiparticle content.

Computations of J_{ij}, L_{ij}, J are extremely cumbersome. However, for a physical theory we do need the integral constant J to evaluate the total charge Q. We shall make only a partial attempt to perform this calculation now. We note that

$$J = I_1 + I_2,$$

$$I_1 \equiv \int_{\mathbb{R}^7} [(p^i\bar{\phi}_{,q^i} - q^i\bar{\phi}_{,p^i}]_{|q^4=0}\, d^3q\, d^4p + (\text{c.c.}),$$

$$I_2 \equiv -\int_{\mathbb{R}^7} \{p^4[\mathscr{L}(..)]_{|..}\}_{|q^4=0}\, d^3q\, d^4p. \tag{7.2.14}$$

We further simplify the situation by putting the antiparticle amplitude $\beta(\mathbf{k}, x) \equiv 0$. In that case

$$\begin{aligned}
I_2 &\equiv (2)^{-1}(2\pi)^{-7} \int_{\mathbb{R}^7}\int_{D_7}\int_{D_7} p^4(\Omega\Omega')^{-1/2} \\
&\quad \{(k'_\alpha k_\alpha + x'_\alpha x_\alpha - \Omega'\Omega - x'_4 x_4 + \mu^2)(\bar{\alpha}'\alpha) \\
&\quad \times \exp[i((k_\alpha - k'_\alpha)q^\alpha + (x_a - x'_\alpha)p^a)]\}\, d^3q\, d^4p\, d^3\mathbf{k}'\, d^4x'\, d^3\mathbf{k}\, d^4x \\
&= (4\pi)^{-1} \int_{\mathbb{R}}\int_{-\omega}^{\omega}\int_{D_7} p^4[\Omega(\mathbf{k}, \mathbf{x}, x'_4)\Omega(\mathbf{k}, \mathbf{x}, x_4)]^{-1/2} \\
&\quad \times \{(k_\alpha k_\alpha + x_\alpha x_\alpha + \mu^2 - \Omega'\Omega - x'_4 x_4)\bar{\alpha}(\mathbf{k}, \mathbf{x}, x'_4) \\
&\quad \times \alpha(\mathbf{k}, \mathbf{x}, x_4)\exp[i(x_4 - x'_4)p^4]\}\, dp^4\, dx'_4\, d^3\mathbf{k}\, d^4x.
\end{aligned}$$

Since I_2 is real we can write

$$\begin{aligned}
I_2 &= (8\pi)^{-1} \int_{\mathbb{R}}\int_{-\omega}^{\omega}\int_{D_7} p^4(\Omega'\Omega)^{-1/2} \\
&\quad \times \{(k_\alpha k_\alpha + x_\alpha x_\alpha + \mu^2 - \Omega'\Omega - x'_4 x_4) \\
&\quad [\bar{\alpha}(\mathbf{k}, \mathbf{x}, x'_4)\alpha(\mathbf{k}, \mathbf{x}, x_4)\exp[i(x_4 - x'_4)p^4] \\
&\quad + \alpha(\mathbf{k}, \mathbf{x}, x'_4)\bar{\alpha}(\mathbf{k}, x, x_4)\exp[-i(x_4 - x'_4)p^4]]\}\, dp_4\, dx'_4\, d^3\mathbf{k}\, d^4x.
\end{aligned} \tag{7.2.15}$$

Let us work out the Cauchy principal value of the integral:

$$\begin{aligned}
&\text{C.P.V.} \int_{-\infty}^{\infty} p^4\{\bar{\alpha}(\mathbf{k}, \mathbf{x}, x'_4)\alpha(\mathbf{k}, \mathbf{x}, x_4)\exp[i(x_4 - x'_4)p^4] \\
&\qquad + \alpha(\mathbf{k}, \mathbf{x}, x'_4)\bar{\alpha}(\mathbf{k}, \mathbf{x}, x_4)\exp[-i(x_4 - x'_4)p^4]\}\, dp^4 \\
&= \begin{cases} 0 \text{ for } x_4 = x'_4, \\ i2(x_4 - x'_4)^{-1}\lim_{L\to\infty}[\alpha(\mathbf{k}, \mathbf{x}, x'_4)\bar{\alpha}(\mathbf{k}, \mathbf{x}, x_4) - (\text{c.c.})] \end{cases} \\
&\quad \times [L\cos((x_4 - x'_4)L) - (x_4 - x'_4)^{-1}\sin(x_4 - x'_4)L)] \\
&\quad \text{for } x_4 \neq x'_4.
\end{aligned}$$

With the further simplifying assumption

$$\alpha(\mathbf{k}, \mathbf{x}, x_4')\bar{\alpha}(\mathbf{k}, \mathbf{x}, x_4) = \bar{\alpha}(\mathbf{k}, \mathbf{x}, x_4')\alpha(\mathbf{k}, \mathbf{x}, x_4), \tag{7.2.16}$$

the integral $I_2 = 0$. The computation of I_1 in (7.2.14) will be done later.

Now, to develop the unified meson field theory, we transform coordinates q^q, p^a of each phase plane to the corresponding cylindrical coordinates by the following equations:

$$\begin{aligned} q^a &= \rho^a \cos\theta^a, \qquad p^a = \rho^a \sin\theta^a; \\ \rho^a &= \sqrt{(q^a)^2 + (p^a)^2}, \qquad \theta^a = \operatorname{arc}(q^a, p^a). \end{aligned} \tag{7.2.17}$$

Here the subscript 'a' is *not* summed, and furthermore coordinates $(\rho^a)^2$, θ^a are a canonically conjugate pair. To make the transformation one-to-one, we restrict $\rho^a > 0$ and $|\theta^a| < \pi$ for each $a \in \{1, 2, 3, 4\}$. We define the new scalar field as $\hat{\phi}(\rho, \theta) \equiv \phi(q, p)$. The generalized Klein–Gordon equation goes over to

$$d^{ij}[\hat{\phi}_{,\rho^i\rho^j} + (\rho^i)^{-1}\hat{\phi}_{,\rho_j} + (\rho^i\rho^j)^{-1}\hat{\phi}_{,\theta^i\theta^j}] = \mu^2\hat{\phi}(\rho, \theta). \tag{7.2.18}$$

Separating variables and demanding that angular functions are (properly) defined, we obtain the basic solutions as

$$\phi^{(t)}(\rho, \theta) = \chi^{(t)}(\rho)\exp[it_a\theta^a], \tag{7.2.19}$$

where $(t) \equiv (t_1, t_2, t_3, t_4)$ and $t_a \in \mathbb{Z}$. The function $\chi^{(t)}$ satisfies the partial differential equation

$$d^{ij}[\chi^{(t)}_{,\rho^i\rho^j} + (\rho^i)^{-1}\chi^{(t)}_{,\rho^j}] = [\mu^2 + d^{ij}(t_it_j/\rho^i\rho^j)]\chi^{(t)}(\rho). \tag{7.2.20}$$

If we *define* the (mass)^2 operator associated with the meson field $\chi^{(t)}(\rho)$ or $\hat{\phi}^{(t)}(\rho, \theta)$ as

$$M^2 \equiv d^{ij}[(\partial^2/\partial\rho^i\partial\rho^j) + (\rho^i)^{-1}(\partial/\partial\rho^j)], \tag{7.2.21}$$

then the equation (7.2.20) implies that

$$M^2\phi^{(t)}(\rho, \theta) = [\mu^2 + (t_1/\rho^1)^2 + (t_2/\rho^2)^2 + (t_3/\rho^3)^2 - (t_4/\rho^4)^2]\phi^{(t)}(\rho, \theta). \tag{7.2.22}$$

The above equation shows some analogy with *Okubo's mass formula* provided t_1, t_2, t_3 are associated with *isotopic* numbers, and $2t_4$ is associated with the *strangeness* quantum number.

The basic solution $\phi^{(t)}(\rho, \theta)$ can be expressed in terms of Bessel functions of the first kind [see the equation (7.2.41-*iv*)]

$$\begin{aligned} \phi^{(t)}(\rho, \theta) &= \left[\prod_{b=1}^{4} J_{t_b}(\kappa_b\rho^b)\right]\exp[it_a\theta^a], \\ \kappa_4 &\equiv \sqrt{\kappa_\alpha\kappa_\alpha + \mu^2}, \end{aligned} \tag{7.2.23}$$

where the index b is *not* summed. The asymptotic version of (7.2.23) for *large* values of ρ^b's is

$$\phi^{(t)}(\rho,\theta) = (2\pi)^{-2}\left[\prod_{b=1}^{4}(\kappa_b\rho^b)^{-1/2}\cos(\kappa_b\rho^b - 2^{-1}\pi t_b - 4^{-1}\pi)\right]\exp[it_a\theta^a] + O((\kappa_b\rho^b)^{-1}). \tag{7.2.23'}$$

Bessel functions of the second kind are *avoided* because of the singular behavior at the origin. A general solution (but *not* the most general solution) of (7.2.18) can be constructed by linear combinations of basic solutions. We can arrive at such a class of general solutions directly from the expression (7.2.7). For that purpose we make the coordinate transformation

$$\kappa_\alpha = \sqrt{k_\alpha^2 + x_\alpha^2}, \qquad \phi^\alpha = \operatorname{arc}(x^\alpha, k_\alpha), \qquad \hat{x}_4 = x_4. \tag{7.2.24}$$

(Please distinguish between the scalar field ϕ and angles $\underline{\phi}$.) The domain of this transformation is D_7 in (7.2.5) (minus some points) and the range is given by

$$\begin{aligned}\hat{D}_7 &\equiv \{(\underline{\kappa},\underline{\phi},\hat{x}_4) \in \mathbb{R}^7 : 0 < \kappa_\alpha, |\phi^\alpha| < \pi, |\hat{x}_4| < \omega\},\\ \omega &= W(k,\mathbf{x}) = \hat{W}(\underline{\kappa}) = \kappa_\alpha\kappa_\alpha + \mu^2.\end{aligned} \tag{7.2.25}$$

Now using (7.2.17), (7.2.24), and Jacobi–Anger's formula (7.2.41-x) for Bessel functions we obtain

$$\begin{aligned}\exp[i(k_1q^1 + x_1p^1)] &= \exp[i\kappa_1\rho^1\cos(\theta^1 - \phi^1)]\\ &= \sum_{t_1=-\infty}^{\infty}(i)^{t_1}J_{t_1}(\kappa_1\rho^1)\exp[it_1(\theta^1 - \phi^1)].\end{aligned} \tag{7.2.26}$$

Utilizing the above and similar expressions, the scalar field in (7.2.7) can be written as

$$\begin{aligned}\hat{\phi}(\underline{\rho},\underline{\theta},q^4,p^4) = {}& (2\pi)^{-7/2}\int_{\hat{D}_7}[2\sqrt{\omega^2 - x_4^2}]^{-1/2}\\ &\times \sum_{t=-\infty}^{\infty}\Bigg\{\hat{\alpha}(\underline{\kappa},\underline{\phi},x_4)\left[\prod_{\beta=1}^{3}(i)^{t_\beta}J_{t_\beta}(\kappa_\beta\rho^\beta)\right]\\ &\times \exp[i(t_\alpha(\theta^\alpha - \phi^\alpha) + k_4q^4 + q^4p^4)]\\ &+ \bar{\hat{\beta}}(\underline{\kappa},\underline{\phi},x_4)\left[\prod_{\beta=1}^{3}(-i)^{t_\beta}J_{t_\beta}(\kappa_\beta\rho^\beta)\right]\\ &\times \exp[-i(t_\alpha(\theta^\alpha - \phi^\alpha) + k_4q^4 + q_4p^4)]\Bigg\}\kappa_1\kappa_2\kappa_3\,d^3\underline{\kappa}\,d^3\underline{\phi}\,dx_4,\end{aligned}$$

$$\begin{aligned}\hat{\phi}(\underline{\rho},\underline{\theta},q^4,p^4) &\equiv \phi(q,p),\\ \hat{\alpha}(\underline{\kappa},\underline{\phi},x_4) = \alpha(\mathbf{k},x), &\qquad \hat{\beta}(\underline{\kappa},\underline{\phi},x_4) \equiv \beta(\mathbf{k},x),\\ k_4 \equiv -\Omega(\mathbf{k},x) &= -\sqrt{\omega^2 - x_4^2}.\end{aligned} \tag{7.2.27}$$

Now we make a coordinate transformation for which the inverse transformation is given by

$$k_4 = -\omega\cos\phi^4, \qquad x_4 = \omega\sin\phi^4,$$
$$\exp[-i(k_4 q^4 + x_4 p^4)] = \sum_{t_4=-\infty}^{\infty} (i)^{t_4} J_{t_4}(\omega p^4)\exp[it_4(\theta^4 + \phi^4)]. \tag{7.2.28}$$

The usual domain of this transformation is the (k_4-x_4)-plane minus the nonnegative k_4 half-line ($\omega > 0$, $|\phi_4| < \pi$). If we insist to keep $k_4 < 0$ (which is mathematically not imperative) then we have to choose $\omega > 0$, $|\phi_4| < \pi/2$. Logically speaking, we can also try multiply connected Riemann-like surfaces for which $\omega > 0$, $|\phi| < n\pi$, $n \in \{2, 3, 4, \ldots\}$. Out of all these choices, the usual choice $\omega > 0$, $|\phi| < \pi$ yields the neatest expressions for the integral constants. On that account, we shall adopt the preceding choice so that the integration has to be performed over the subset

$$D_7^{\#} \equiv \{(\underline{\kappa}, \phi) \in \mathbb{R}^7 : 0 < \kappa_\alpha, |\phi^j| < \pi\}. \tag{7.2.29}$$

The scalar field in (7.2.27) goes over to

$$\begin{aligned}\phi^{\#}(\rho,\theta) = {} & (2\pi)^{-7/2}\int_{D_7^{\#}} [(\omega\cos\phi^4)/2]^{1/2} \\ & \times \sum_{t=-\infty}^{\infty}\left\{\hat{\alpha}(\underline{\kappa}, \underline{\phi}, \omega\sin\phi^4)\left[\prod_{\beta=1}^{3}(i)^{t_\beta}J_{t_\beta}(\kappa_\beta\rho^\beta)\right]\right. \\ & \times (-i)^{t_4}J_{t_4}(\omega\rho^4)\exp[i(t_\alpha(\theta^\alpha - \phi^\alpha) - t_4(\theta^4 + \phi^4))] \\ & + \bar{\hat{\beta}}(.,.,.)\left[\prod_{\beta=1}^{3}(-i)^{t_\beta}J_{t_\beta}(\kappa_\beta\rho^\beta)\right](i)^{t_4}J_{t_4}(\omega\rho^4) \\ & \left.\times \exp[-i(t_\alpha(\theta^\alpha - \phi^\alpha) + t_4(\theta^4 + \phi^4))]\right\}\kappa_1\kappa_2\kappa_3\, d^3\underline{\kappa}\, d^4\phi,\end{aligned} \tag{7.2.30}$$

$$\phi^{\#}(\rho,\theta) \equiv \hat{\phi}(\underline{\rho}, \underline{\theta}, q^4, p^4).$$

Let us consider the integrals

$$\begin{aligned}\alpha^{\#}(\underline{\kappa}, \underline{\phi}, t_4) &\equiv (4\pi)^{-1/2}\int_{-\pi}^{\pi} \hat{\alpha}(\underline{\kappa}, \underline{\phi}, \omega\sin\phi^4) \\ &\qquad \times (\omega\cos\phi^4)^{1/2}\exp[-it_4\phi^4]\, d\phi^4, \\ \beta^{\#}(.,.,.) &\equiv (4\pi)^{-1/2}\int_{-\pi}^{\pi} \bar{\hat{\beta}}(.,.,.) \\ &\qquad \times (\omega\cos\phi^4)^{1/2}\exp[it_4\phi^4]\, d\phi^4.\end{aligned} \tag{7.2.31}$$

This equation can be inverted (under usual conditions) by the theory of

Fourier series to obtain

$$\begin{aligned}
&[(\omega\cos\phi^4)/2]^{1/2}\hat{\alpha}(\underline{\kappa},\underline{\phi},\omega\sin\phi^4)\\
&\quad = (2\pi)^{-1/2}\sum_{t_4=-\infty}^{\infty}\alpha^{\#}(\underline{\kappa},\underline{\phi},t_4)\exp[it_4\phi^4],\\
&[(\omega\cos\phi^4)/2]^{1/2}\hat{\bar{\beta}}(.,.,.)\\
&\quad = (2\pi)^{-1/2}\sum_{t_4=-\infty}^{\infty}\bar{\beta}^{\#}(.,.,.)\exp[-it_4\phi^4].
\end{aligned} \tag{7.2.32}$$

Now we shall define the following functions:

$$\begin{aligned}
\gamma(\underline{\kappa},t) &\equiv 2^{1/2}(2\pi)^{-3/2}\int_{-\pi}^{\pi}\cdot\int_{-\pi}^{\pi}\alpha^{\#}(\underline{\kappa},\underline{\phi},t_4)\exp[-it_\beta\phi^\beta]\,d^3\underline{\phi},\\
\bar{\nu}(\underline{\kappa},t) &\equiv 2^{1/2}(2\pi)^{-3/2}\int_{-\pi}^{\pi}\cdot\int_{-\pi}^{\pi}\bar{\beta}^{\#}(\underline{\kappa},\underline{\phi},t_4)\exp[it_\beta\phi^\beta]\,d^3\underline{\phi}.
\end{aligned} \tag{7.2.33}$$

These formulae can be inverted to obtain

$$\begin{aligned}
\alpha^{\#}(\underline{\kappa},\underline{\phi},t_4) &= 2^{-1/2}(2\pi)^{-3/2}\sum_{t=-\infty}^{\infty}\gamma(..)\exp[it_\beta\phi^\beta],\\
\bar{\beta}^{\#}(.,.,.) &= 2^{-1/2}(2\pi)^{-3/2}\sum_{t=-\infty}^{\infty}\bar{\nu}(.,.)\exp[-it_\beta\phi^\beta],
\end{aligned} \tag{7.2.34}$$

Let us denote $D_6^{\#} \equiv \{(\underline{\kappa},\underline{\phi}) \in \mathbb{R}^6 : 0 < \kappa_\alpha,\ |\phi^\beta| < \pi\}$. Then utilizing (7.2.31), (7.2.33), the equation (7.2.$\bar{3}$0) (assuming that interchange of integrations and summations is valid) yields

$$\begin{aligned}
\phi^{\#}(\rho,\theta) = (2\pi)^{-3}\sum_t\int_{D_6^{\#}}\Big\{&\alpha^{\#}(\underline{\kappa},\underline{\phi},t_4)\left[\prod_{\beta=1}^{3}(i)^{t_\beta}J_{t_\beta}(\kappa_\beta\rho^\beta)\right]\\
&\times\exp[it_\alpha(\theta^\alpha-\phi^\alpha)](-i)^{t_4}J_{t_4}(\omega\rho^4)\exp[-it_4\theta^4]\\
&+\bar{\beta}^{\#}(\underline{\kappa},\underline{\phi},t_4)\left[\prod_{\beta=1}^{3}(-i)^{t_\beta}J_{t_\beta}(\kappa_\beta\rho^\beta)\right]\\
&\times\exp[-it_\alpha(\theta^\alpha-\phi^\alpha)](i)^{t_4}J_{t_4}(\omega\rho^4)\exp[it_4\theta^4]\Big\}\\
&\times\kappa_1\kappa_2\kappa_3\,d^3\underline{\kappa}\,d^3\underline{\phi}\\
= \sum_t \phi^{(t)}(\rho,\theta),&\\
\phi^{(t)}(\rho,\theta) \equiv 2^{-1/2}(2\pi)^{-3/2}\int_{0^+}^{\infty}\cdot\cdot\Big\{&\gamma(\underline{\kappa},t)\left[\prod_{\beta=1}^{3}(i)^{t_\beta}J_{t_\beta}(\kappa_\beta\rho^\beta)\right]\\
&\times(-i)^{t_4}J_{t_4}(\omega\rho^4)\exp[i(t_\alpha\theta^\alpha-t_4\theta^4)]\\
&+\bar{\nu}(\underline{\kappa},t)\left[\prod_{\beta=1}^{3}(-i)^{t_\beta}J_{t_\beta}(\kappa_\beta\rho^\beta)\right](i)^{t_4}J_{t_4}(\omega\rho^4)\\
&\times\exp[-i(t_\alpha\theta^\alpha-t_4\theta^4)]\Big\}\kappa_1\kappa_2\kappa_3\,d^3\underline{\kappa}.
\end{aligned} \tag{7.2.35}$$

Assuming that above integral converge uniformly and differentiations under integral signs are permitted, the fields $\phi^{\#}(\rho, \theta)$ and $\phi^{(t)}(\rho, \theta)$ satisfy partial differential equations (7.2.18) and (7.2.22) respectively. We postulate that $\phi^{(t)}(\rho, \theta)$ represents a particular meson field. In that sense the equation $\phi^{\#}(\rho, \theta) = \sum_{t=-\infty}^{\infty} \phi^{(t)}(\rho, \theta)$ stands for the *unified meson field*.

Now we shall compute many constants of motion in terms of meson fields $\phi^{(t)}(\rho, \theta)$. In these calculations we shall put antiparticle amplitude $\beta(\mathbf{k}, \mathbf{x}) \equiv 0$ for the sake of simplicity. Consider the constant N in the equation (7.2.10) (with $\beta \equiv 0$). It is given by

$$\begin{aligned} N &= \int_{D_7} |\alpha(\mathbf{k}, x)|^2 \, d^3\mathbf{k} \, d^4x \\ &= \int_{\hat{D}_7} |\hat{\alpha}(\underline{\kappa}, \underline{\phi}, x_4)|^2 \kappa_1 \kappa_2 \kappa_3 \, d^3\underline{\kappa} \, d^3\underline{\phi} \, dx_4 \\ &= \int_{D_7^{\#}} |\hat{\alpha}(\underline{\kappa}, \underline{\phi}, \omega \sin \phi^4)|^2 \omega \cos \phi^4 \kappa_1 \kappa_2 \kappa_3 \, d^3\underline{\kappa} \, d^4\phi. \end{aligned} \tag{7.2.36}$$

Now using (7.2.32), the integral

$$\begin{aligned} 2\int_{-\pi}^{\pi} &|[\omega \cos \phi^4)/2]^{1/2} \hat{\alpha}(\underline{\kappa}, \underline{\phi}, \omega \sin \phi^4)|^2 \, d\phi^4 \\ &= (\pi)^{-1} \sum_{t_4} \sum_{\hat{t}_4} \left\{ \alpha^{\#}(\underline{\kappa}, \underline{\phi}, t_4) \bar{\alpha}^{\#}(\underline{\kappa}, \underline{\phi}, \hat{t}_4) \right. \\ &\qquad \left. \times \left[\int_{-\pi}^{\pi} \exp[i(t_4 - \hat{t}_4)\phi^4] \, d\phi^4 \right] \right\} \\ &= 2 \sum_{t_4} |\alpha^{\#}(\underline{\kappa}, \underline{\phi}, t_4)|^2. \end{aligned}$$

Therefore, from (7.2.36) we obtain

$$\begin{aligned} N &= 2 \sum_{t_4} \int_{D_6^{\#}} |\alpha^{\#}(\underline{\kappa}, \underline{\phi}, t_4)|^2 \kappa_1 \kappa_2 \kappa_3 \, d^3\underline{\kappa} \, d^3\underline{\phi} \\ &= \sum_{t_4} \sum_{\mathbf{t}} \sum_{\hat{\mathbf{t}}} \int_{0^+}^{\infty} \cdot\cdot \left\{ \gamma(\underline{\kappa}, \mathbf{t}, t_4) \bar{\gamma}(\underline{\kappa}, \hat{\mathbf{t}}, t_4) \right. \\ &\qquad \left. \times \left[(2\pi)^{-3} \int_{-\pi}^{\pi} \exp[i(t_\beta - \hat{t}_\beta)\phi^\beta] \, d^3\underline{\phi} \right] \right\} \kappa_1 \kappa_2 \kappa_3 \, d^3\underline{\kappa} \\ &= \sum_{t} \int_{0^+}^{\infty} \cdot\cdot |\gamma(\underline{\kappa}, t)|^2 \kappa_1 \kappa_2 \kappa_3 \, d^3\underline{\kappa}. \end{aligned} \tag{7.2.37}$$

Let us consider the total energy H from the equation (7.2.12) (with $\beta \equiv 0$):

$$H = \int_{D_7} |\alpha(\mathbf{k}, x)|^2 \Omega(\mathbf{k}, x)\, d^3\mathbf{k}\, d^4x$$
$$= \int_{\hat{D}_7} |\hat{\alpha}(\underline{\kappa}, \underline{\phi}, x_4)|^2 (\omega^2 - x_4^2)^{1/2} \kappa_1 \kappa_2 \kappa_3\, d^3\underline{\kappa}\, d^3\underline{\phi}\, dx_4$$
$$= \int_{D_7^\#} |\hat{\alpha}(\underline{\kappa}, \underline{\phi}, \omega \sin\phi^4)|^2 (\omega \cos\phi^4)^2 \kappa_1 \kappa_2 \kappa_3\, d^3\underline{\kappa}\, d^4\phi. \qquad (7.2.38)$$

Now the integral (with the equation (7.2.32))

$$\omega^2 \int_{-\pi}^{\pi} |\hat{\alpha}(\underline{k}, \underline{\phi}, \omega \sin\phi^4|^2 \cos^2\phi^4\, d\phi^4$$
$$= 2\omega \int_{-\pi}^{\pi} |[(\omega\cos\phi^4)/2]^{1/2} \hat{\alpha}(\underline{\kappa}, \underline{\phi}, \omega\sin\phi^4)|^2 \cos\phi^4\, d\phi^4$$
$$= \omega \sum_{t_4} \sum_{\hat{t}_4} \Bigg\{ \bar{\alpha}^\#(\underline{\kappa}, \underline{\phi}, t_4) \alpha^\#(\underline{\kappa}, \underline{\phi}, \hat{t}_4)$$
$$\times \Bigg[(2\pi)^{-1} \int_{-\pi}^{\pi} (\exp[i(\hat{t}_4 - t_4 + 1)\phi^4]$$
$$+ \exp[i(\hat{t}_4 - t_4 - 1)]\phi^4)\, d\phi^4 \Bigg] \Bigg\}$$
$$= \omega \sum_{t_4} \{ \bar{\alpha}^\#(\underline{\kappa}, \underline{\phi}, t_4) [\alpha^\#(\underline{\kappa}, \underline{\phi}, t_4 - 1) + \alpha^\#(\underline{\kappa}, \underline{\phi}, t_4 + 1)] \}.$$

Therefore, from (7.2.38), (7.2.33) we have

$$H = \sum_{t_4} \int_{D_6^\#} \{ \bar{\alpha}^\#(\underline{\kappa}, \underline{\phi}, t_4) [\alpha^\#(\underline{\kappa}, \underline{\phi}, t_4 - 1)$$
$$+ \alpha^\#(\underline{\kappa}, \underline{\phi}, t_4 + 1)] \} \omega \kappa_1 \kappa_2 \kappa_3\, d^3\underline{\kappa}\, d^3\underline{\phi}$$
$$= (2)^{-1} \sum_{t_4} \sum_{\mathbf{t}} \sum_{\hat{\mathbf{t}}} \int_{0^+}^{\infty} \cdot\cdot \{ \bar{\gamma}(\underline{\kappa}, \hat{\mathbf{t}}, t_4) [\gamma(\underline{\kappa}, \mathbf{t}, t_4 - 1) + \gamma(\underline{\kappa}, \mathbf{t}, t_4 + 1)] \}$$
$$\times \left\{ (2\pi)^{-3} \int_{-\pi}^{\pi} \cdot\cdot \exp[i(t_\beta - \hat{t}_\beta)\phi^\beta]\, d^3\underline{\phi} \right\} \omega \kappa_1 \kappa_2 \kappa_3\, d^3\underline{\kappa}$$
$$= (2)^{-1} \sum_{t} \int_{0^+}^{\infty} \cdot\cdot \{ \bar{\gamma}(\underline{\kappa}, t) [\gamma(\underline{\kappa}, \mathbf{t}, t_4 - 1)$$
$$+ \gamma(\underline{\kappa}, \mathbf{t}, t_4 + 1)] \} \omega \kappa_1 \kappa_2 \kappa_3\, d^3\underline{\kappa}. \qquad (7.2.39)$$

The reality of the right-hand side can be proved.

Now we shall compute the integral constant J from the equation (7.2.14) with assumptions $\beta \equiv 0 \equiv \nu$ and equation (7.2.16). In that case

$$J = I_1 = \int_{\mathbb{R}^7} [(p^j \bar{\phi}_{,q^j} - q^j \bar{\phi}_{,p_j}) \phi_{,q^4}]_{|q^4=0}\, d^3q\, d^4p + (\text{c.c.}). \qquad (7.2.40)$$

We shall compute J using equation (7.2.35) involving Bessel functions. For that reason, we digress to list some relevant formulae for Bessel functions, namely,

$$J_{-t}(\rho) = (-1)^t J_t(\rho), \tag{7.2.41-i}$$

$$(d/d\rho)J_t(\rho) = (1/2)[J_{t-1}(\rho) - J_{t+1}(\rho)], \tag{7.2.41-ii}$$

$$J_{t-1}(\rho) + J_{t+1}(\rho) = (2t/\rho)J_t(\rho), \tag{7.2.41-iii}$$

$$[(d^2/d\rho^2) + (1/\rho)(d/d\rho) + 1 - (t^2/\rho^2)]J_t(\rho) = 0, \tag{7.2.41-iv}$$

$$\begin{aligned}(\partial/\partial q)[J_t(k\rho)e^{\pm it\theta}] &= (k/2)[J_{t-1}(\cdot)e^{\mp i\theta} - J_{t+1}(\cdot)e^{\pm i\theta}]e^{\pm it\theta} \\ &= [(1/2)k\cos\theta(J_{t-1} - J_{t+1}) \mp (it/\rho)\sin\theta J_t]e^{\pm it\theta},\end{aligned} \tag{7.2.41-v}$$

$$\begin{aligned}(\partial/\partial p)[J_t(k\rho)e^{\pm it\theta}] &= \pm(ik/2)[J_{t-1}(\cdot)e^{\mp i\theta} + J_{t+1}(\cdot)]e^{\pm it\theta} \\ &= [(1/2)k\sin\theta(J_{t-1} - J_{t+1}) \pm (it/\rho)\cos\theta J_t]e^{\pm it\theta},\end{aligned} \tag{7.2.41-vi}$$

$$\int_{0^+}^{\infty} J_t(k\rho)J_t(\hat{k}\rho)\rho\, d\rho = (k)^{-1}\delta(k - \hat{k}), \tag{7.2.41-vii}$$

$$\int_{0^+}^{\infty} [J_t(\omega\rho)]^2(\rho)^{-1}\, d\rho = (2t)^{-1}, \tag{7.2.41-viii}$$

$$\hat{t}\int_{0^+}^{\infty} J_t(\rho)J_{\hat{t}}(\rho)(\rho)^{-1}\, d\rho$$

$$= S(t,\hat{t}) := \begin{cases} 0 \quad \text{for } \hat{t} = 0, \\ [\operatorname{sgn}(\hat{t})]^{|\hat{t}|}[\sin(|\hat{t}|\pi/2)](\hat{t}\pi/2) \quad \text{for } t = 0, \hat{t} \neq 0, \\ 2^{-1}\operatorname{sgn}(t) \quad \text{for } \hat{t} \neq 0, |t| = |\hat{t}|, \\ 2(\pi)^{-1}[\operatorname{sgn}(t)]^{|t|}[\operatorname{sgn}(\hat{t})]^{|\hat{t}|}\hat{t}[|\hat{t}|^2 - |t|^2]^{-1}\sin[(|\hat{t}| - |t|)\pi/2] \\ \qquad\qquad \text{for } \hat{t} \neq 0, t \neq 0, |t| \neq |\hat{t}|, \end{cases} \tag{7.2.41-ix}$$

$$\exp[i\rho\cos\theta] = \sum_{t=-\infty}^{\infty} (i)^t J_t(\rho)\exp[it\theta]. \tag{7.2.41-x}$$

We use now the equation (7.2.35) (with $v \equiv 0$) and equations (7.2.41-v), (7.2.41-vi) to obtain

$$\bar{\phi}^{\#}(\rho,\theta) = 2^{-1/2}(2\pi)^{-3/2}\sum_{t}\int_{0_+}^{\infty}\cdot\cdot\bar{\gamma}(\underline{\kappa},t)\left[\prod_{\beta=1}^{3}(-i)^{t_\beta}J_{t_\beta}(\kappa_\beta\rho^\beta)\right](i)^{t_4}J_{t_4}(\omega\rho^4)$$
$$\times \exp[-i(t_\alpha\theta^\alpha - t_4\theta^4)]\kappa_1\kappa_2\kappa_3\, d^3\underline{\kappa},$$

$$\phi^{\#}_{,q^4} = 2^{-1/2}(2\pi)^{-3/2} \sum_t \int_{0^+}^{\infty} \cdot\cdot \gamma(.,.) \left[\prod_{\beta=1}^{3} (i)^{t_\beta} J_{t_\beta}(\cdot) \right] (-i)^{t_4}$$
$$\times [(\omega/2)(\cos\theta^4)(J_{t_4-1}(\cdot) - J_{t_4+1}(\cdot))$$
$$+ it_4(\rho^4)^{-1}(\sin\theta^4)J_{t_4}(\cdot)]\exp[i(t_\alpha\theta^\alpha - t_4\theta^4)]\kappa_1\kappa_2\kappa_3\, d^3\underline{\kappa},$$

$$q^j\bar{\phi}_{,p^j} - p^j\bar{\phi}_{,q^j} = \rho^j\cos\theta^j[(\sin\theta^j)\phi^{\#}_{,\rho^j} + (\cos\theta^j)(\rho^j)^{-1}\bar{\phi}^{\#}_{,\theta^j}]$$
$$- \rho^j\sin\theta^j[(\cos\theta^j)\bar{\phi}^{\#}_{,\rho^j} - (\sin\theta^j)(\rho^j)^{-1}\bar{\phi}^{\#}_{,\theta^j}]$$
$$= \sum_{j=1}^{4} \bar{\phi}^{\#}_{,\theta^j} = -i2^{-1/2}(2\pi)^{-3/2}$$
$$\times \sum_t (t_1 + t_2 + t_3 - t_4) \int_{0^+}^{\infty} \cdot\cdot \bar{\gamma}(.,.) \left[\prod_{\beta=1}^{3} (-i)^{t_\beta} J_{t_\beta}(\cdot) \right]$$
$$\times (i)^{t_4} J_{t_4}(\cdot)\exp[-i(t_\alpha\theta^\alpha - t_4\theta^4)]\kappa_1\kappa_2\kappa_3\, d^3\underline{\kappa}. \tag{7.2.42}$$

Now, $q^4 = 0$ implies that either $\theta^4 = \pi/2$ or else $\theta^4 = -\pi/2$ [by equation (7.2.17)]. For $\theta^4 = \pi/2$ we have $p^4 > 0$, whereas for $\theta^4 = -\pi/2$ we get $p^4 < 0$. Let us compute a part I_1^+ of J from (7.2.14), (7.2.42) at $\theta^4 = \pi/2$.

$$I_1^+ = -(2)^{-1}(2\pi)^{-3} \sum_t \sum_{\hat{t}} (t_1 + t_2 + t_3 - t_4)\hat{t}_4$$
$$\times \int_{D_7^{\#}} \int_{0^+}^{\infty} \cdots \bar{\gamma}(\underline{\kappa}, t)\gamma(\underline{\hat{\kappa}}, \hat{t}) \left[\prod_{\beta=1}^{3} (-i)^{t_\beta}(i)^{\hat{t}_\beta} J_{t_\beta}(\cdot) J_{\hat{t}_\beta}(\cdot) \right]$$
$$\times (i)^{t_4}(-i)^{\hat{t}_4} J_{t_4}(\omega\rho^4) J_{\hat{t}_4}(\hat{\omega}\rho^4)\exp[i((\hat{t}_\alpha - t_\alpha)\theta^\alpha - (\hat{t}_4 - t_4)(\pi/2))]$$
$$\times \rho^1\rho^2\rho^3(\rho^4)^{-1}\kappa_1\kappa_2\kappa_3\hat{\kappa}_1\hat{\kappa}_2\hat{\kappa}_3\, d^4\rho\, d^3\underline{\theta}\, d^3\underline{\kappa}\, d^3\underline{\hat{\kappa}} + \text{(c.c.)}. \tag{7.2.43}$$

Using the θ-integrations and the equation (7.2.41-vii), (7.2.41-ix) the above equation yields

$$I_1^+ = -(2)^{-1} \sum_{\mathbf{t}} \sum_{t_4} \sum_{\hat{t}_4} (-1)^{t_4+\hat{t}_4}(t_1 + t_2 + t_3 - t_4)\hat{t}_4$$
$$\times \int_{0^+}^{\infty} \int_{0^+}^{\infty} \cdot\cdot \bar{\gamma}(\underline{\kappa}, \mathbf{t}, t_4)\gamma(\underline{\kappa}, t, \hat{t}_4) J_{t_4}(\omega\rho^4) J_{\hat{t}_4}(\omega\rho^4)(\rho^4)^{-1}$$
$$\times \kappa_1\kappa_2\kappa_3\, d\rho^4\, d^3\underline{\kappa} + \text{(c.c.)}$$
$$= -(2)^{-1} \sum_{\mathbf{t}} \sum_{t_4} \sum_{\hat{t}_4} (-1)^{t_4+\hat{t}_4}(t_1 + t_2 + t_3 - t_4)$$
$$\times \int_{0^+}^{\infty} \cdot\cdot \bar{\gamma}(\underline{\kappa}, t)\gamma(\underline{\kappa}, \mathbf{t}, \hat{t}_4) S(t_4, \hat{t}_4)\kappa_1\kappa_2\kappa_3\, d^3\underline{\kappa} + \text{(c.c.)}. \tag{7.2.44}$$

Similar computation yields for $\theta = -\pi/2$,

$$I_1^- = (2)^{-1} \sum_{\mathbf{t}} \sum_{t_4} \sum_{\hat{t}_4} (t_1 + t_2 + t_3 - t_4)$$
$$\times \int_{0^+}^{\infty} \cdot\cdot \bar{\gamma}(\underline{\kappa}, t)\gamma(\underline{\kappa}, \mathbf{t}, \hat{t}_4) S(t_4, \hat{t}_4)\kappa_1\kappa_2\kappa_3\, d^3\underline{\kappa} + \text{(c.c.)}. \tag{7.2.45}$$

Therefore, the integral constant

$$J = I_1^+ + I_1^- = 2^{-1} \sum_t \sum_{\hat{t}_4} [1 - (-1)^{t_4+\hat{t}_4}](t_1 + t_2 + t_3 - t_4)$$
$$\times \int_{0^+}^{\infty} \bar{\gamma}(\underline{\kappa}, t)\gamma(\underline{\kappa}, t, \hat{t}_4) S(t_4, \hat{t}_4)\kappa_1\kappa_2\kappa_3\, d^3\underline{\kappa} + (\text{c.c.}). \tag{7.2.46}$$

We summarize many constants of motion in the following:

$$2K_\alpha = 2^{-1} \sum_t \int_{0^+}^{\infty} \cdot\cdot \{\bar{\gamma}(\underline{\kappa}, t)[\gamma(\underline{\kappa}, \ldots, t_\alpha - 1, \ldots, t_4) + \gamma(\underline{\kappa}, \ldots, t_\alpha + 1, \ldots, t_4)]\}$$
$$\times\, \kappa_\alpha\kappa_1\kappa_2\kappa_3\, d^3\underline{\kappa} + (\text{c.c.}),$$

$$2H = 2^{-1} \sum_t \int_{0^+}^{\infty} \cdot\cdot \{\bar{\gamma}(\underline{\kappa}, t)[\gamma(\underline{\kappa}, t, t_4 - 1) + \gamma(\underline{\kappa}, t, t_4 + 1)]\}$$
$$\times\, \omega\kappa_1\kappa_2\kappa_3\, d^3\underline{\kappa} + (\text{c.c.}),$$

$$2X_\alpha = (i2)^{-1} \sum_t \int_{0^+}^{\infty} \cdot\cdot \{\bar{\gamma}(\underline{\kappa}, t)[\gamma(\underline{k}, \ldots, t_\alpha - 1, \ldots, t_4) - \gamma(\underline{\kappa}, \ldots, t_\alpha + 1, \ldots, t_4)]\}$$
$$\times\, \kappa_\alpha\kappa_1\kappa_2\kappa_3\, d^3\underline{\kappa} + (\text{c.c.}),$$

$$B \equiv 0,$$

$$N = \sum_t \int_{0^+}^{\infty} \cdot\cdot |\gamma(\underline{\kappa}, t)|^2 \kappa_1\kappa_2\kappa_3\, d^3\underline{\kappa},$$

$$Q = e_0 J = 2^{-1} e_0 \sum_t \sum_{\hat{t}_4} [1 - (-1)^{t_4+\hat{t}_4}](t_1 + t_2 + t_3 - t_4) S(t_4, \hat{t}_4)$$
$$\times \int_{0^+}^{\infty} \cdot\cdot \bar{\gamma}(\underline{\kappa}, t)\gamma(\underline{\kappa}, t, \hat{t}_4)\kappa_1\kappa_2\kappa_3\, d^3\underline{\kappa} + (\text{c.c.}). \tag{7.2.47}$$

Suppose that we consider a mixture of four kinds of elemental meson fields $\phi^{(T_3, T_4)}(\rho, \theta)$, $\phi^{(T_3, T_4+1)}(\rho, \theta)$, $\phi^{(T_3+1, T_4)}(\rho, \theta)$, $\phi^{(T_3, T_4+1)}(\rho, \theta)$, $\phi^{(T_3+1, T_4+1)}(\rho, \theta)$. Furthermore, let the meson field amplitude $\gamma(\underline{\kappa}, t)$ satisfy the special conditions characterized by

$$(2\kappa_1\kappa_2\kappa_3)^{1/2}\gamma(\underline{\kappa}, t) \equiv \begin{cases} \sqrt{F(\kappa, T_3, T_4)} \exp[ig(\underline{\kappa}, T_3, T)] \\ \text{for } t_1 = t_2 = 0;\ t_3 = T_3, T_3 + 1;\ t_4 = T_4, T_4 + 1; \\ 0 \quad \text{otherwise.} \end{cases} \tag{7.2.48}$$

This condition physically means that the special four modes are equally excited and other modes are absent. In this case, the constants of motion (7.2.47) reduce to

$$K_1 = K_2 = 0,$$

$$K_3 = \int_{0^+}^{\infty} \ldots F(\underline{\kappa}, T_3, T_4)\kappa_3 \, d^3\underline{\kappa},$$

$$H = \int_{0^+}^{\infty} \ldots F(\underline{\kappa}, T_3, T_4)\omega \, d\underline{\kappa},$$

$$X_1 = X_2 = X_3 = 0,$$

$$B = 0,$$

$$N = 2\int_{0^+}^{\infty} \ldots F(\underline{\kappa}, T_3, T_4)\, d^3\underline{\kappa},$$

$$Q = 4(\pi)^{-1}e_0(T_3 - T_4)\int_{0^+}^{\infty} \ldots F(\underline{\kappa}, T_3, T_4)\, d^3k = e(T_3 - T_4)N,$$

$$e \equiv 2(\pi)^{-1}e_0. \tag{7.2.49}$$

In this case F should be interpreted as a statistical distribution of mesons in "momentum and charge" space. The expression for the charge resembles Gellman–Nishijima's formula for the meson charge provided T_3 is identified with isotropic quantum number and $2T_4$ is identified with strangeness quantum number. In (7.2.49) each of the momentum component is integrated over $\mathbb{R}^+$ instead of $\mathbb{R}$ in the usual theory. This difference can be reconciled by noting that for an absolutely integrable, piecewise differentiable real-valued function f, the Fourier integral is

$$f(x) = \int_0^{\infty} [a(k)\cos kx + b(k)\sin kx]\, dk = \int_0^{\infty} [\gamma(k)e^{ikx} + \bar{\gamma}(k)e^{-ikx}]\, dk,$$

$$\gamma(k) \equiv 2^{-1}[a(k) - ib(k)].$$

By extending the domain of definition for γ from $\mathbb{R}^+$ into $\mathbb{R}$ by $\gamma(-k) \equiv \overline{\gamma(k)}$, the usual formula $\int_{-\infty}^{\infty} \gamma(k)e^{ikx}\, dk$ results. In this model $K_1 = K_2 = 0$ in (7.2.49). To obtain all nonzero components we need a mixture of *sixteen* elemental meson fields satisfying

$$(8\kappa_1\kappa_2\kappa_3)^{1/2}\gamma(\underline{\kappa}, t) \equiv \begin{cases} \sqrt{F(\underline{k}, T)}\exp[ig(\underline{k}, T)] & \text{for } t_a = T_a \text{ and } T_a + 1, \\ 0 & \text{otherwise.} \end{cases} \tag{7.2.50}$$

7.3. Spin-$\frac{1}{2}$ Fields in the Extended Phase Space

The relevant matrix algebra for a spin-$\frac{1}{2}$ in the extended phase space is the Clifford algebra of eight generators Γ^A satisfying

$$\Gamma^A\Gamma^B + \Gamma^B\Gamma^A = 2D^{AB}I, \tag{7.3.1}$$

where I stands for the unit element. A convenient basis set for this 256-dimensional algebra is given by

$$e^{n_1 n_2 \cdots n_8} = (\Gamma^1)^{n_1}(\Gamma^2)^{n_2}\cdots(\Gamma^8)^{n_8}. \tag{7.3.2}$$

Here $n_A \in \{0, 1\}$. The only irreducible representation of this algebra is sixteen dimensional. A possible 16×16 irreducible representation of the generators is furnished by

$$\begin{aligned}
&\Gamma^1 = \sigma^1 \times I \times I \times I = (\Gamma^1)^\dagger, && \Gamma^2 = \sigma^3 \times \sigma^1 \times I \times I = (\Gamma^2)^\dagger,\\
&\Gamma^3 = \sigma^3 \times \sigma^3 \times \sigma^1 \times I = (\Gamma^3)^\dagger, && \Gamma^4 = i\sigma^3 \times \sigma^3 \times \sigma^3 \times \sigma^1 = -(\Gamma^4)^\dagger,\\
&\Gamma^5 = \sigma^2 \times I \times I \times I = (\Gamma^5)^\dagger, && \Gamma^6 = \sigma^3 \times \sigma^2 \times I \times I = (\Gamma^6)^\dagger,\\
&\Gamma^7 = \sigma^3 \times \sigma^3 \times \sigma^2 \times I = (\Gamma^7)^\dagger, && \Gamma^8 = i\sigma^3 \times \sigma^3 \times \sigma^3 \times \sigma^2 = -(\Gamma^8)^\dagger.
\end{aligned} \tag{7.3.3}$$

Here, σ^a are the Pauli matrices, I is the 2×2 unit matrix, and the $\times$ denotes the Kronecker product [equation (3.3.2)]. There exists a nonsingular hermitian matrix Ω such that

$$\Omega \equiv \sigma^3 \times \sigma^3 \times \sigma^3 \times I \equiv \Omega^\dagger, \qquad (\Gamma^A)^\dagger = -\Omega\Gamma^A\Omega^{-1}. \tag{7.3.4}$$

The Lagrangian of a spin-$\frac{1}{2}$ field $\psi(\xi)$ is given by

$$\begin{aligned}
\mathscr{L}(..)_{|..} &= \tfrac{1}{2}[\tilde{\psi}(\xi)\Gamma^A\psi_{,A} - \tilde{\psi}_{,A}\Gamma^A\psi(\xi)] + m\tilde{\psi}(\xi)\psi(\xi),\\
\tilde{\psi}(\xi) &\equiv [\psi(\xi)]^\dagger\Omega.
\end{aligned} \tag{7.3.5}$$

Here, $\psi(\xi)$ is a 16×1 column vector field. The corresponding Euler–Lagrange equations yield

$$\begin{aligned}
\Gamma^A\psi_{,A} + m\psi(\xi) &= \mathbf{0},\\
\tilde{\psi}_{,A}\Gamma^A - m\tilde{\psi}(\xi) &= \mathbf{0}^\dagger,
\end{aligned} \tag{7.3.6}$$

where the $\mathbf{0}$ is the 16×1 zero column vector.

The various tensor fields obeying differential conservation laws can be constructed from the Lagrangian in (7.3.5). These are given by

$$\begin{aligned}
T^A{}_B(\xi) &= (1/2)\tilde{\psi}_{,B}\Gamma^A\psi(\xi) + (\text{h.c.}),\\
j^C{}_{AB}(\xi) &= (1/2)(\tilde{\psi}_{,A}\Gamma^C\psi - \tilde{\psi}\Gamma^C\psi_{,A})\xi_B - (1/2)(\tilde{\psi}_{,B}\Gamma^C\psi - \tilde{\psi}\Gamma^C\psi_{,B})\xi_A\\
&\quad + (1/8)[\tilde{\psi}\Gamma^C(\Gamma_A\Gamma_B - \Gamma_B\Gamma_A)\psi + (\text{h.c.})],\\
j^{(A)}(\xi) &= T^{(A)}(\xi) + (1/2)B^{(A)}(\xi)\\
&= (1/2)[(q^j\tilde{\psi}_{,p^j} - p^j\tilde{\psi}_{,q^j})\Gamma^A\psi - \tilde{\psi}\Gamma^A(q^j\psi_{,p^j} - p^j\psi_{,q^j})]\\
&\quad - (1/16)[\tilde{\psi}\Gamma^A d^{ij}(\Gamma_i\Gamma_{j+4} - \Gamma_{j+4}\Gamma_i - \Gamma_{i+4}\Gamma_j + \Gamma_j\Gamma_{i+4})\psi + (\text{h.c.})],\\
v^A(\xi) &= -i\tilde{\psi}(\xi)\Gamma^A\psi(\xi).
\end{aligned} \tag{7.3.7}$$

Here, (h.c.) stands for the hermitian conjugation of the preceding terms.

Some of the constants of motion, which can be constructed out of corresponding densities in (7.3.7), are furnished as:

$$
\begin{aligned}
K_\alpha &= (1/2)\int_{\mathbb{R}_7} [\tilde{\psi}_{,q^\alpha}\Gamma^4\psi + (\text{h.c.})]_{|q^4=0}\, d^3q\, d^4p, \\
H &= -K_4 = (1/2)\int_{\mathbb{R}_7} [\tilde{\psi}\Gamma^4\psi_{,q^4} + (\text{h.c.})]_{|..}\, d^3q\, d^4p, \\
X_a &= (1/2)\int_{\mathbb{R}_7} [\tilde{\psi}_{,p^a}\Gamma^4\psi + (\text{h.c.})]_{|..}\, d^3q\, d^4p, \\
Q &= e_0[T + (1/2)B] \\
&= (1/2)\int_{\mathbb{R}_7} \{q^j\tilde{\psi}_{,p^j} - p^j\tilde{\psi}_{,q^j})\Gamma^4\psi \\
&\qquad - (1/8)[\tilde{\psi}\Gamma^4 d^{ab}(\Gamma_a\Gamma_{b+4} - \Gamma_{b+4}\Gamma_a - \Gamma_{a+4}\Gamma_b + \Gamma_b\Gamma_{a+4})\psi] \\
&\qquad + (\text{h.c.})\}_{|..}\, d^3q\, d^4p, \\
N &= -i\int_{\mathbb{R}_7} (\tilde{\psi}\Gamma^4\psi)_{|..}\, d^3q\, d^4p,
\end{aligned}
\tag{7.3.8}
$$

where e_0 is the charge parameter.

We can write the first-order matrix differential equation (7.3.6) in the form

$$
\begin{aligned}
v^a\psi_{,q^a} + a^b\psi_{,p^b} + m\psi(q,p) &= 0, \\
v^a \equiv \Gamma^a, \qquad a^b &\equiv \Gamma^{b+4}, \\
v^a v^b + v^b v^a = 2d^{ab}I &= a^a a^b + a^b a^a, \\
v^a a^b + a^b v^a &= 0.
\end{aligned}
\tag{7.3.9}
$$

The v^a matrices have the significance of the four-velocity components and a^b matrices have the meaning of the four-acceleration components. The equation (7.3.9) has an analogy to the Boltzmann's transport equation for the statistical distribution function in the extended phase space. We shall call this equation the Boltzmann–Dirac–Yukawa or, in short, B.D.Y. equation.

Now we shall derive the unified spin-$\frac{1}{2}$ fields out of the B.D.Y. equation (7.3.9). For that purpose we introduce polar coordinates [also see equation (7.2.17)]

$$
q^b = \rho^b \cos\theta^b, \qquad p^b = \rho^b \sin\theta^b,
$$

where b is not summed. We recall that

$$
\begin{aligned}
\frac{\partial}{\partial q^b} &= \cos\theta^b \frac{\partial}{\partial \rho^b} - (\rho^b)^{-1}\sin\theta^b \frac{\partial}{\partial \theta^b}, \\
\frac{\partial}{\partial p^b} &= \sin\theta^b \frac{\partial}{\partial \rho^b} + (\rho^b)^{-1}\cos\theta^b \frac{\partial}{\partial \theta^b}, \\
\hat{\psi}(\rho,\theta) &\equiv \psi(q,p).
\end{aligned}
$$

Therefore, the B.D.Y. equation goes over into

$$
\begin{aligned}
&\alpha^{b+}[e^{-i\theta^b}(\hat{\psi}_{,\rho^b} - i(\rho^b)^{-1}\hat{\psi}_{,\theta^b})] \\
&\qquad + \alpha^{b-}[e^{i\theta^b}(\hat{\psi}_{,\rho^b} + i(\rho^b)^{-1}\hat{\psi}_{,\theta^b})] + m\hat{\psi}(\rho,\theta) = 0, \qquad (7.3.10a) \\
&\alpha^{b+} \equiv (1/2)(\Gamma^b + i\Gamma^{b+4}), \qquad \alpha^{b-} \equiv (1/2)(\Gamma^b - i\Gamma^{b+4}).
\end{aligned}
$$

We can more explicitly write 16×16 $\alpha^{b\pm}$ matrices in the following way:

$$
\alpha^{1+} = \begin{bmatrix}
 & & & & I & & & \\
 & & & & & I & & \\
 & & & & & & I & \\
 & & & & & & & I \\
 & & & & & & & \\
 & & & & & & & \\
 & & & & & & & \\
 & & & & & & &
\end{bmatrix}, \quad
\alpha^{1-} = \begin{bmatrix}
 & & & & & & & \\
 & & & & & & & \\
 & & & & & & & \\
 & & & & & & & \\
I & & & & & & & \\
 & I & & & & & & \\
 & & I & & & & & \\
 & & & I & & & &
\end{bmatrix},
$$

$$
\alpha^{2+} = \begin{bmatrix}
 & & I & & & & & \\
 & & & I & & & & \\
 & & & & & & & \\
 & & & & & & & \\
 & & & & & & -I & \\
 & & & & & & & -I \\
 & & & & & & & \\
 & & & & & & &
\end{bmatrix}, \quad
\alpha^{2-} = \begin{bmatrix}
 & & & & & & & \\
 & & & & & & & \\
I & & & & & & & \\
 & I & & & & & & \\
 & & & & & & & \\
 & & & & & & & \\
 & & & & -I & & & \\
 & & & & & -I & &
\end{bmatrix},
$$

$$
\alpha^{3+} = \begin{bmatrix}
 & I & & & & & & \\
 & & & & & & & \\
 & & & -I & & & & \\
 & & & & & & & \\
 & & & & & -I & & \\
 & & & & & & & \\
 & & & & & & & I \\
 & & & & & & &
\end{bmatrix}, \quad
\alpha^{3-} = \begin{bmatrix}
 & & & & & & & \\
I & & & & & & & \\
 & & & & & & & \\
 & & -I & & & & & \\
 & & & & & & & \\
 & & & & -I & & & \\
 & & & & & & & \\
 & & & & & & I &
\end{bmatrix},
$$

$$
2\alpha^{4+} = \begin{bmatrix}
-(\sigma_2 - i\sigma_1) & & & & & & & \\
 & \sigma_2 - i\sigma_1 & & & & & & \\
 & & \sigma_2 - i\sigma_1 & & & & & \\
 & & & -(\sigma_2 - i\sigma_1) & & & & \\
 & & & & \sigma_2 - i\sigma_1 & & & \\
 & & & & & -(\sigma_2 - i\sigma_1) & & \\
 & & & & & & -(\sigma_2 - i\sigma_1) & \\
 & & & & & & & \sigma_2 - i\sigma_1
\end{bmatrix},
$$

$$2\alpha^{4-} = \begin{bmatrix} \sigma_2+i\sigma_1 & & & & & & & \\ & -(\sigma_2+i\sigma_1) & & & & & & \\ & & -(\sigma_2+i\sigma_1) & & & & & \\ & & & (\sigma_2+i\sigma_1) & & & & \\ & & & & -(\sigma_2+i\sigma_1) & & & \\ & & & & & \sigma_2+i\sigma_1 & & \\ & & & & & & \sigma_2+i\sigma_1 & \\ & & & & & & & -(\sigma_2+i\sigma_1) \end{bmatrix}. \tag{7.3.10b}$$

where I stands for the 2×2 identity matrix and blank spaces contain zeros only.

We shall seek basic, separable solutions of the polar B.D.Y. equation (7.3.10a) by putting the spin-$\frac{1}{2}$ field components as

$$\begin{aligned} \hat{\psi}^L(\rho,\theta) &= \chi^L(\rho)\exp(it^L_b\theta^b), \\ L &\in \{1,2,\ldots,16\}. \end{aligned} \tag{7.3.11}$$

In the sequel, L, M, $N \in \{1,2,\ldots,16\}$, and summation convention will be carried out between a pair of dummy superscript and subscript. According to our convention, the index L in (7.3.11) is *not* summed.

Substituting (7.3.11) into (7.3.10) we can derive the following system of linear partial differential equations:

$$\begin{aligned}
&[\chi^9_{,\rho^1} + (t^9_1/\rho^1)\chi^9]\exp[i(t^9_b\theta^b - \theta^1)] + [\chi^5_{,\rho^2} + (t^5_2/\rho^2)\chi^5]\exp[i(t^5_b\theta^b - \theta^2)] \\
&\quad + [\chi^3_{,\rho^3} + (t^3_3/\rho^3)\chi^3]\exp[i(t^3_b\theta^b - \theta^3)] \\
&\quad + i[\chi^2_{,\rho^4} + (t^2_4/\rho^4)\chi^2]\exp[i(t^2_b\theta^b - \theta^4)] + m\chi^1\exp[it^1_b\theta^b] = 0, \\
&[\chi^{10}_{,\rho^1} + (t^{10}_1/\rho^1)\chi^{10}]\exp[i(t^{10}_b\theta^b - \theta^1)] + [\chi^6_{,\rho^2} + (t^6_2/\rho^2)\chi^6]\exp[i(t^6_b\theta^b - \theta^2)] \\
&\quad + [\chi^4_{,\rho^3} + (t^4_3/\rho^3)\chi^4]\exp[i(t^4_b\theta^b - \theta^3)] + i[\chi^1_{,\rho^4} - (t^1_4/\rho^4)\chi^1] \\
&\quad \times \exp[i(t^1_b\theta^b + \theta^4)] + m\chi^2\exp[it^2_b\theta^b] = 0, \\
&[\chi^{11}_{,\rho^1} + (t^{11}_1/\rho^1)\chi^{11}]\exp[i(t^{11}_b\theta^b - \theta^1)] + [\chi^7_{,\rho^2} + (t^7_2/\rho^2)\chi^7]\exp[i(t^7_b\theta^b - \theta^2)] \\
&\quad + [\chi^1_{,\rho^3} - (t^1_3/\rho^3)\chi^1]\exp[i(t^1_b\theta^b + \theta^3)] \\
&\quad - i[\chi^4_{,\rho^4} + (t^4_4/\rho^4)\chi^4]\exp[i(t^4_b\theta^b - \theta^4)] + m\chi^3\exp[it^3_b\theta^b] = 0, \\
&[\chi^{12}_{,\rho^1} + (t^{12}_1/\rho^1)\chi^{12}]\exp[i(t^{12}_b\theta^b - \theta^1)] + [\chi^8_{,\rho^2} + (t^8_2/\rho^2)\chi^8]\exp[i(t^8_b\theta^b - \theta^2)] \\
&\quad + [\chi^2_{,\rho^3} - (t^2_3/\rho^3)\chi^2]\exp[i(t^2_b\theta^b + \theta^3)] \\
&\quad - i[\chi^3_{,\rho^4} - (t^3_4/\rho^4)\chi^3]\exp[i(t^3_b\theta^b + \theta^4)] + m\chi^4\exp[it^4_b\theta^b] = 0, \\
&[\chi^{13}_{,\rho^1} + (t^{13}_1/\rho^1)\chi^{13}]\exp[i(t^{13}_b\theta^b - \theta^1)] + [\chi^1_{,\rho^2} - (t^1_2/\rho^2)\chi^1]\exp[i(t^1_b\theta^b + \theta^2)] \\
&\quad - [\chi^7_{,\rho^3} + (t^7_3/\rho^3)\chi^7]\exp[i(t^7_b\theta^b - \theta^3)] \\
&\quad - i[\chi^6_{,\rho^4} + (t^6_4/\rho^4)\chi^6]\exp[i(t^6_b\theta^b - \theta^4)] + m\chi^5\exp[it^5_b\theta^b] = 0,
\end{aligned}$$

$$\begin{aligned}
&[\chi^{14}_{,\rho^1} + (t_1^{14}/\rho^1)\chi^{14}]\exp[i(t_b^{14}\theta^b - \theta^1)] + [\chi^{2}_{,\rho^2} - (t_2^{2}/\rho^2)\chi^{2}]\exp[i(t_b^{2}\theta^b + \theta^2)] \\
&\quad - [\chi^{8}_{,\rho^3} + (t_3^{8}/\rho^3)\chi^{8}]\exp[i(t_b^{8}\theta^b - \theta^3)] \\
&\quad - i[\chi^{5}_{,\rho^4} - (t_4^{5}/\rho^4)\chi^{5}]\exp[i(t_b^{5}\theta^b + \theta^4)] + m\chi^6\exp[it_b^6\theta^b] = 0,
\end{aligned}$$

$$\begin{aligned}
&[\chi^{15}_{,\rho^1} + (t_1^{15}/\rho^1)\chi^{15}]\exp[i(t_b^{15}\theta^b - \theta^1)] + [\chi^{3}_{,\rho^2} - (t_2^{3}/\rho^2)\chi^{3}]\exp[i(t_b^{3}\theta^b + \theta^2)] \\
&\quad - [\chi^{5}_{,\rho^3} - (t_3^{5}/\rho^3)\chi^{5}]\exp[i(t_b^{5}\theta^b + \theta^3)] \\
&\quad + i[\chi^{8}_{,\rho^4} + (t_4^{8}/\rho^4)\chi^{8}]\exp[i(t_b^{8}\theta^b - \theta^4)] + m\chi^7\exp[it_b^7\theta^b] = 0,
\end{aligned}$$

$$\begin{aligned}
&[\chi^{16}_{,\rho^1} + (t_1^{16}/\rho^1)\chi^{16}]\exp[i(t_b^{16}\theta^b - \theta^1)] + [\chi^{4}_{,\rho^2} - (t_2^{4}/\rho^2)\chi^{4}]\exp[i(t_b^{4}\theta^b + \theta^2)] \\
&\quad - [\chi^{6}_{,\rho^3} - (t_3^{6}/\rho^3)\chi^{6}]\exp[i(t_b^{6}\theta^b + \theta^3)] \\
&\quad + i[\chi^{7}_{,\rho^4} - (t_4^{7}/\rho^4)\chi^{7}]\exp[i(t_b^{7}\theta^b + \theta^4)] + m\chi^8\exp[it_b^8\theta^b] = 0,
\end{aligned}$$

$$\begin{aligned}
&[\chi^{1}_{,\rho^1} - (t_1^{1}/\rho^3)\chi^{1}]\exp[i(t_b^{1}\theta^b + \theta^1)] - [\chi^{13}_{,\rho^2} + (t_2^{13}/\rho^2)\chi^{13}]\exp[i(t_b^{13}\theta^b - \theta^4)] \\
&\quad - [\chi^{11}_{,\rho^3} + (t_3^{11}/\rho^3)\chi^{11}]\exp[i(t_b^{11}\theta^b - \theta^3)] \\
&\quad - i[\chi^{10}_{,\rho^4} + (t_4^{10}/\rho^4)\chi^{10}]\exp[i(t_b^{10}\theta^b - \theta^4)] + m\chi^9\exp[it_b^9\theta^b] = 0,
\end{aligned}$$

$$\begin{aligned}
&[\chi^{2}_{,\rho^1} - (t_1^{2}/\rho^1)\chi^{2}]\exp[i(t_b^{2}\theta^b + \theta^1)] - [\chi^{14}_{,\rho^2} + (t_2^{14}/\rho^2)\chi^{14}]\exp[i(t_b^{14}\theta^b - \theta^2)] \\
&\quad - [\chi^{12}_{,\rho^3} + (t_3^{12}/\rho^3)\chi^{12}]\exp[i(t_b^{12}\theta^b - \theta^3)] \\
&\quad - i[\chi^{9}_{,\rho^4} - (t_4^{9}/\rho^4)\chi^{9}]\exp[i(t_b^{9}\theta^b + \theta^4)] + m\chi^{10}\exp[it_b^{10}\theta^b] = 0,
\end{aligned}$$

$$\begin{aligned}
&[\chi^{3}_{,\rho^1} - (t_1^{3}/\rho^1)\chi^{3}]\exp[i(t_b^{3}\theta^b + \theta^1)] - [\chi^{15}_{,\rho^2} + (t_2^{15}/\rho^2)\chi^{15}]\exp[i(t_b^{15}\theta^b - \theta^2)] \\
&\quad - [\chi^{9}_{,\rho^3} - (t_3^{9}/\rho^3)\chi^{9}]\exp[i(t_b^{9}\theta^b + \theta^3)] \\
&\quad + i[\chi^{12}_{,\rho^4} + (t_4^{12}/\rho^4)\chi^{12}]\exp[i(t_b^{12}\theta^b - \theta^4)] + m\chi^{11}\exp[it_b^{11}\theta^b] = 0,
\end{aligned}$$

$$\begin{aligned}
&[\chi^{4}_{,\rho^1} - (t_1^{4}/\rho^1)\chi^{4}]\exp[i(t_b^{4}\theta^b + \theta^1)] - [\chi^{16}_{,\rho^2} + (t_2^{16}/\rho^2)\chi^{16}]\exp[i(t_b^{16}\theta^b - \theta^2)] \\
&\quad - [\chi^{10}_{,\rho^3} - (t_3^{10}/\rho^3)\chi^{10}]\exp[i(t_b^{10}\theta^b + \theta^3)] \\
&\quad + i[\chi^{11}_{,\rho^4} - (t_4^{11}/\rho^4)\chi^{11}]\exp[i(t_b^{11}\theta^b + \theta^4)] + m\chi^{12}\exp[it_b^{12}\theta^b] = 0,
\end{aligned}$$

$$\begin{aligned}
&[\chi^{5}_{,\rho^1} - (t_1^{5}/\rho^1)\chi^{5}]\exp[i(t_b^{5}\theta^b + \theta^1)] - [\chi^{9}_{,\rho^2} - (t_2^{9}/\rho^2)\chi^{9}]\exp[i(t_b^{9}\theta^b + \theta^2)] \\
&\quad + [\chi^{15}_{,\rho^3} + (t_3^{15}/\rho^3)\chi^{15}]\exp[i(t_b^{15}\theta^b - \theta^3)] \\
&\quad + i[\chi^{14}_{,\rho^4} + (t_4^{14}/\rho^4)\chi^{14}]\exp[i(t_b^{14}\theta^b - \theta^4)] + m\chi^{13}\exp[it_b^{13}\theta^b] = 0,
\end{aligned}$$

$$\begin{aligned}
&[\chi^{6}_{,\rho^1} - (t_1^{6}/\rho^1)\chi^{6}]\exp[i(t_b^{6}\theta^b + \theta^1)] - [\chi^{10}_{,\rho^2} - (t_2^{10}/\rho^2)\chi^{10}]\exp[i(t_b^{10}\theta^b + \theta^2)] \\
&\quad + [\chi^{16}_{,\rho^3} + (t_3^{16}/\rho^3)\chi^{16}]\exp[i(t_b^{16}\theta^b - \theta^3)] \\
&\quad + i[\chi^{13}_{,\rho^4} - (t_4^{13}/\rho^4)\chi^{13}]\exp[i(t_b^{13}\theta^b + \theta^4)] + m\chi^{14}\exp[it_b^{14}\theta^b] = 0,
\end{aligned}$$

$$\begin{aligned}
&[\chi^{7}_{,\rho^1} - (t_1^{7}/\rho^1)\chi^{7}]\exp[i(t_b^{7}\theta^b + \theta^1)] - [\chi^{11}_{,\rho^2} - (t_2^{11}/\rho^2)\chi^{11}]\exp[i(t_b^{11}\theta^b + \theta^2)] \\
&\quad + [\chi^{13}_{,\rho^3} - (t_3^{13}/\rho^3)\chi^{13}]\exp[i(t_b^{13}\theta^b + \theta^3)] \\
&\quad - i[\chi^{16}_{,\rho^4} + (t_4^{16}/\rho^4)\chi^{16}]\exp[i(t_b^{16}\theta^b - \theta^4)] + m\chi^{15}\exp[it_b^{15}\theta^b] = 0,
\end{aligned}$$

$$[\chi^8_{,\rho^1} - (t^8_1/\rho^1)\chi^8]\exp[i(t^8_b\theta^b + \theta^1)] - [\chi^{12}_{,\rho^2} - (t^{12}_2/\rho^2)\chi^{12}]\exp[i(t^{12}_b\theta^b + \theta^2)]$$
$$+ [\chi^{14}_{,\rho^3} - (t^{14}_3/\rho^3)\chi^{14}]\exp[i(t^{14}_b\theta^b + \theta^3)]$$
$$- i[\chi^{15}_{,\rho^4} - (t^{15}_4/\rho^4)\chi^{15}]\exp[i(t^{15}_b\theta^b + \theta^4)] + m\chi^{16}\exp[it^{16}_b\theta^b] = 0. \tag{7.3.12}$$

Let us analyze the first of these equations. It can be written in the form

$$G_1(\rho;\theta^2,\theta^3,\theta^4)\exp[i(t^9_1 - 1 - t^1_1)\theta^1]$$
$$+ G_2(\rho;\theta^2,\theta^3,\theta^4)\exp[i(t^5_1 - t^1_1)\theta^1] + G_3(\rho;\theta^2,\theta^3,\theta^4)\exp[i(t^3_1 - t^1_1)\theta^1]$$
$$+ G_4(\rho;\theta^2,\theta^3,\theta^4)\exp[i(t^2_1 - t^1_1)\theta^1]$$
$$+ G_5(\rho;\theta^2,\theta^3,\theta^4) = 0. \tag{7.3.13}$$

Here it is assumed that at least one of the $G_a(..) \not\equiv 0$ (say, $G_1(..) \not\equiv 0$). Otherwise, the equation becomes an identity. We differentiate the left-hand side of (7.3.13) with respect to θ^1 and divide by $i\exp[i(t^2_1 - t^1_1)\theta^1]$. Then again we differentiate with respect to θ^1 and divide by $i\exp[i(t^3_1 - t^2_1)\theta^1]$. Next we differentiate and divide out $i\exp[i(t^5_1 - t^3_1)\theta^1]$. The resulting equation yields (since $G_1(..) \not\equiv 0$)

$$(t^9_1 - 1 - t^1_1)(t^9_1 - 1 - t^2_1)(t^9_1 - 1 - t^3_1)(t^9_1 - 1 - t^5_1) = 0. \tag{7.3.14}$$

Similarly, many more equations can be derived from the sixteen equations of (7.3.12). Invoking separation of variables, the solution of these sixteen equations is

$$t^1_1 = t^2_1 = t^3_1 = t^4_1 = t^5_1 = t^6_1 = t^7_1 = t^8_1 = t^9_1 - 1$$
$$= t^{10}_1 - 1 = t^{11}_1 - 1 = t^{12}_1 - 1 = t^{13}_1 - 1 = t^{14}_1 - 1$$
$$= t^{15}_1 - 1 = t^{16}_1 - 1. \tag{7.3.14a}$$

Similarly, from differentiations of θ^2, θ^3, θ^4 the following relations emerge:

$$t^1_2 = t^2_2 = t^3_2 = t^4_2 = t^5_2 - 1 = t^6_2 - 1 = t^7_2 - 1$$
$$= t^8_2 - 1 = t^9_2 = t^{10}_2 = t^{11}_2 = t^{12}_2 = t^{13}_2 - 1 = t^{14}_2 - 1$$
$$= t^{15}_2 - 1 = t^{16}_2 - 1,$$
$$t^1_3 = t^2_3 = t^3_3 - 1 = t^4_3 - 1 = t^5_3 = t^6_3 = t^7_3 - 1$$
$$= t^8_3 - 1 = t^9_3 = t^{10}_3 = t^{11}_3 - 1 = t^{12}_3 - 1 = t^{13}_3 = t^{14}_3$$
$$= t^{15}_3 - 1 = t^{16}_3 - 1,$$
$$t^1_4 = t^2_4 - 1 = t^3_4 = t^4_4 - 1 = t^5_4 = t^6_4 - 1 = t^7_4$$
$$= t^8_4 - 1 = t^9_4 = t^{10}_4 - 1 = t^{11}_4 = t^{12}_4 - 1 = t^{13}_4 = t^{14}_4 - 1$$
$$= t^{15}_4 = t^{16}_1 - 1. \tag{7.3.14b}$$

The above sixty linear equations hold among sixty-four integers t_b^L. It is clear that four of t_b^L's can be chosen arbitrarily and that the remaining sixty can be obtained through (7.3.14*a*,*b*). The following t_b^L are chosen freely:

$$t_1 \equiv t_1^1, \qquad t_2 \equiv t_2^2, \qquad t_3 \equiv t_3^3 - 1, \qquad t_4 \equiv t_4^4 - 1. \tag{7.3.15}$$

With these choices of t_b^L, the functions of θ^j's are all cancelled and the polar B.D.Y. equation (7.3.10*a*) or (7.3.12) goes over to

$$\begin{aligned}(\alpha^{b+})^N_L[\chi^L_{,\rho^b} + (t_b^L/\rho^b)\chi^L] + (\alpha^{b-})^N_L[\chi^L_{,\rho^b} - (t_b^L/\rho^b)\chi^L]& \\ + m\chi^N(\rho) = 0.&\end{aligned} \tag{7.3.16}$$

We want to solve the above system of equations. We notice that, if we operate from the left by $\Gamma^A\partial/\partial\xi^A - mI$ the first of equations (7.3.6), then we obtain the decoupled second-order system $D^{AB}\psi_{,AB} - m^2\psi(\xi) = 0$. Similarly, the system (7.3.16) yields (under the assumption that the χ^L's are twice differentiable) the decoupled system:

$$\begin{aligned}M^2\chi^L(\rho) &\equiv d^{ab}[\chi^L_{,\rho^a\rho^b} + (\rho^a)^{-1}\chi^L_{,\rho^b}] \\ &= [m^2 + (d^{ab}t_a^L t_b^L/\rho^a\rho^b)]\chi^L(\rho).\end{aligned} \tag{7.3.17}$$

(Here L is not summed.) This equation indicates an Okubo-like mass splitting [compare the equation (7.2.22)]. From the knowledge of Sec. 2, we try to solve (7.3.16) in terms of Bessel functions of the first kind. Let us assume for separable solutions

$$\chi^L(\rho) = u^L(\kappa)\prod_{j=1}^{4}[J_{t_j^L}(\kappa_j\rho^j)]. \tag{7.3.18}$$

Here j is not summed, t_b^L are given by (7.3.14*a*,*b*), and $u^L(\kappa)$ are undetermined amplitude functions. We substitute (7.3.18) into (7.3.16). The first equation (for $N = 1$) yields

$$\begin{aligned}&[J_{t_1^9,\rho^1} + (t_1^9/\rho^1)J_{t_1^9}]J_{t_2^9}J_{t_3^9}J_{t_4^9}u^9 + [J_{t_2^5,\rho^2} + (t_2^5/\rho^2)J_{t_2^5}]J_{t_1^5}J_{t_3^5}J_{t_4^5}u^5 \\ &\quad + [J_{t_3^3,\rho^3} + (t_3^3/\rho^3)J_{t_3^3}]J_{t_1^3}J_{t_2^3}J_{t_4^3}u^3 \\ &\quad + i[J_{t_4^2,\rho^4} + (t_4^2/\rho^4)J_{t_4^2}]J_{t_1^2}J_{t_2^2}J_{t_3^2}u^2 + mJ_{t_1^1}J_{t_2^1}J_{t_3^1}J_{t_4^1}u^1 = 0.\end{aligned} \tag{7.3.19}$$

Using the rules $[\partial_\rho \mp (t/\rho)]J_t(\kappa\rho) = \mp\kappa J_{t\pm 1}(\kappa\rho)$ of the equations (7.2.41-ii,iii) and the equations (7.3.14*a*,*b*), we observe that Bessel functions in (7.3.19) can all be *cancelled* out. Therefore, (7.3.19) yields the algebraic equation

$$\kappa_1 u^9 + \kappa_2 u^5 + \kappa_3 u^3 + i\kappa_4 u^2 + mu^1 = 0. \tag{7.3.20}$$

Similar simplifications occur in other fifteen equations of (7.3.16). The resulting algebraic system of equations can be neatly summarized into

$$[i\kappa_b a^b + mI]u(\kappa) = \mathbf{0}. \tag{7.3.21}$$

Here the 16×16 matrices a^b have been defined in the equation (7.3.9) and

$u(\kappa)$ is the unknown 16×1 column vector with components $u^L(\kappa)$. Multiplying the above equation from the left by $[-i\kappa_c a^c + mI]$ we obtain

$$[d^{bc}\kappa_b\kappa_c + m^2]u(\kappa) = 0.$$

For a nontrivial solution we must have the necessary condition:

$$\begin{aligned} d^{bc}\kappa_b\kappa_c + m^2 &= 0, \\ \kappa_4 &= \pm E(\underline{\kappa}), \\ E(\underline{\kappa}) &\equiv \sqrt{\kappa_\alpha\kappa_\alpha + m^2}. \end{aligned} \tag{7.3.22}$$

The present representation of the a^b matrices in (7.3.9) is too complicated to solve (7.3.21) neatly. We notice, however, that $a_b a_c + a_c a_b = 2d_{bc}I$, which implies that the a^b-matrices are reducible to four direct sums of Dirac matrices by a similarity transformation. Such a similarity matrix, which is unitary as well, is explicitly furnished:

$$4S \equiv \begin{bmatrix} I+\sigma^2 & I-\sigma^2 & \sigma^3+i\sigma^1 & \sigma^3-i\sigma^1 & i(I-\sigma^2) & i(I+\sigma^2) & -(\sigma^1+i\sigma^3) & \sigma^1-i\sigma^3 \\ -(\sigma^3+i\sigma^1) & \sigma^3-i\sigma^1 & I+\sigma^2 & -(I-\sigma^2) & \sigma^1+i\sigma^3 & \sigma^1-i\sigma^3 & i(I-\sigma^2) & -i(I+\sigma^2) \\ \sigma^3-i\sigma^1 & \sigma^3+i\sigma^1 & -(I-\sigma^2) & -(I+\sigma^2) & \sigma^1-i\sigma^3 & -(\sigma^1+i\sigma^3) & -i(I+\sigma^2) & -i(I-\sigma^2) \\ -(I-\sigma^2) & I+\sigma^2 & -(\sigma^3-i\sigma^1) & \sigma^3+i\sigma^1 & -i(I+\sigma^2) & i(I-\sigma^2) & -(\sigma^1-i\sigma^3) & -(\sigma^1+i\sigma^3) \\ -i(I+\sigma^2) & i(I-\sigma^2) & -(\sigma^1-i\sigma^3) & -(\sigma^1+i\sigma^3) & -(I-\sigma^2) & I+\sigma^2 & -(\sigma^3-i\sigma^1) & \sigma^3+i\sigma^1 \\ \sigma^1-i\sigma^3 & -(\sigma^1+i\sigma^3) & -i(I+\sigma^2) & -i(I-\sigma^2) & \sigma^3-i\sigma^1 & \sigma^3+i\sigma^1 & -(I-\sigma^2) & -(I+\sigma^2) \\ -(\sigma^1+i\sigma^3) & -(\sigma^1-i\sigma^3) & -i(I-\sigma^2) & i(I+\sigma^2) & \sigma^3+i\sigma^1 & -(\sigma^3-i\sigma^1) & -(I+\sigma^2) & I-\sigma^2 \\ -i(I-\sigma^2) & -i(I+\sigma^2) & \sigma^1+i\sigma^3 & -(\sigma^1-i\sigma^3) & -(I+\sigma^2) & -(I-\sigma^2) & -(\sigma^3+i\sigma^1) & -(\sigma^3-i\sigma^1) \end{bmatrix} \tag{7.3.23}$$

We can verify by tedious computations (which involve block by block matrix multiplication) that

$$\begin{aligned} S^\dagger S &= I, & Sa^bS^\dagger &= I \times \hat{\gamma}^b, \\ \hat{\gamma}^\alpha &= \sigma^2 \times \sigma^\alpha, & \hat{\gamma}^4 &= i\sigma^3 \times I. \end{aligned} \tag{7.3.24}$$

Here the same symbol I has been used for unit matrices of different sizes. The $\hat{\gamma}^b$ matrices are representations of Dirac matrices that are different from those in either (6.5.4) or (6.5.6). Multiplying the equation (7.3.21) by S from the left and using (7.3.24), we can derive

$$\begin{aligned} [I \times (i\kappa_b\hat{\gamma}^b + mI)]\hat{u} &= 0, \\ \hat{u} = Su, \qquad u &= S^\dagger\hat{u}. \end{aligned} \tag{7.3.25}$$

If we put

$$\hat{u} = \hat{I} \times u_D, \qquad [\hat{I}]^T \equiv (1, 1, 1, 1), \tag{7.3.26}$$

where u_D is a 4×1 column vector, then (7.3.25) reduces to

$$\begin{aligned} I \times [(i\kappa_b\hat{\gamma}^b + mI)u_D] &= 0, \\ [i\kappa_b\hat{\gamma}^b + mI]u_D &= 0. \end{aligned} \tag{7.3.27}$$

The last equation is equivalent to (6.5.8), the Dirac equation for the plane wave amplitude. Four basic solutions of (7.3.27), which are analogous to those given in (6.5.12*a,b*), are listed in the following:

$$\begin{aligned} \nu[e_1]^{\mathrm{T}} &= (1, 0, (E+m)^{-1}\kappa^3, (E+m)^{-1}(\kappa_1 + i\kappa_2)), \\ \nu[e_2]^{\mathrm{T}} &= (0, 1, (E+m)^{-1}(\kappa_1 - i\kappa_2), -(E+m)^{-1}\kappa_3), \\ \nu[e_1']^{\mathrm{T}} &= (-(E+m)^{-1}\kappa_3, -(E+m)^{-1}(\kappa_1 + i\kappa_2), 1, 0), \\ \nu[e_2']^{\mathrm{T}} &= (-(E+m)^{-1}(\kappa_1 - i\kappa_2), (E+m)^{-1}\kappa_3, 0, 1), \\ \nu^{-1} &\equiv \sqrt{(E+m)/2m}. \end{aligned} \tag{7.3.28}$$

Here T denotes the transposition. We can construct *sixteen basic solutions* of (7.3.25) by using (7.3.28). These are furnished by

$$\begin{aligned} \hat{u}^{\mathrm{T}}_{(11)} &= [e_1^{\mathrm{T}}, 0, 0, 0], & \hat{u}^{\mathrm{T}}_{(12)} &= [e_2^{\mathrm{T}}, 0, 0, 0], \\ \hat{u}^{\mathrm{T}}_{(21)} &= [0, e_1^{\mathrm{T}}, 0, 0], & \hat{u}^{\mathrm{T}}_{(22)} &= [0, e_2^{\mathrm{T}}, 0, 0], \\ \hat{u}^{\mathrm{T}}_{(31)} &= [0, 0, e_1^{\mathrm{T}}, 0], & \hat{u}^{\mathrm{T}}_{(32)} &= [0, 0, e_2^{\mathrm{T}}, 0], \\ \hat{u}^{\mathrm{T}}_{(41)} &= [0, 0, 0, e_1^{\mathrm{T}}], & \hat{u}^{\mathrm{T}}_{(42)} &= [0, 0, 0, e_2^{\mathrm{T}}], \\ \hat{v}^{\mathrm{T}}_{(11)} &= [e_1'^{\mathrm{T}}, 0, 0, 0], & \hat{v}^{\mathrm{T}}_{(12)} &= [e_2'^{\mathrm{T}}, 0, 0, 0], \\ \hat{v}^{\mathrm{T}}_{(21)} &= [0, e_1'^{\mathrm{T}}, 0, 0], & \hat{v}^{\mathrm{T}}_{(22)} &= [0, e_2'^{\mathrm{T}}, 0, 0], \\ \hat{v}^{\mathrm{T}}_{(31)} &= [0, 0, e_1'^{\mathrm{T}}, 0], & \hat{v}^{\mathrm{T}}_{(32)} &= [0, 0, e_2'^{\mathrm{T}}, 0], \\ \hat{v}^{\mathrm{T}}_{(41)} &= [0, 0, 0, e_1'^{\mathrm{T}}], & \hat{v}^{\mathrm{T}}_{(42)} &= [0, 0, 0, e_2'^{\mathrm{T}}]. \end{aligned} \tag{7.3.29}$$

Here, each entry represents an 1×4 row matrix. By the linear independence of e_1, e_2, e_1', e_2', the linear independence of $\hat{u}_{(a\alpha)}$, $\hat{v}_{(b\beta)}$ (where α, $\beta \in \{1, 2\}$) follows. The sixteen (linearly independent) basic solutions of (7.3.21) will be given by [using equations (7.3.29), (7.3.25), (7.3.24)]

$$u_{(a\alpha)} \equiv S^{\dagger}\hat{u}_{(a\alpha)}, \qquad v_{(b\beta)} \equiv S^{\dagger}\hat{v}_{(b\beta)}. \tag{7.3.30}$$

The orthogonality and normalization conditions among basic solutions can be computed directly, and they are listed as follows:

$$\begin{aligned} \hat{u}^{\dagger}_{(a\alpha)}(\underline{\kappa})\hat{u}_{(b\beta)}(\underline{\kappa}) &= \hat{v}^{\dagger}_{(a\alpha)}(\underline{\kappa})\hat{v}_{(b\beta)}(\underline{\kappa}) = u^{\dagger}_{(a\alpha)}(\underline{\kappa})u_{(b\beta)}(\underline{\kappa}) \\ &= v^{\dagger}_{(a\alpha)}(\underline{\kappa})v_{(b\beta)}(\underline{\kappa}) = (E/m)\delta_{(ab)}\delta_{(\alpha\beta)}, \\ \hat{v}^{\dagger}_{(a\alpha)}(\underline{\kappa})\hat{u}_{(b\beta)}(\underline{\kappa}) &= \hat{u}^{\dagger}_{(a\alpha)}(\underline{\kappa})\hat{v}_{(b\beta)}(\underline{\kappa}) = v^{\dagger}_{(a\alpha)}(\underline{\kappa})u_{b\beta}(\underline{\kappa}) \\ &= u^{\dagger}_{(a\alpha)}(\underline{\kappa})v_{(b\beta)}(\underline{\kappa}) \equiv 0. \end{aligned} \tag{7.3.31}$$

Every solution of (7.3.21) can be obtained as a linear combination (or superposition) of basic solutions in (7.3.30). Therefore,

$$u^L(\underline{\kappa}) = \sum_{r=1}^{4} \sum_{\alpha=1}^{2} [\beta^{(r\alpha)}(\underline{\kappa})u^L_{(r\alpha)}(\underline{\kappa}) + \bar{\gamma}^{(r\alpha)}(\underline{\kappa})v^L_{(r\alpha)}(\underline{\kappa})] \tag{7.3.32}$$

for some complex coefficients $\beta^{(r\alpha)}(\underline{\kappa})$, $\gamma^{(r\alpha)}(\underline{\kappa})$. A general class of solutions of the polar B.D.Y. equation (7.3.10) is given by the Fourier–Bessel integral [the equations (7.3.11), (7.3.14*a,b*), (7.3.18), (7.3.22), (7.3.32) are assumed]:

$$\begin{aligned}\hat{\psi}^L(\rho,\theta) &= (2\pi)^{-3/2}\sqrt{m}\\ &\quad\times\sum_{r=1}^{4}\sum_{\alpha=1}^{2}\sum_{t=-\infty}^{\infty}\Bigg\{\int_{0^+}^{\infty}\cdots[\beta^{(r\alpha)}(\underline{\kappa},t)u^L_{(r\alpha)}(\underline{\kappa})\\ &\qquad+\bar{\gamma}^{(r\alpha)}(\underline{\kappa},t)v^L_{(r\alpha)}(\underline{\kappa})]\\ &\qquad\times\left[\prod_{j=1}^{4}J_{t_j^L}(\kappa_j\rho^j)\right]\exp[i(t_b^L\theta^b)]\kappa_1\kappa_2\kappa_3\,d^3\underline{\kappa}\Bigg\}.\end{aligned} \tag{7.3.33}$$

Here, t_b^L satisfy (7.3.14*a,b*) and $\kappa_4 \equiv E(\underline{\kappa})$. Furthermore, we assume that the above improper integral converges uniformly and partial differentiations with respect to ρ^a, θ^b are permitted under the integral signs. Otherwise, $\beta^{(r\alpha)}$, $\gamma^{(r\alpha)}$ are *arbitrary* complex-valued functions. The spin-$\frac{1}{2}$ field in (7.3.33) can also be expressed as

$$\begin{aligned}\hat{\psi}^L(\rho,\theta) &= \sum_{t=-\infty}^{\infty}\hat{\psi}^L(\rho,\theta;t_1,t_2,t_3,t_4),\\ \hat{\psi}^L(\rho,\theta,t) &= (2\pi)^{-3/2}\sqrt{m}\\ &\quad\times\sum_{r=1}^{4}\sum_{\alpha=1}^{2}\Bigg\{\int_{0^+}^{\infty}\cdots[\beta^{(r\alpha)}(\underline{\kappa},t)u^L_{(r\alpha)}(\underline{\kappa})+\bar{\gamma}^{(r\alpha)}(\underline{\kappa},t)v^L_{(r\alpha)}(\underline{\kappa})]\\ &\qquad\times\left[\prod_{j=1}^{4}J_{t_j^L}(\kappa_j\rho^j)\right]\exp[i(t_b^L\theta^b)]\kappa_1\kappa_2\kappa_3\,d^3\underline{\kappa}\Bigg\}.\end{aligned} \tag{7.3.34}$$

This equation resembles the unified meson field equation (7.2.35). The constants of motion in (7.3.8) have been calculated in our paper (referred after this chapter,) using a *single* t-mode $\psi^L(\rho,\theta;t_1,t_2,t_3,t_4)$. But those results may not be physically reliable, since the *complete* set of separable solutions in (7.3.33) has not been used. Such a calculation with the complete set will be extremely difficult.

The functions $\beta^{(r\alpha)}(\underline{\kappa},t)$ and $\gamma^{(r\alpha)}(\underline{\kappa},t)$ in (7.3.34) represent the particle and antiparticle amplitudes. The index $\alpha \in \{1,2\}$ represents the usual two degrees of freedom for a spin-$\frac{1}{2}$ wave field. The index $r \in \{1,2,3,4\}$ stands for *four families* of spin-$\frac{1}{2}$ wave fields. Furthermore, each family can have fourfold infinitely many flavorlike quantum numbers $t_1, t_2, t_3, t_4 \in \{1,2,3,\ldots\}$. Thus, the equation (7.3.34) stands for the unified spin-$\frac{1}{2}$ fields on the extended phase space M_8.

References

1. M. Born, Proc. Roy. Soc. London Ser. A **165** (1938), 291; **116** (1938), 552; Rev. Mod. Phys. **21** (1949), 463.
2. J. A. Brooke and E. Prugovecki, Nuovo Cimento *A* **78** (1984), 17.

3. H. Brandt, Nucl. Phys. B **6** (Proc. Suppl.) (1989), 367.
4. E. R. Caianiello, Lett. Nuovo Cim. **25** (1979), 225.
5. E. R. Caianiello, M. Gasperini, and G. Scarpetta, Nuovo Cim. B **105** (1990), 259.
6. A. Das, J. Math. Phys. **7** (1966), 45; **21** (1980), 1506, 1513, 1521; Nucl. Phys. B **6** (Proc. Suppl.) (1989); Z. Phys. C. **41** (1988), 505.
7. M. Gell-Mann, Cal-Tech. Lab. Report CTSL-20, 1961.
8. Y. S. Kim and E. P. Wigner, Phys. Rev. A **36** (1987), 1293.
9. S. Okubo, Prog. Theor. Phys. (Kyoto) **27** (1962), 949.
10. G. N. Watson, *A treatise on the theory of Bessel functions*, 2nd ed., Cambridge Univ. Press, London and New York, 1966, pp. 248.
11. E. P. Wigner, Ann. of Math. **40** (1939), 149; in *Proceedings of the First International Conference on the Physics of Phase Space* (Y. S. Kim and W. W. Zachary, eds.), Springer, Verlag, Heielberg, 1987.
12. H. Yukawa, Phys. Rev. **76** (1949), 300; **77** (1950), 899; **80** (1950), 1047.

Answers and Hints to Selected Exercises

EXERCISES 1.1

1 (ii). $\sigma(\mathbf{a}) = \sqrt{2}$, $\sigma(\mathbf{b}) = \sigma(\mathbf{c}) = 0$, $\sigma(\mathbf{d}) = 1$.

EXERCISES 1.3

2 (i). Hint: Note that $\tau_{ij} \equiv (1/2)(\tau_{ij} + \tau_{ji}) + (1/2)(\tau_{ij} - \tau_{ji})$.

EXERCISES 2.1

2. $\mathscr{X}^i(t) = \pm\sin(t - c^i)$.

EXERCISES 2.2

1. $S(\gamma) = 2$.

EXERCISES 2.4

1. Hint: $\tau^{i_1 \cdots i_r}{}_{j_1 \cdots j_s}(x) = a^{i_1}{}_{k_1} \cdots a^{i_r}{}_{k_r} \ell^{m_1}{}_{j_1} \cdots \ell^{m_s}{}_{j_s} \hat{\tau}^{k_1 \cdots k_r}{}_{m_1 \cdots m_s}(\hat{x})$.

EXERCISES 3.2

2. Hint: Note that the 4×4 identity matrix does not belong to this subset.

EXERCISES 3.4

3(i). Hint: $\dfrac{\partial}{\partial x} = \dfrac{\partial}{\partial v} + \dfrac{\partial}{\partial u}, \dfrac{\partial}{\partial t} = \dfrac{\partial}{\partial v} - \dfrac{\partial}{\partial u}$.

EXERCISES 4.1

1. Hint: Assuming $\hat{x}^\alpha = -x^\alpha$ prove that $UPU^\dagger + P = 0$, or $U\sigma_\alpha + \sigma_\alpha U = 0$. (The last equation leads to a contradiction.)

EXERCISES 4.2

1(i). Hint: Compare the problem 2(ii) of Exercises 1.3.

EXERCISES 4.3

2. $\mathscr{T} = \gamma^4\gamma^5$.

EXERCISES 5.1

1. $H(\theta, p_\theta) = (p_\theta^2/2m\ell^2) + mg\ell(1 - \cos\theta)$.

EXERCISES 5.3

1. $M^\alpha{}_\beta(v) = m(1 - \delta_{\mu\nu}v^\mu v^\nu)^{-1/2}[\delta^\alpha{}_\beta + (1 - \delta_{\rho\sigma}v^\rho v^\sigma)^{-1}v^\alpha v_\beta]$.

EXERCISES 5.4

2. Hint: $$\frac{\partial\Lambda(x,u)}{\partial x^j} = \frac{\partial\hat{X}^k(x)}{\partial x^j}\frac{\partial\hat{\Lambda}(\hat{x},\hat{u})}{\partial\hat{x}^k} + u^b\frac{\partial^2\hat{X}^k(x)}{\partial x^b\partial x^j}\frac{\partial\hat{\Lambda}(\hat{x},\hat{u})}{\partial\hat{u}^k},$$

$$\frac{\partial\Lambda(x,u)}{\partial u^j} = \frac{\partial\hat{X}^k(x)}{\partial x^j}\frac{\partial\hat{\Lambda}(\hat{x},\hat{u})}{\partial\hat{u}^k}.$$

EXERCISES 6.1

1. Hint: $\mathscr{L}(x, \zeta, \bar{\zeta}, \zeta_k, \bar{\zeta}_k) = -d^{kj}\zeta_k\bar{\zeta}_j + k\cos\zeta$.

EXERCISES 6.2

4. Hint: $$\frac{\partial\psi}{\partial x^1} = (\cos\alpha)\frac{\partial h}{\partial\omega}, \quad \frac{\partial\psi}{\partial x^2} = \left[(\sin\alpha)\frac{\partial h}{\partial\omega} + \frac{\partial h}{\partial\xi}\right],$$

$$\frac{\partial\psi}{\partial x^3} = i\left[\frac{\partial h}{\partial\omega} + (\sin\alpha)\frac{\partial h}{\partial\xi}\right], \quad \frac{\partial\psi}{\partial x^4} = (\cos\alpha)\frac{\partial h}{\partial\xi}.$$

EXERCISES 6.3

2. Hint: Use special Minkowski coordinates such that
$$F_{34}(x) = F_{12}(x) \equiv 0.$$

Index of Symbols

$\mathscr{A}(F)$ action functional of a relativistic field
$A_k(x)$ electromagnetic field potential
$\mathbf{a} \cdot \mathbf{b}$ inner product between two vectors
$A \times B$ Cartesian product of two sets
$M \times N$ Kronecker product of two matrices
$B^A{}_i(x)$ gauge potential components
$\mathscr{C}^2(D; \mathbb{R})$; $\mathscr{C}^2(D; S)$ the set of twice differentiable functions from D into $\mathbb{R}$, S respectively
$C^A{}_{BD}$ structure constants of a Lie algebra
$\chi(p) = x = (x^1, x^2, x^3, x^4)$ local coordinates of a point p in the manifold
(χ, U) chart in a differentiable manifold
$D = [d_{ij}] \equiv \mathrm{diag}[1, 1, 1, -1]$ Lorentz metric; a domain $D \subset \mathbb{R}^n$
∂D boundary of a domain D
$\square$ d'Alembertian or wave operator, also denotes the completion of an *example*
$\blacksquare$ Q.E.D.
$D_a \psi^K(x)$ gauge covariant derivative
$\delta(k - k')$ Dirac delta function
$\delta^a{}_b$ Kronecker delta
$\Delta\tilde{\omega}$ perihelion shift per revolution
$\partial(\hat{x}^1, \hat{x}^2, \hat{x}^3, \hat{x}^4)/\partial(x^1, x^2, x^3, x^4)$ Jacobian of a coordinate transformation
$\mathbb{E}_3$ three-dimensional Euclidean space
$\{\mathbf{e}_i\}_1^4 \equiv \{\mathbf{e}_1, \mathbf{e}_2, \mathbf{e}_3, \mathbf{e}_4\}$ a basis set or tetrad
$\mathbf{e}_i \cdot \mathbf{e}_j = d_{ij}$ an M-orthonormal tetrad
$E(\mathbf{p})$ frequency associated with Dirac plane wave
$\in$ belongs to
ε_{ijkl} totally antisymmetric permutation symbol (Levi–Civita)
η_{ijkl} totally antisymmetric pseudotensor (Levi–Civita)
$F^i(x, u)$ four-force components
F_{ij} electromagnetic tensor components (field)
F^*_{ij} Hodge dual of electromagnetic tensor components
$F^A{}_{ij}(x)$ gauge field components

ϕ^A spinor (field) components
G a group
g_{ij} metric tensor components
$g_{ij}(x, u)$ Finsler metric components
$\gamma: x^i = \mathscr{X}^i(t)$ a parameterized curve on the manifold
γ^k Dirac matrices
$\gamma^{ku}{}_v$ entries of Dirac matrices, also numerical bispinor-tensor components
$\gamma_{ab}(x)$ linearized gravitational ten-potential components
$H(t, \mathbf{x}, \mathbf{p})$ (prerelativistic) Hamiltonian function
$I = [\delta^i{}_j]$ the identity matrix
$J[\gamma]$ prerelativistic action functional
$J_h[\tilde{\gamma}]$ prerelativistic canonical action functional
$J_L[\gamma]$ relativistic action functional
$J_H[\tilde{\gamma}]$ relativistic canonical action functional
$\mathscr{J}^i{}_{ab}(x)$ relativistic angular momentum tensor density
$j^k(x)$ charge-current four-vector density
J_{ab} the total relativistic angular momentum tensor
$\mathbf{L}$, $L \equiv [l^i{}_j]$ a linear (or Lorentz) mapping and the corresponding matrix
L^T the transposed matrix
$L(t, \mathbf{x}, \mathbf{v})$ prerelativistic Lagrangian
$L'(t, \mathbf{x}, t', \mathbf{x}')$ prerelativisitc Lagrangian in space *and* time
$\mathscr{L}(x, y^A, y^A{}_i)$ the Lagrangian of a field
λ a scalar (a real or complex number)
$\Lambda(x, u)$ relativistic Lagrangian of a particle
$\mathscr{L}_4$ general Lorentz group
$\mathscr{L}_{4+}^+$ proper, orthochronous subgroup of $\mathscr{L}_4$
$\mathbf{M}_4$ (flat) Minkowski space–time manifold
$\mathscr{N}_{x_0}$ Null cone with vertex at x_0
$\nu(\mathbf{k})$ frequency associated with electromagnetic plane wave
$\omega(\mathbf{k})$ frequency associated with Klein–Gordon plane wave
$\Omega(x, p) = 0$ constraining mass hypersurface, Super-Hamiltonian
$\mathbf{P}$ space reflection mapping
P_a the total four-momentum vector
p_k four-momentum components
$\mathscr{P}_4$ Poincaré group
$\mathscr{Q}$ the total charge
$\mathbb{R}$ the set of real numbers
$\mathbb{R}^n \equiv \mathbb{R} \times \cdots \times \mathbb{R}$ (n factors) Cartesian product of the set $\mathbb{R}$
$S(\gamma)$ separation functional on the curve γ
$SU(N)$ the set of unitary unimodular matrices
$\sigma(\mathbf{v})$ separation number of the vector $\mathbf{v}$
σ_α Pauli matrices
$\sigma^{k\bar{A}B}$ numerical spinor-tensor
$\psi(x)$ Klein–Gordon field
ψ^u, $\psi^u(x)$ bispinor, Dirac field components

$\Sigma_1, \Sigma_2, \Sigma_3$ 1-flat, 2-flat, 3-flat submanifolds
T time reversal mapping
$T^i{}_j(x)$ canonical energy-momentum-stress tensor density
$\tau^{i_1\cdots i_r}_{j_1\cdots j_s}$ Minkowski tensor components
$\tau^{i_1\cdots i_r}_{j_1\cdots j_s}(x)$ Minkowski tensor field components
$\theta_{ij}(x)$ symmetrized energy-momentum-stress tensor density
$\tilde{u}$, u_i covariant vector (or covector) and components
$u^i(s)$ four-velocity components
$V(\mathbf{x})$ the potential energy
$\mathbf{V}_4$ Minkowski vector space
$\mathbf{v}$, v^i Minkowski vector, components
$z \equiv x^1 + ix^2$ a complex variable
$\bar{z} \equiv x^1 - ix^2$ the complex conjugate variable
$\mathbb{Z}$ the set of integers
$\mathbb{Z}^+$ the set of positive integers

Subject Index

Universitext *(continued)*

Sagan: Space-Filling Curves
Samelson: Notes on Lie Algebras
Schiff: Normal Families
Shapiro: Composition Operators and Classical Function Theory
Smith: Power Series From a Computational Point of View
Smoryński: Self-Reference and Modal Logic
Stillwell: Geometry of Surfaces
Stroock: An Introduction to the Theory of Large Deviations
Sunder: An Invitation to von Neumann Algebras
Tondeur: Foliations on Riemannian Manifolds

9 780387 940427